North America's Natural Wonders

Geologic Tours of the World

Gary L. Prost

North America's Natural Wonders, 2-Volume Set

North America's Natural Wonders
Canadian Rockies, California, The Southwest, Great Basin, Tetons-Yellowstone Country

North America's Natural Wonders
Appalachians, Colorado Rockies, Austin-Big Bend Country, Sierra Madre

For more information about this series, please visit: *www.crcpress.com/Geologic-Tours-of-the-World-Series/book-series/CRCREMSENAPP*

North America's Natural Wonders

Appalachians, Colorado Rockies, Austin-Big Bend Country, Sierra Madre

Gary L. Prost

CRC Press is an imprint of the
Taylor & Francis Group, an **informa** business

CRC Press
Taylor & Francis Group
6000 Broken Sound Parkway NW, Suite 300
Boca Raton, FL 33487-2742

Printed on acid-free paper

International Standard Book Number-13: 978-0-367-82122-7 (Paperback)
International Standard Book Number-13: 978-0-367-85944-2 (Hardback)

Visit the Taylor & Francis Web site at
http://www.taylorandfrancis.com

and the CRC Press Web site at
http://www.crcpress.com

These volumes are dedicated to my parents, Ruth and Paul Baldwin, for inspiring in me a sense of exploration and wonder, and to Nancy, Adam, Benjamin, and Elizabeth, who accompanied me on family vacations and endured my endless rambling on about the geology.

Contents

Preface

People come from all around the world to see the unique and spectacular parks and landscapes of North America. Most folks have no idea how the rugged peaks of the Rocky Mountains, caverns of Texas and New Mexico, or forest-mantled vistas of the Appalachians formed and what they are made of. As a young man I became fascinated by the stories the rocks had to tell. As a career geologist I learned about many of these areas, mostly by working there or from technical field guides.

It occurred to me that I have been on many geological field trips and have led a few myself, but they were all directed toward professional geoscientists. Professional field trip guidebooks provided examples and gave technical details to those working on some narrow, focused aspect of geology. That is great as far as it goes. But there are lots of people that are curious about the rocks beneath their feet, or about the landscape they are touring through.

The purpose of *this* guidebook is to inspire and inform. Inspire you to think "I want to go there and see that"; to inform you about what you are seeing, about the geologic history and the human story; and to answer questions such as what are those rocks? How old is that mountain? How did that canyon (or rock, natural arch, or geyser) form? Those questions are answered as we travel on these geological expeditions. Together, we will unlock the secrets behind the beauty of the landscape.

This self-guided geological tour book is a collection of classic excursions across some of the best known natural wonders in central and eastern North America. We begin in the Appalachian Mountains, the original blue ridges that separated early settlers from the vast interior of the continent. From there, we move to the Colorado Rocky Mountains with their awesome crags and breath-taking beauty. We continue across the mid-continent from the tranquil Gulf Coast to the Texas Hill Country and Edwards Plateau. Much of the geology buried beneath the Gulf Coast is exposed as you drive west across Texas and into New Mexico. We see ancient seafloors and reefs lifted high on mountaintops, vast dunes of shifting gypsum sand, recent lava flows, and the Rio Grande Rift, where the continent is literally tearing itself apart. We end our explorations in the majestic and mysterious Sierra Madre of Mexico. During these tours we visit Shenandoah National Park, Florissant Fossil Beds National Monument, Big Bend National Park, Carlsbad Caverns National Park, Guadalupe Mountains National Park, White Sands National Monument, Organ Rock National Monument, and Prehistoric Pathways National Monument, among many others. Taken together, the six tours in Volume 2 total 5,055 km (3,130 mi), roughly the distance from Mexico City, Mexico, to Quebec City, Canada.

My objective in this guidebook is to explain to the curious layman, the interested general public, the rock hound, the student, and the geologist what they are seeing when they look out across the Appalachian Plateau, at a volcanic dike or a cave in the Texas Hill Country. In simple terms, you will learn what you are looking at, how it came to be, and why it is important. Along the way, we add historical context, say a word or two about the resources of an area, and point out interesting fossil and mineral locations.

Written in easy to understand language, and richly illustrated, these geological tour guides to classic and not-so-well-known areas are meant to educate and entertain. A fuller understanding of natural formations should enhance the traveler's experience beyond simple sightseeing.

These trips can be run in either direction, but most start and stop at a major town where people have easy access. Some transects, such as the Appalachians or Colorado Rockies, are short enough to be traveled in 1 or 2 days. Others, such as the Texas-New Mexico trip, can take several days to a week or more. The trips can easily be broken into multiple separate tours. Times and distances are provided, but are meant as a guide only and will change depending on traffic and weather conditions. Each stop has, in addition to the distance from the previous stop, GPS coordinates in latitude and longitude, and when possible, mileage keyed to roadside mileposts. Although described stops are provided, there are many more stops that can be made along these routes: do not feel constrained to use only those stops in the guidebook. Interesting side trips are provided for those with the time to explore. Obviously not every point of interest could be included: maybe they will be in future editions. Some routes are closed in winter, or are unpaved, and I have tried to note these in the text.

Please be careful when pulling off the road, especially when there are narrow shoulders or curves in the road. Be aware of traffic at all times. Safety first!

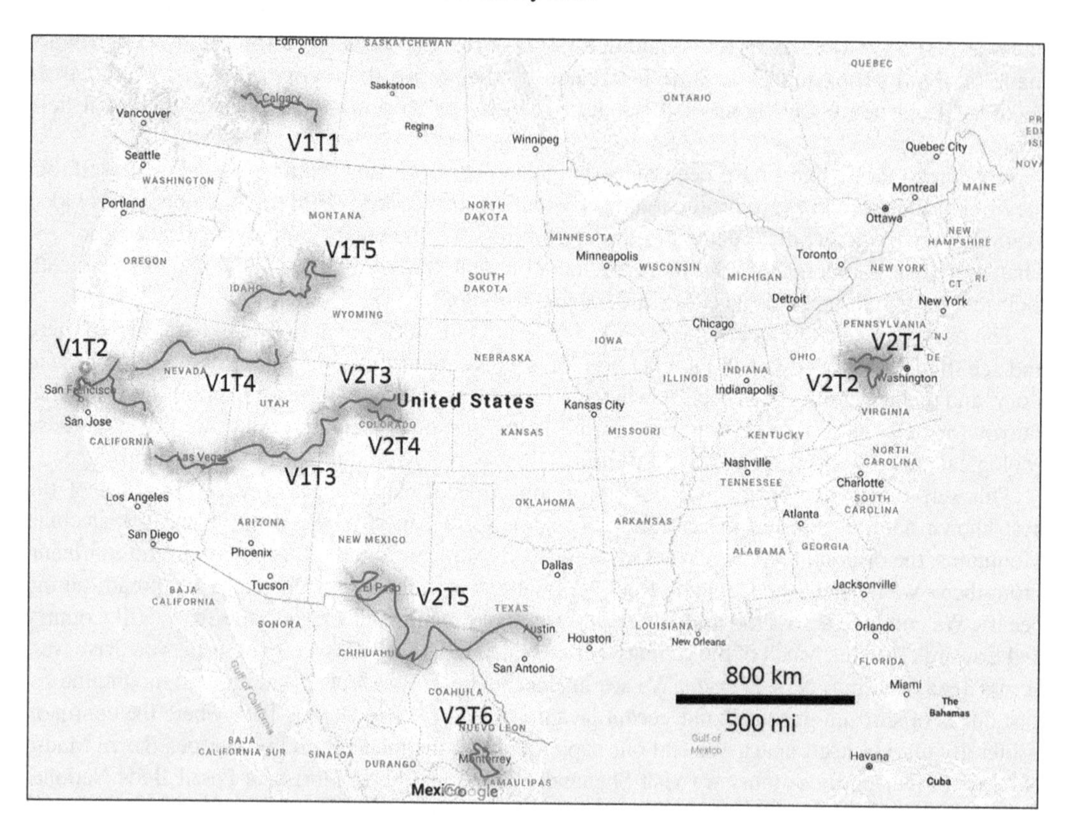

Overview of geologic transects in North America Volumes 1 (V1T1-5) and 2 (V2T1-6).

Charts are provided to show what rock strata exist in each area, how old the layers are, and what they are made of. Maps and cross sections reveal the three-dimensional geology. Each region is placed in a plate-tectonic context, so that you know how the mountains and basins formed and when the rocks were folded and faulted. Geologic influences on our economy are discussed so that you know, for instance, that the Appalachians contain important economic coal deposits that influenced America's steel industry and that the Denver Basin is an important source of oil and gas.

Units are mostly provided metric first and then their American English equivalent.

Abbreviations are explained and then used. For example, millions of years is abbreviated to Ma (for mega-annum), Ga is billions of years, Fm is used for Formation, Gp for Group, mbr for member, and bbl stands for barrels of oil.

Road descriptions, pullouts, travel times, mine and other special tours, contact names, websites, and emails are current as of the time of writing. For current hours and entry fees of a park or museum, you should visit their website.

Each geo-tour was scouted with different companions, and their contributions are gratefully acknowledged in each section.

A geologic time scale is provided here as a useful reference for those not familiar with the different geologic periods referred to in the trip descriptions.

EON	ERA	PERIOD		EPOCH		Ma
Phanerozoic	Cenozoic	Quaternary		Holocene		0.01
				Pleistocene	Late	0.8
					Early	1.8
		Tertiary	Neogene	Pliocene	Late	3.6
					Early	5.3
				Miocene	Late	11.2
					Middle	16.4
					Early	23.7
			Paleogene	Oligocene	Late	28.5
					Early	33.7
				Eocene	Late	41.3
					Middle	49.0
					Early	54.8
				Paleocene	Late	61.0
					Early	65.0
	Mesozoic	Cretaceous		Late		99.0
				Early		144
		Jurassic		Late		159
				Middle		180
				Early		206
		Triassic		Late		227
				Middle		242
				Early		248
	Paleozoic	Permian		Late		256
				Early		290
		Pennsylvanian				323
		Mississippian				354
		Devonian		Late		370
				Middle		391
				Early		417
		Silurian		Late		423
				Early		443
		Ordovician		Late		458
				Middle		470
				Early		490
		Cambrian		D		500
				C		512
				B		520
				A		543
Precambrian	Proterozoic	Late				900
		Middle				1600
		Early				2500
	Archean	Late				3000
		Middle				3400
		Early				3800?

Geologic time scale. Occasionally "Carboniferous" is used for the period of time that includes Mississippian and Pennsylvanian.

Author

Gary L. Prost obtained his BSc in geology from Northern Arizona University in 1973 and an MSc (1975) and PhD (1986) in geology at Colorado School of Mines. Over the past 45 years, he has worked for Norandex (mineral exploration), Shell U.S.A. (petroleum exploration worldwide), the U.S. Geological Survey (geologic mapping, coal), the Superior Oil Company (mineral and oil exploration), Amoco Production Company (worldwide oil exploration, remote sensing, and structural geology), Gulf Canada (international new ventures), and ConocoPhillips Canada (Canadian Arctic exploration, gas field development, oil sands development, and reservoir characterization). He spent over 20 years working as a satellite image analyst in the search for hydrocarbons and minerals in more than 30 countries. During this time, he applied structural geology and remote sensing to exploration, development, and environmental projects. The second half of his career was spent working on regional studies, new ventures and frontier exploration, and oil and gas field development. His most recent work has been in public outreach, leading field trips and educating the public on topics of geological interest. He is the principal geologist for G.L. Prost GeoConsulting of El Cerrito, California, and has been a registered professional geologist in Wyoming (United States) and in Alberta and the Northwest Territories (Canada). Previously, he has published three previous books, namely, *The Geology Companion: Essentials for Understanding the Earth* (Taylor and Francis, 2018), *Remote Sensing for Geoscientists: Image Analysis and Integration* (third edition, Taylor & Francis, 2013), and *The English–Spanish and Spanish–English Glossary of Geoscience Terms* (Taylor & Francis, 1997). He is already at work on his next volume, *Geologic Tours of the World – South America's Natural Wonders.*

1 Central Appalachians

Harpers Ferry, West Virginia, to Terra Alta, West Virginia

Folding in the Silurian McKenzie Formation, railroad cut at Pinto, Maryland.

OVERVIEW

Our visit to the Appalachians starts at Harpers Ferry in the hinterland of the Pennsylvanian-Permian Alleghanian Orogeny and traverses the Central Appalachian Fold-Thrust Belt. We cross Virginia and West Virginia from east to west. Along the way, we examine the rocks and structures that make this area unique. We review the geologic history of an area that has been the site of multiple continent-continent collisions and discuss how this geology led to development of the geosynclinal theory that was widely accepted for almost 100 years before yielding to the concept of plate tectonics. The influence of geology on American history is examined.

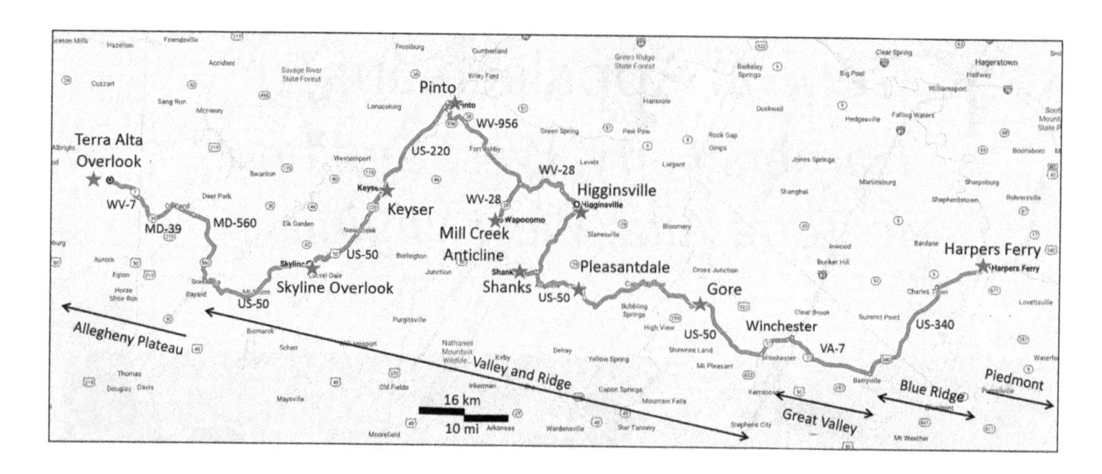

ITINERARY

Stop 1 Harpers Ferry
Stop 2 Gore Roadcut
Stop 3 Pleasantdale Fault Zone
Stop 4 Shanks Roadcut
Stop 5 Higginsville
Stop 6 Mill Creek Anticline at Hanging Rocks
Stop 7 Pinto Railroad Cut
Stop 8 Kaiser Cut
Stop 9 Skyline Scenic Overlook, Allegheny Structural Front
Stop 10 Terra Alta Overlook of the Allegheny Plateau

GEOLOGIC SETTING OF THE CENTRAL APPALACHIAN MOUNTAINS

Geologic History

The broad sweep of geologic history in the Appalachians includes Precambrian rifting and sedimentation followed by a series of continental collisions that added new terrain to eastern North America. These collisions compressed existing sedimentary rocks to form folded and thrusted mountains. An orogeny is a mountain-building event. The Ordovician Taconic Orogeny was caused by the collision of Laurentia (which contained part of what is now North America) with Baltica (which now underlies much of Scandinavia). Another mountain-building episode occurred during the Devonian Acadian Orogeny as a result of the collision of Laurentia with Proto-(southern) Europe that formed a new continent, Euramerica. A period of quiet sedimentation was followed by mountain building during the Late Pennsylvanian to Permian Alleghanian Orogeny. This episode records the collision of Euramerica with Gondwana (South America-Africa-Arabia-India-Australia-Antarctica). Deformation was mainly west-directed thrusting and folding inclined (verging) to the west. Thrusts emerge from a master detachment surface in Early Cambrian to Devonian shales. The Alleghanian episode was followed in the Triassic by rifting related to opening of the Atlantic. Sedimentation along the relatively quiet east coast of North America ('passive margin' sedimentation) has been ongoing since the Jurassic. Some good recent regional studies on the Appalachians include Hatcher et al. (2007) and Tollo et al. (2010).

The oldest units in this area are 1.2 billion-year-old granitic rocks exposed in the Blue Ridge, and we will see them along Skyline Drive (Chapter 2). These rocks cored a mountain system that formed during the Grenville Orogeny, long before the Appalachians.

A long period of erosion wore away these early mountains and exposed the granite below. The next oldest rocks are 600–700 million-year-old (Ma) basaltic lava flows that filled in valleys in the granite. These lavas have since been buried and metamorphosed to greenstone.

Opening (rifting) of the Iapetus Ocean between Laurentia and Gondwana led to deposition of early Cambrian sedimentary rocks. Between 600 and 400 million years ago, the Iapetus Ocean was south and east of Laurentia.

The entire Appalachian region was covered by ocean between 600 and 260 million years ago. During this time, almost 30.5 km (19 mi) of sediments were deposited, mostly in a shallow marine setting, which means the area was slowly subsiding. The pile of sediments include two carbonate (limestone and dolomite) sequences up to 3,000 m (10,000 ft) thick.

As the sea began to retreat around 350 Ma, it was being filled by sediments deposited in swamps and on river floodplains. Plants in the swamps died, accumulated, and formed peat deposits that turned to coal after they were buried.

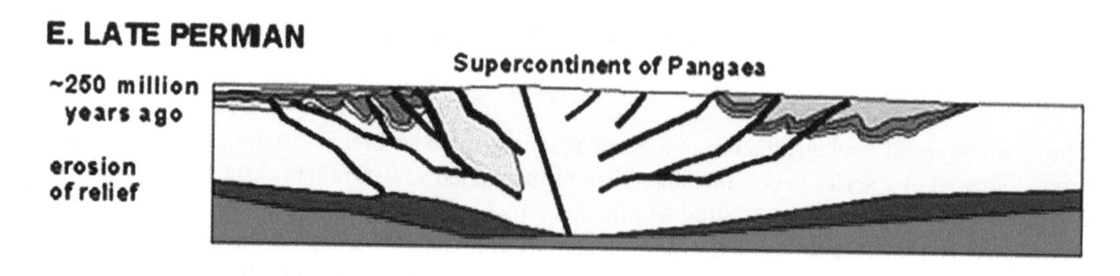

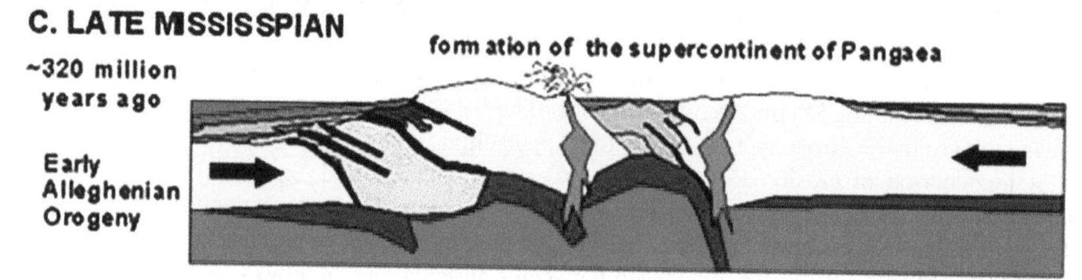

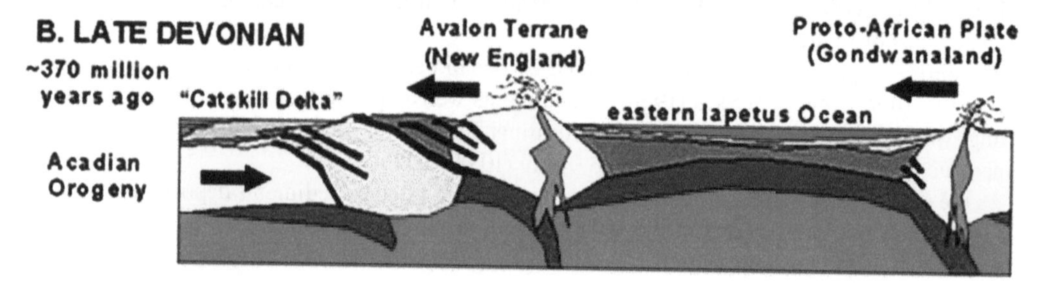

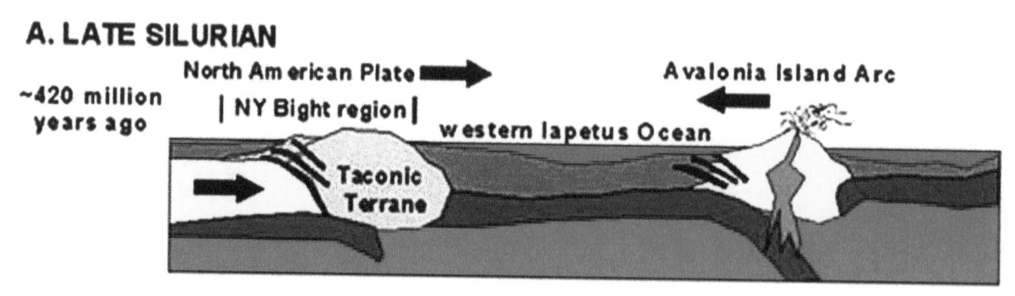

The Appalachian Mountains are a result of three separate continental collisions involving the Proto-North American Continent with the Taconic and Avalon terranes, and finally the collision of the African and North American continents during the Alleghanian Orogeny at the end of the Paleozoic (US Geological Survey, https://commons.wikimedia.org/wiki/File:Appalachian_orogeny.jpg).

Between 325 and 260 Ma, the Iapetus Ocean closed because the African tectonic plate collided with the Laurentian tectonic plate to form the supercontinent Pangea. 'Tectonics' are processes that affect the Earth's crust; 'tectonic plates' are slabs of the Earth's crust that move over the deeper mantle and, where they collide, raise up mountains. In this case, the collision, the Alleghanian Orogeny, caused crumpling and thrusting that formed the Appalachian Mountains. During their heyday, these mountains were every bit as high and rugged as the Canadian Rockies and Alps. Mountain building here ended around 250 million years ago.

Around 200 Ma (Late Triassic-Early Jurassic), Pangea began to split apart. The rifting created the Tethys Ocean, which evolved into the Atlantic Ocean, between the new continents of Laurasia (North America-Europe-Asia) and Gondwana (Africa-South America-India-Australia-Antarctica). The east coast of North America has largely been a passive margin since that time.

There is some research that suggests that the Piedmont and Coastal Plain physiographic provinces were subject to episodic transpressional faulting and uplift from Middle Jurassic to possibly Miocene time (Brown et al., 1977; Mixon and Newell, 1977; Prowell, 1988; McLaurin and Harris, 2001). Transpression is a combination of compression (shortening) and translation (sideways slip), and transpressional faulting combines thrust and strike-slip movement on the same fault plane.

Since Jurassic time, the region has been subject mostly to erosion and karst development. Erosion has exposed the bones of the mountains, mainly the resistant sandstone layers that form most of the high ridges. Erosion and deposition formed the valleys, which are underlain by easily eroded shale and limestone. Karst is the slow dissolving of limestone by groundwater that leads to caverns and sinkholes (collapsed caverns). Both caverns and sinkholes are common in the local limestones.

Geosynclinal Theory Development

The Appalachian Mountains were key to developing early ideas of how mountains form. For over 100 years, from the mid-1800s until the 1960s, most North American and European geologists accepted that continents were fixed in place on the surface of the earth and grew from a central core, or craton, by adding mountain ranges around their margins. The American geologists James Hall and James Dana studied the Appalachian Mountains and found that a thick accumulation of sedimentary rocks had been crumpled and uplifted. In 1857, Hall presented his idea that mountain belts are the result of thick wedges of sediment deposited in deep basins or troughs along the margins of continents that are later compressed and uplifted. He published his "geosynclinal theory" in 1882. Dana, in his *Manual of Geology* (1895), defined a geosyncline as a belt of thick, deformed sediments, metamorphic rock, and granitic intrusions along the margin of a continent. He proposed that a great thickness of sediments must have loaded and depressed the crust until the sediments at the bottom got extremely hot. At that point, the hot, deformed sediments and associated metamorphic rocks began to melt, whereupon they rose up buoyantly to form mountain belts.

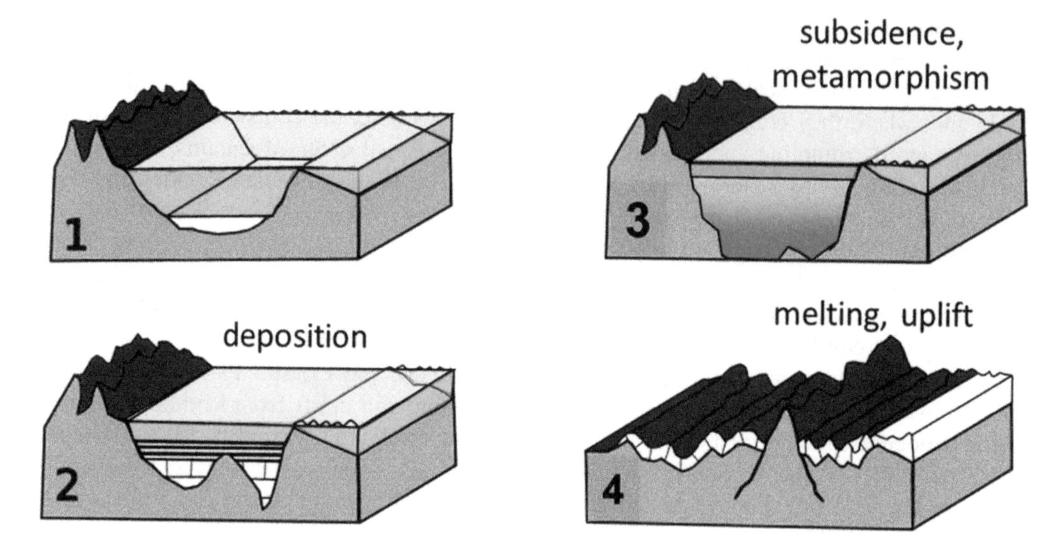

Development of a geosyncline. (Modified after Núñez González, 2016; https://commons.wikimedia.org/wiki/File:Geosinclinal,_2016.svg.)

The geosynclinal theory stated that mountain chains developed along the margins of continents. The theory seemed to work well in eastern North America, but it didn't explain why some continents don't have folded sedimentary mountain ranges along their margins. Nor did it explain why some folded mountain ranges appear deep within continents. Clearly, this theory had some kinks that needed to be worked out.

Modern Plate Tectonics Theory

Fast forward to the 1960s. Workers now had an Earth model with rigid crustal plates above a ductile, mobile mantle. And a mechanism, mantle convection, had been proposed to move continents. Multiple lines of evidence merged into the theory of plate tectonics. Plate tectonics theory states that Earth is covered by a number of brittle plates that move over a ductile interior. Most of the deformation that builds mountains takes place at or near plate margins. These plates are created at and move away from mid-ocean ridges (e.g., Mid-Atlantic Ridge, East Pacific Rise). Convection currents in the mantle move up to the surface at mid-ocean ridges, and then push or pull the ocean plates away from the spreading centers. The plates plunge back into the mantle and are consumed in subduction zones (convergent margins) along oceanic trenches (e.g., Aleutian Trench, Japan Trench, Cascadia Trench). Subduction zones are the site of volcanic arcs and deep earthquakes that result from downward-moving and melting oceanic crustal plates. The plates slide past each other at transform margins (e.g., San Andreas Fault, Queen Charlotte Fault, Dead Sea Transform). Where continents collide, the two plates and the sediments between them crumple and form thrusted and folded mountain belts (e.g., Himalayas, Alps, Appalachians). Excellent reviews of plate tectonics theory can be found in Wilson (1972), Redfern (2001), Frisch (2010), Saunders (2011), Molnar (2015), Oreskes (2018), Prost and Prost (2018), and Sykes (2019).

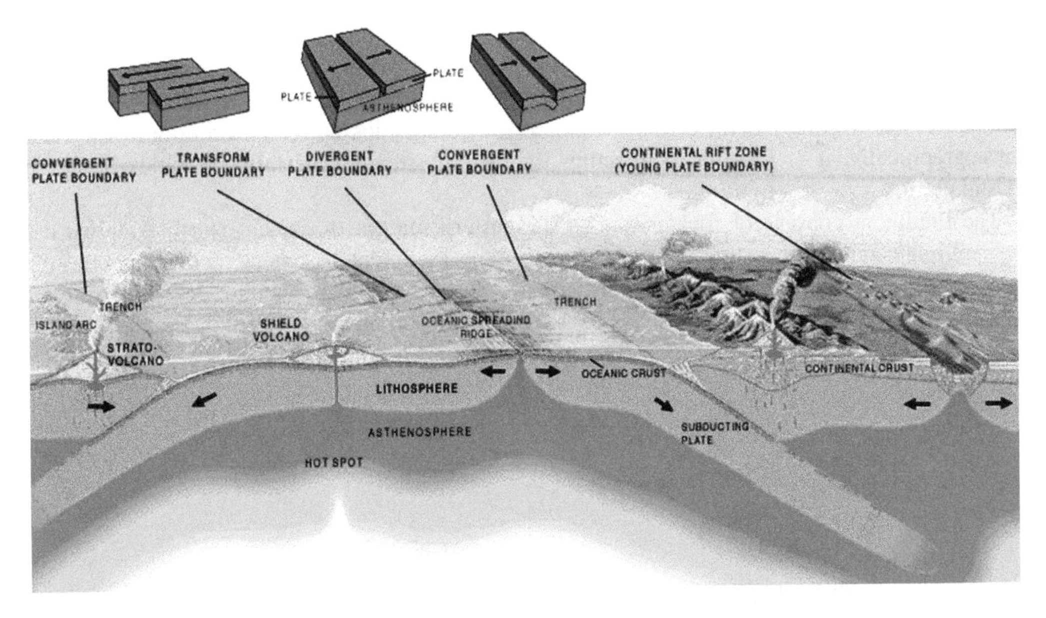

Artist's cross section illustrating the main types of plate boundaries. (Cross section by José F. Vigil from This Dynamic Planet – a wall map produced jointly by the U.S. Geological Survey, the Smithsonian Institution, and the U.S. Naval Research Laboratory.) http://pubs.usgs.gov/gip/dynamic/Vigil.html.

Appalachian Provinces

Six major physiographic provinces each have characteristic strata, levels of metamorphism, and structures along the central Appalachians of North America. Each landscape is the result of unique rock types, deformation, and erosion that has affected the area.

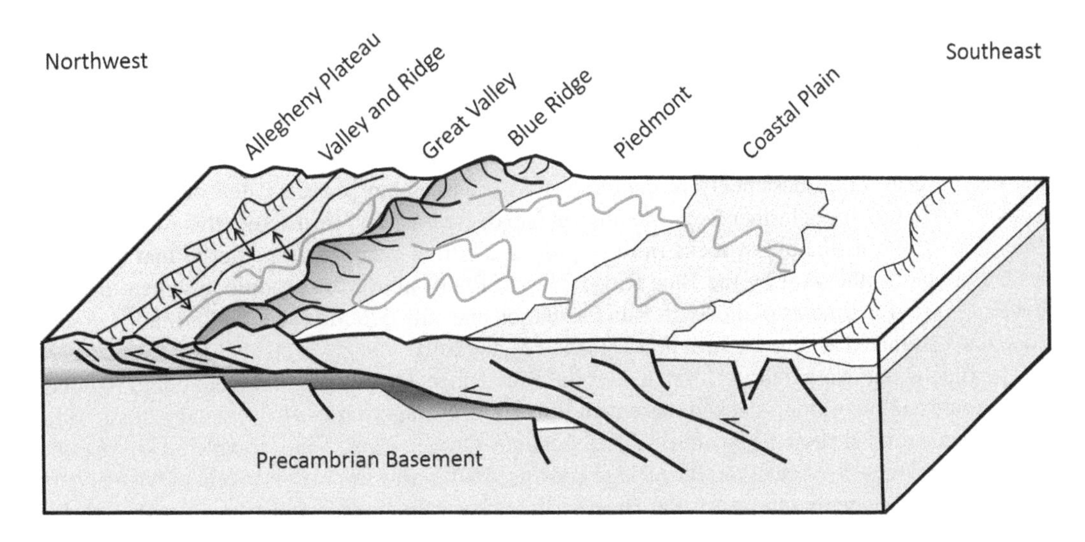

Schematic cross section through the Appalachians in northern Virginia and West Virginia showing the physiographic provinces and major structures that underpin them. (Modified after Pohn et al., 1985.)

The Coastal Plain or Tidewater Province contains an eastward thickening wedge of Cretaceous to Recent sediments. These sediments are deposited over granitic and metamorphic basement rocks. The western boundary, known as the "Fall Line," is the beginning of the Piedmont province. This was originally defined as the limit of sea-going ships on rivers draining to the coast. Physiographically, it is the line connecting the first upstream waterfalls or rapids along the coastal plain.

The Piedmont consists of rock pasted to the eastern margin of ancient North America during multiple mountain-building episodes. Most of the Piedmont basement rock is underlain by metamorphosed Late Precambrian and Paleozoic sedimentary and igneous rocks. These rocks were overprinted by Mesozoic pull-apart, or rift basins that contain continental, lake, and marine sediments reflecting early stages of opening of the Atlantic.

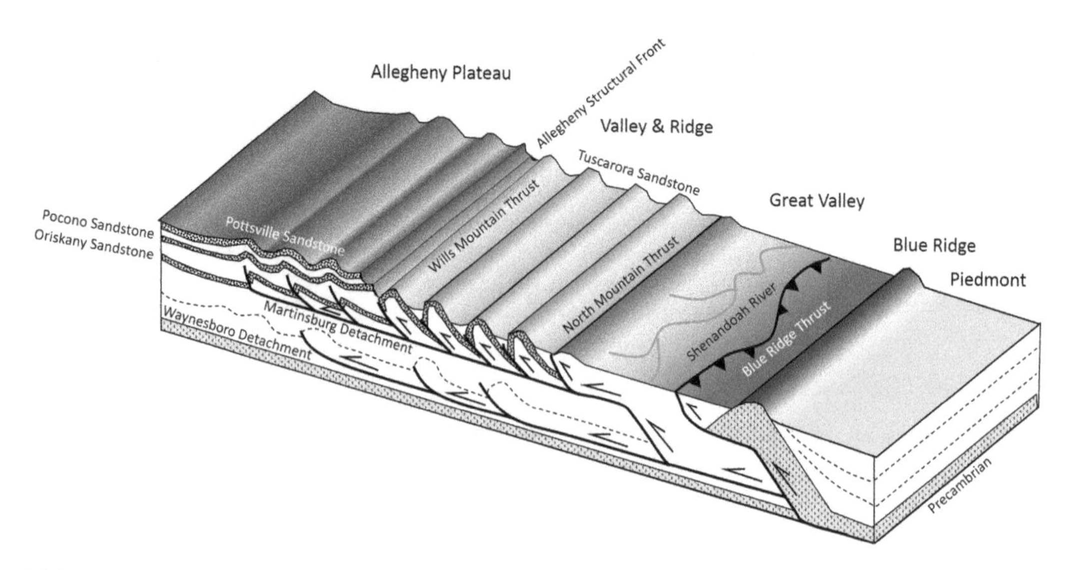

Major thrusts across the central Appalachian Mountains of Virginia and West Virginia. (Modified after Renton, 2018.)

West of the Piedmont is the Blue Ridge Province. Metamorphosed Cambrian sedimentary rocks that overlie crystalline basement are continuous with unmetamorphosed Paleozoic sedimentary rocks found in provinces farther west. Metamorphism decreases westward across this province. The Blue Ridge contains the oldest rocks in the region, 1.2 billion-year-old granodiorite that has been pushed up and to the west by the Blue Ridge Thrust. Rocks above the granitic basement include metamorphosed sediments of the Swift Run Formation and 550–570 Ma Catoctin Formation greenstone that originated as basaltic lava flows (Fichter et al., 2010).

The Blue Ridge Anticlinorium, carried on the Blue Ridge Thrust, is a basement-cored regional uplift containing multiple stacked basement thrusts and superimposed secondary folds. The anticlinorium and thrusts began during the Acadian Orogeny and were completed during the Alleghanian Orogeny. Most of the Blue Ridge Province in this area lies in or adjacent to Shenandoah National Park. We begin our transect at Harpers Ferry, near the boundary between the Blue Ridge and Great Valley.

The Great Valley Province contains thick (up to 3,500 m, or 11,500 ft) Cambrian and Ordovician carbonate platform deposits similar to those seen in the Bahamas today. The "Great American Carbonate Bank," as it has been called (Derby et al., 2012), was much more than a fringing reef, as it extended around Laurentia from present-day Virginia to Nevada, and from present-day Texas to South Dakota. These rocks were folded and thrust from east to west during the Alleghanian Orogeny, with the intensity of folding and faulting decreasing westward. The Great Valley Province is characterized by karst (sinkhole) topography, a result of rainfall and snowmelt dissolving the limestone. These limestones contain abundant cave systems.

A key landmark of the Great Valley is Massanutten Mountain, an 80 km (50 mi) ridge within the Shenandoah Valley. It is folded into a U-shaped fold (syncline) and supported by resistant sandstone. The valley floor, at an elevation around 300 m (1,000 ft), consists of easily eroded limestone and shale. The limestone dissolves to form caverns and, when they collapse, sinkholes. The province is bounded on the east by the Blue Ridge Thrust system and on the west by the North Mountain-Little North Mountain Thrust (NMT-LNMT). The NMT-LNMT was emplaced after the Blue Ridge Thrust and moved the Cambrian-Ordovician rocks carried by these thrusts 60 km (36 mi) to the west (Evans, 1989).

After emplacement of the NMT-LNMT, the Cambrian and Ordovician carbonates below the thrust were deformed into low-amplitude folds detached in the Cambrian Waynesboro Formation. This is known as the Lower Carbonate Duplex. Continued displacement, thrusting, and folding formed the western Valley and Ridge structures.

West of the Great Valley is the Valley and Ridge Province which, as the name implies, is a series of ridges and valleys that make up the Allegheny Mountains. The eastern boundary in this area is the trace of the NMT-LNMT. West of this thrust, the rocks consist mainly of folded and faulted Cambrian to Devonian sandstone, shale, and carbonate. These rocks originally came from highlands to the east and were deposited at or near the shore of an ocean that covered most of Laurentia/Euramerica at the time. The rocks were folded and thrust during the Alleghanian Orogeny. Detachments occur in weak, shale-dominated layers. As we discussed earlier, thrusting records the collision of the African part of Gondwana with Laurentia (ancestral North America). Major Valley and Ridge structures, from east to west, include the Cacapon Mountain Anticlinorium, the Broad Top Synclinorium (an intricately folded, regional syncline-like structure), and the Wills Mountain Anticline.

The western margin of the Valley and Ridge is the Allegheny Structural Front along the Wills Mountain Thrust. The front marks the transition from tight folds of the Valley and Ridge to broad folds and flat-lying rocks of the Allegheny Plateau.

The Allegheny Plateau (in some places called the Appalachian Plateau) contains mainly Mississippian, Pennsylvanian, and Permian strata (roughly 350–250 million years old), the youngest we've seen. The bluffs characteristic of the Plateau are formed by resistant, flat sandstone layers. Gentle, wide open folds in the Plateau are thought to overlie buried thrusts that originate in the Valley and Ridge Province to the east but never quite reach the surface.

Estimates of the amount of shortening across the Central Appalachian fold-thrust belt vary widely: they range from 77 km (48 mi; Sak et al., 2012) to as much as 300 km (186 mi; Mitchell, 2017).

Effect of Geology on American History

The first settlers occupied the Tidewater, coastal plain areas next to bays and estuaries and subject to ocean tides. With the development of tobacco-based agriculture in the 1600s, more land was needed and farming quickly moved into the Piedmont, the area of low rolling hills and valleys between the Blue Ridge and the Tidewater. By the early 1700s, a few adventurous souls had crossed the Blue Ridge, comprising the easternmost thrusts and folds, and settled in the valley of the Shenandoah River. Further westward expansion was held back by the ranges of the Valley and Ridge Province, whose ridges consisted of resistant sandstone layers outlining anticlines or capping synclines. By the mid-1700s, settlers found their way through these ridges at water gaps (Cumberland Gap, Keys Gap, Swift Run Gap, for example). Gaps were formed by steams cutting down across the ridges, often where a ridge had been cut by a fault that made the rock more susceptible to erosion. The less-rugged topography through the gap made it a preferred route for the original natives as well as the pioneers on their way west to the Ohio country. After the American

Revolution until the mid-1800s, the Cumberland Road was a primary starting point for emigrant trails to the west. Eventually, the original native track or game trail became a wagon road, a railroad, and today a highway.

Likewise, geology played a role in the Civil War. Ridges were used to screen troop movements, as in the movement of Southern troops through the Shenandoah Valley west of the Blue Ridge in the lead up to the battle of Gettysburg. Both sides understood the importance of controlling the high ground to see and fire upon an attacking army. Union troops took advantage of the high ground on Seminary Ridge, Cemetery Ridge, and Little Round Top to turn the tide at Gettysburg. At Gettysburg the ridges consist of resistant diabase dikes intruded into the Triassic Gettysburg Sandstone. Wide, flat valleys, with their lack of cover and wide-open fields of fire, were formed by erosion of limestone layers and are where many of the bloodiest battles, such as Antietam, occurred.

Resources

Carbonates

Limestone and dolomite are quarried for building stone, for cement, and for the agriculture and steel industries. Limestone and dolomite underlie the Great Valley and create, through erosion and weathering, extremely rich farming soils. The Great Valley, extending from Virginia through Maryland and into Pennsylvania, was and is the main agricultural area west of the Atlantic Coastal Plain.

Coal

The Appalachian Basin is over 1,600 km (1,000 mi) long and as much as 560 km (350 mi) wide. Rocks in the basin range from Early Cambrian to Early Permian age. The major sources for much of the sediment in the basin were mountains to the east. Sandstone and shale deposited by rivers or in shallow marine settings are concentrated on the east side of the basin. Carbonate rocks dominate the western part of the basin. Prodigious amounts of coal (the remains of trees) accumulated in Mississippian and Pennsylvanian non-marine rocks (de Witt, 1993).

During the Pennsylvanian period, from 323 to 299 Ma, the area of Virginia and West Virginia was close to sea level and covered by a vast swamp. For tens of millions of years, the trees and plants in these swamps died, piled up, and were buried before they could rot away. The peat that formed in this way gradually was converted to coal by the earth's heat and pressure.

Coal in Virginia occurs in three provinces: relatively minor Mesozoic coal in the rift basins/rift valleys of the Piedmont province, relatively minor Early Mississippian coal in the Valley and Ridge Province, and extensive Pennsylvanian coal in the Appalachian Plateau. All of the coal mined today is bituminous coal from the Appalachian Plateau. The value of coal produced in Virginia in 1996 was $974 million. About 20.4 million tonnes (22.5 million short tons) was produced from 110 mines around 2015 (Virginiaplaces.org).

Coal was discovered in Virginia (and West Virginia) as early as 1742. There is coal in the Valley and Ridge, but it is not mined today because mining is difficult and costly in these folded rocks. Two-thirds of the coal is in the relatively flat-lying Appalachian Plateau, and this area has all the mineable coal. Coal is produced from both surface and underground mines.

In 2011, central Appalachian Basin coal production accounted for approximately 77% of all U.S. metallurgical (or coking) coal and 29% of total U.S. production. In the last few years, West Virginia has produced around 122 million tonnes (135 million short tons) and was the second largest producer of coal in the United States. Reserves (discovered resource still in the ground) are around 106 billion tonnes (117 billion short tons), which gives it the fourth largest reserves in the United States. Because the shallow and thicker coals are mined first, much of what remains is thin and deep, thus more expensive to mine.

West Virginia coal is Early Pennsylvanian to Permian mostly low-volatile bituminous coal (which happens to be the state rock). Low-volatile coal is considered metallurgical grade and is used to make coke for the steel industry. Higher volatile coals are used to generate steam for electricity production.

Petroleum

Oil seeps were first discovered and used commercially in western New York (1821) and Ohio and West Virginia (mid-1830s). Large-scale production in the United States began in 1859 with Edwin Drake's discovery of oil near Titusville in western Pennsylvania. Between 1860 and 1989, the Appalachian Basin produced more than 397 million m^3 (2.5 billion barrels) of oil and more than 850 billion m^3 (30 trillion ft^3) of gas from more than 500,000 wells (de Witt and Milici, 1989). Although both oil and gas continue to be produced in the Appalachian Basin, most new wells in the region are drilled in organic-rich shales to produce natural gas (Ruppert and Ryder, 2014).

The Appalachian Basin contains extensive marine shale layers rich in organic matter that blanketed large swaths of the basin during the Ordovician and Devonian Periods. These shales are the source of most of the hydrocarbons produced in the basin and are currently being targeted for natural gas production (Coleman et al., 2014) In addition to rich shale source rocks, Mississippian and Pennsylvanian coal may have been a source for gas. The thermal maturity (the amount of heating) of the rocks increases eastward across the basin, leading to more dry gas accumulations in the central and eastern basin. Oil and associated (wet) gas occur mainly in the western part of the basin (de Witt, 1993). Most natural gas and all oil come from the Allegheny Plateau region.

The major oil and gas reservoirs in the Appalachian Basin are the Upper Devonian sandstones, Lower Devonian Oriskany Formation sandstone, Lower Silurian sandstones, Ordovician shelf carbonates, and Upper Cambrian-Lower Ordovician Knox Formation carbonates. The main shale gas reservoirs include the upper Ordovician Utica Formation and Devonian Marcellus Formation. Gas is also found in Pennsylvanian coal beds. Source rocks, the organic-rich layers that the oil and gas originated in, are found in Ordovician and Devonian strata with minor source rock intervals present in Cambrian and Silurian strata in this region (Coleman et al., 2014).

Age	Orogeny	Group	Formation	Thickness (m)	Lithology
Pennsylvanian	Alleghenian	Conemaugh Gp	Glenshaw	91-111	red and light to dark gray sandstone, siltstone, shale, coal, and lenticular non-marine limestones
			Allegheny	46-61	Cyclothemic sequences of coal, sandstone, shale, limestone
					Unconformity
		Pottsville Gp	Kanawha	110	Gray conglomerate; fine to coarse-grained sandstone; minor limestone, siltstone, shale, and coal
			New River		
			Pocahontas		
Miss.			Mauch Chunk	330	Sandstone, siltstone, shale, coal; plant fossils
	Calm		Greenbrier	0-120	Limestone
	Acadian		Pocono	90-520	Medium to light gray sandstone, conglomerate, thin black carbonaceous shales
Devonian			Hampshire	600	Red sandstone, siltstone, shale
		Greenland Gap	Foreknobs	600	Red and green sandstone, siltstone
			Scheer		
			Brallier	457-518	Shale and thin siltstones
		Millboro	Harrel	360-420	Dark gray fine sandstone and siltstone
			Mahantango		
			Marcellus	100-150	Black shale
			Needmore	30-160	Olive gray fine sandstone, siltstone, shale; fossils abundant in places
					Unconformity
	Calm		Oriskany	3-40	Resistant white, gray, or tan sandstone; abundant fossils
		Helderberg	Licking Creek	15-27	Mainly limestone, often chert-bearing, interbedded with shale and sandstone; abundant fossils
			Mandata	0-33	
			New Scotland	20-45	
			New Creek	5-15	
			Keyser	20-180	
Silurian			Tonoloway	15-75	Medium-dark gray limestone
		Bloomsburg	Wills Creek	0-120	Red, fine-grained sandstone, siltstone, and shale
			Williamsport		
			McKenzie	0-23	Buff to gray calcareous shale with fossils
			Rochester	12-15	Shale
	Taconic	Clinton	Keefer	21	Sandstone
			Rose Hill	200	Red fine to coarse sandstone and shale
			Tuscarora	15-76	Resistant sandstone
Ordovician			Juniata/Oswego	0-114	Red crossbedded sandstone/gray-white coarse crossbedded sandstone; red shale
			Martinsburg	915	Olive-green shale; brown sandstone; fossiliferous
			Edinburg	130-182	Massive black limestone and interbedded thin shales
	Divergent Continental Margin		Lincolnshire	8-52	Light gray to black limestone
			New Market	12-76	Medium to dark gray, shaly and partly dolomitic limestone; abundant fossils
					Unconformity
			Beekmantown	760	Thick-bedded dolomite; black chert
			Stonehenge	152	Blue-gray thick-bedded limestone
Cambrian			Conococheague	760	Limestone, dolomite, sandstone
			Elbrook	610	Blue-gray limestone, dolomite
			Rome	610	Red-green shale, dolomite, limestone
			Shady	488	Mainly dolomite; limestone at top and base
		Chilhowee	Antietam	150-457	Buff crossbedded sandstone
			Harpers	610	Thin-bedded gray phyllite and quartzite
			Weverton	244	Crossbedded coarse sandstone, conglomeratic sandstone, green-gray phyllite
Precambrian					Unconformity
			Catoctin	0-610	Greenstone (altered basalt flows)
			Swift Run	0-61	Dark green-brown to light gray conglomeratic quartzite and phyllite
					Unconformity
			Pedlar		Granodiorite, granite gneiss

Stratigraphy of the central Appalachian Mountains, northern Virginia, and West Virginia.

The first oil discovery in Virginia was at the Early Grove Field near Bristol in 1931. Since then, over 9,580 wells have been drilled through 2013, with 124 wells drilled in 2013 alone. Natural gas is currently produced from 7,940 wells, and 57 wells produce oil (Milici and Upchurch, 2014).

The first oil in West Virginia was produced just before the Civil War, and peak production of 2.5 million m^3 (16 million barrels) per year was reached in 1900. After many years of decline, production has recently been increasing rapidly, and in 2016, West Virginia wells produced about 1.27 million m^3 (8 million barrels) of oil. Recent drilling in the Marcellus Shale discovered oil and natural gas liquids. Most of the increased production has come from the northern part of the state. West Virginia is the ninth largest natural gas-producing state in the nation. Natural gas from shale wells exceeded 28.3 billion m^3 (1 trillion ft^3)/year by 2014. In 2015, the state's shale gas reserves exceeded 538 billion m^3 (19 trillion ft^3). Natural gas resources also occur in West Virginia's conventional gas fields and in many coal fields as coalbed methane. Coalbed methane production and proved reserves are small (US Energy Information Administration, 2017).

We begin our transect at the eastern margin of the Blue Ridge and proceed west from older, highly deformed and metamorphosed strata to younger, less deformed formations of the Allegheny Plateau. This is equivalent to crossing from the heart of the deformed belt to the ancient continental craton.

Begin—Stop 1, Shenandoah Street Parking Area *(39.321896, −77.742828) in Harpers Ferry, West Virginia.*

STOP 1 HARPERS FERRY

Harpers Ferry is a town in West Virginia at the confluence of the Potomac and Shenandoah rivers. In 1733, Peter Stephens started a ferry across the Potomac here. In 1747, Robert Harper paid Stephens for his squatting rights, since the land actually belonged to Lord Fairfax, and began the first regular ferry service. The ferry ended in 1824, when a wooden bridge was built across the river.

In 1833, the Chesapeake and Ohio Canal reached Harpers Ferry, running alongside the Potomac to avoid its rapids. A year later, the Baltimore and Ohio (B&O) Railroad began service through the town.

Harpers Ferry is best known for John Brown's raid. In 1796, the federal government bought 125 ac (0.5 km^2) from the Harper family. An armory and arsenal was built, and the town became an industrial center. Between 1801 and 1861, the armory produced over 600,000 muskets, rifles, and pistols. [John Hall invented interchangeable firearms parts at the armory during the 1820s and 1830s.]

The abolitionist John Brown wanted to use the Armory as the base for a slave revolt. On October 16, 1859, Brown led 21 men in a raid on the arsenal. He attacked and captured several buildings hoping to arm the slaves and start a revolt across the south. Ironically, the only man killed was Heyward Shepherd, a free black man who was a night baggage porter for the B&O Railroad. The shot alerted a neighbor who rallied the local militia. Brown's men were quickly surrounded and took refuge next to the armory. Lieutenant Colonel Robert E. Lee took a force of 86 marines to the town, arriving on October 18. After negotiations failed, they stormed the house and captured most of the raiders. Brown was tried for treason and hanged. The raid and trial became a cause célèbre, and roused northern abolitionists like nothing else just prior to the Civil War.

During the Civil War, Harpers Ferry changed sides eight times because of its strategic location. Union troops were stationed there when General Lee decided to invade Maryland in September 1862. To protect his supply lines, Lee ordered Stonewall Jackson to take the town. The Confederates took the heights around town and began a bombardment. After 2 days, the Federals surrendered. With 12,419 Federal troops captured by Jackson, the surrender at Harpers Ferry was the largest surrender of U.S. troops until the Battle of Bataan in World War II. Because of the delay in capturing Harpers Ferry, Lee was forced to regroup. The delay allowed the Union Army to confront Lee at Antietam, the bloodiest single day in U.S. military history. Lee retreated through the area, and by July 1864, the Union again controlled the town.

In 1944, Congress authorized the establishment of Harpers Ferry National Historical Park, to take in most of the town and be administered by the National Park Service. Today, the Appalachian Trail passes directly through town.

Geologically, Harpers Ferry sits in a valley that contains lightly metamorphosed phyllite (a mica-rich slate) and quartzite of the Lower Cambrian Harpers Formation. This unit is exposed on the overturned west limb of the Blue Ridge Anticlinorium, meaning the rocks are upside down, inclined to the east, and younger layers are down section and to the west. The rocks have a pervasive slaty cleavage, or alignment of platy minerals such as mica, that is parallel to the northeast-trending axial plane of the anticlinorium.

Overturned, east-dipping Lower Cambrian Harpers Formation quartzite near Shenandoah Street public parking.

Lower Cambrian Harpers Formation crenulated phyllite near Shenandoah Street public parking. Crenulation is a metamorphic fabric caused by two or more stress directions.

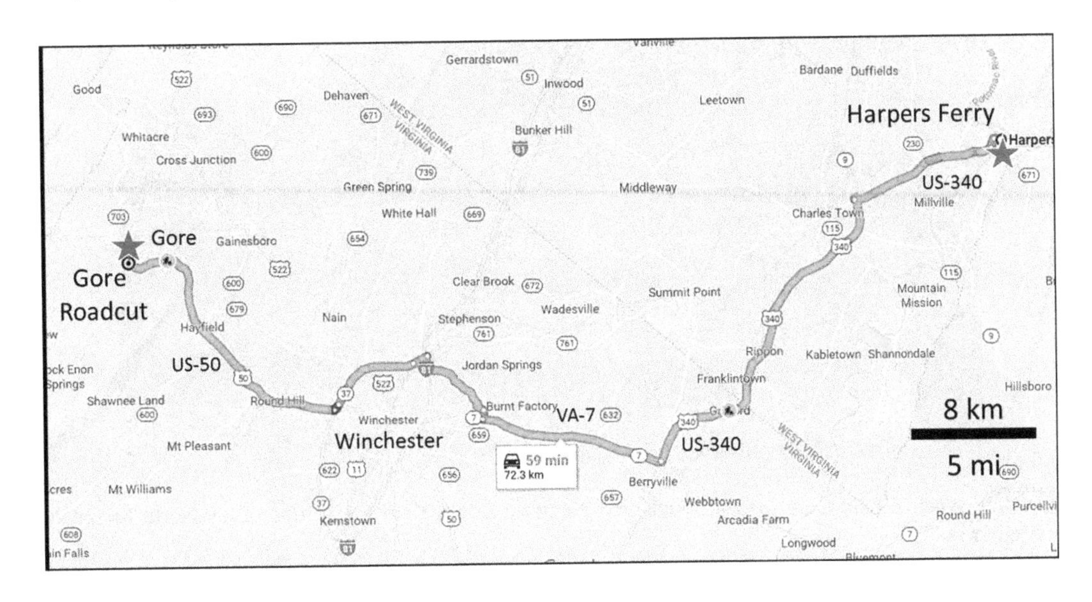

Harpers Ferry to Gore Roadcut: *From Washington Street in Harpers Ferry drive west to Union Street; turn left (south) on Union St and drive to US-340 S/William Wilson Freeway; turn right (west) on US-340 and drive to the intersection with WV-9; continue on US-340 S to Berryville; turn right (west) onto VA-7 W; continue on VA-7 to VA-660 outside of Winchester; turn right (northwest) onto VA-660 and drive to US-11 S; turn left (west) onto US-11; continue on US-11/VA-37 west to US-50; take US-50 west toward Romney; pull over on right at Roys Lane just west of Gore. This is* ***Stop 2, Gore Roadcut*** *(39.267713, −78.340249), for a total of 72.3 km (44.9 mi; 59 min).*

STOP 2 GORE ROADCUT

We have crossed the Great Valley and are near the eastern margin of the Valley and Ridge Province. Folded Devonian Brallier Shale and Greenland Gap Group occur as disharmonic folds with weak axial plane cleavage (mineral alignment, or cleavage, that is parallel to the fold axis). This stop is on the northwest limb of the North Mountain Anticline. The overall fold is more-or-less symmetrical; subsidiary folds on this limb verge (are inclined) to the southeast.

About 1,000 m (3,000 ft) of Middle to Upper Devonian rocks are exposed in the Gore roadcut. The subsidiary folds seen here are on the northwest flank of the North Mountain Anticlinorium and verge to the southeast, as they should if they developed as part of the larger structure. Different deformation styles result from changing rock types, especially at the northwest end of the roadcut.

Anticline-syncline pair, Gore roadcut. Folds verge southeast on the northwest limb of the North Mountain Anticlinorium.

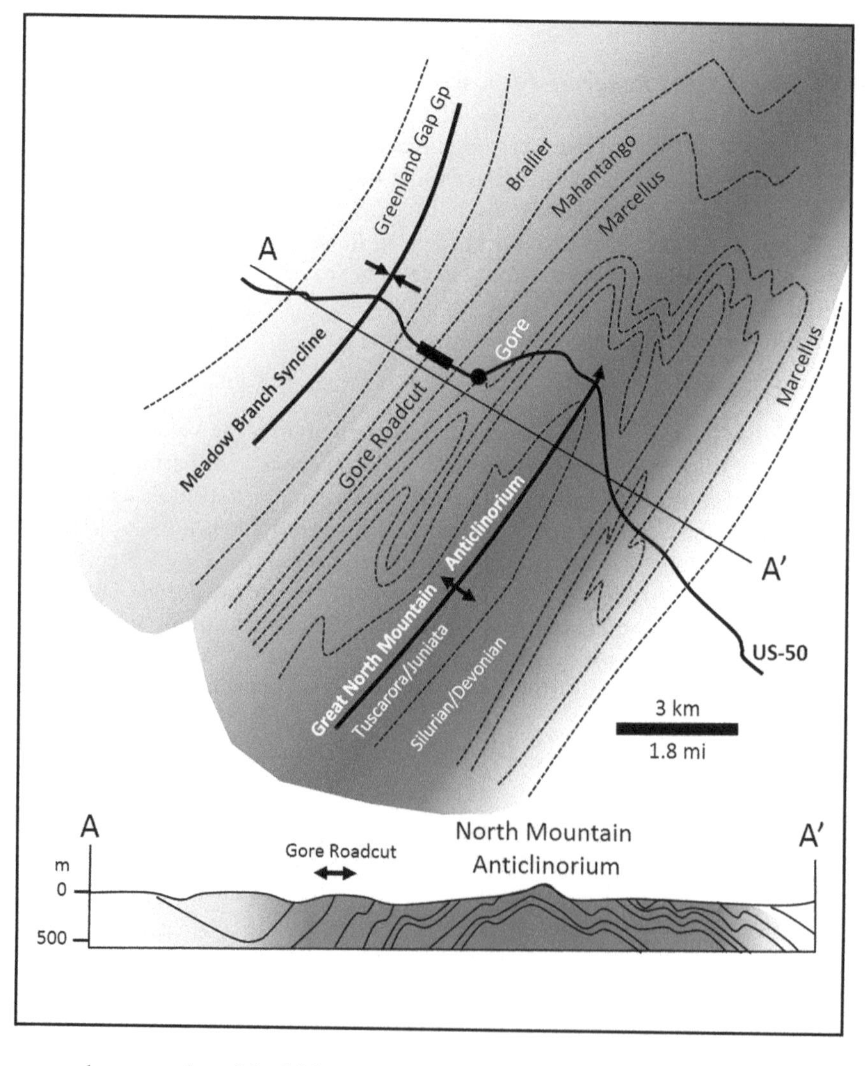

Geologic map and cross section of the folding near Gore, VA. (Modified after Rader et al., 1996.)

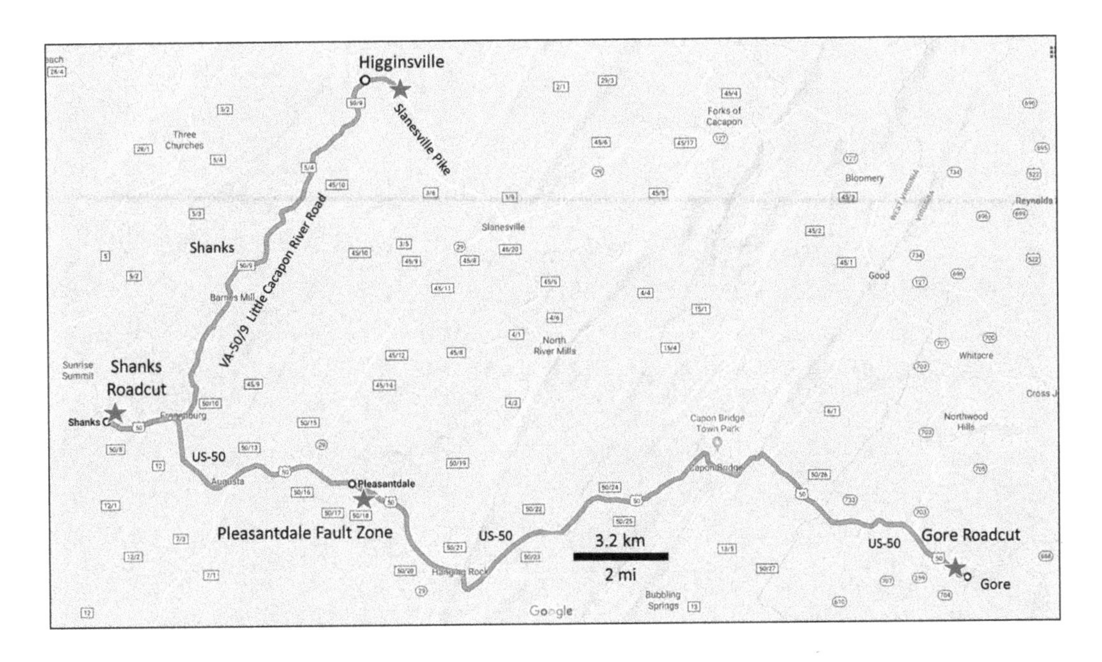

***Gore to Pleasantdale:** From Gore roadcut, continue west on US-50 to Morris Brothers Garage and Texaco Station on the left approximately 300 m (1000 ft) past Rebel Hill Drive, at the west end of the roadcut. This is **Stop 3, Pleasantdale Fault Zone** (39.286852, −78.566259) for a total of 25.5 km (15.9 mi; 21 min). Walk east along the roadcut.*

STOP 3 PLEASANTDALE FAULT ZONE

This outcrop exposes the Pleasantdale Fault Zone within the Broadtop Synclinorium. The Lower Devonian Hampshire Formation and upper Greenland Gap Formation are cut mostly by southeast-inclined thrust faults that stack a series of synclinal folds. Slickensides indicate shear surfaces (fault planes). This fault zone cuts through the overturned to near-vertical west flank of the Hanging Rock Anticline and the east limb of the Sideling Hill Syncline.

Rare, east-directed thrust in the Greenland Gap Formation, Pleasantdale Fault Zone, south side of highway.

Pleasantdale to Shanks: *Continue west on US-50 to Johnson's Auto Repair (on right) ~300m (1000ft) west of Frenchburg Estates Drive, on the eastern outskirts of Shanks. This is* ***Stop 4, Shanks Roadcut*** *(39.311944, −78.680918), for a total of 12.3km (7.7 mi; 9min).*

STOP 4 SHANKS ROADCUT

These folds are developed in the Middle-Upper Devonian Brallier Formation as part of the Broadtop Synclinorium. From a symmetrical fold in the center, they verge outward as if pushed out of the syncline due to lack of room. These are thought to be parasitic buckle folds on a small syncline along the southeast back limb of the regional Whip Cove Anticline. The axis of the syncline is near the west end of the roadcut.

Shanks roadcut, view north.

Shanks to Higginsville: *From Shanks roadcut backtrack east on US-50 to Frenchburg; turn left (northeast) on VA-50/9, Little Cacapon River Road and drive to Higginsville; turn right (southeast) on Slanesville Pike and drive 1.6 km (1 mi) east of Higginsville (39.415006, −78.565492). Pull off on the right shoulder (be careful; it is narrow): this is* ***Stop 5, Higginsville****, for a total of 19.4 km (12.1 mi; 25 min). Walk 100 m (330 ft) north along roadcut.*

STOP 5 HIGGINSVILLE

At this stop, you are looking at folded and faulted Brallier Shale. This highly disturbed zone is part of a thrust on the upper limb of the West Whip Cove Anticline. Starting at the east end of the roadcut, you can see a thrust-cut anticline, then as you walk to the west you go into a highly faulted and folded shale-siltstone sequence deformed into tight, west-verging anticlines and synclines. Axial planar cleavage is present throughout the outcrop.

Deformed Brallier Shale at Higginsville. View is northeast at the roadcut.

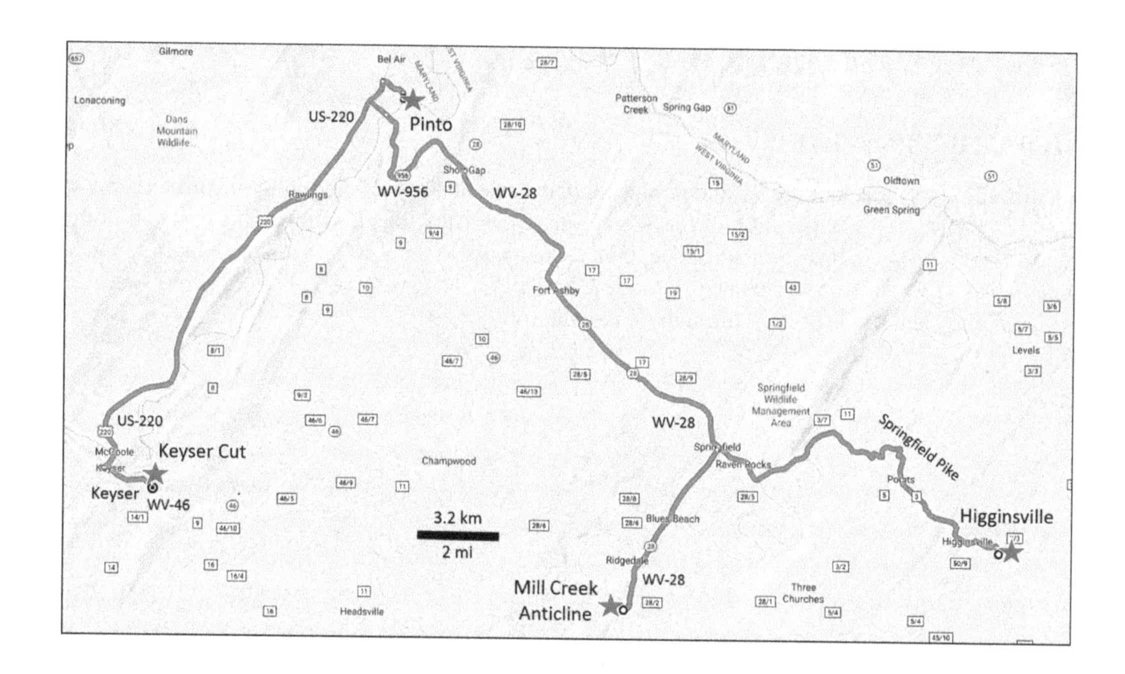

Higginsville to Mill Creek Anticline: *Drive northwest on Slanesville Pike past Higginsville; continue straight onto Springfield Pike and drive to Springfield; turn left (south) on WV-28/ Cumberland Road and drive to Rox Road/Hanging Rock Campground on the right. This is* ***Stop 6, Mill Creek Anticline*** *(39.394552, −78.736547), for a total of 23.8 km (14.8 mi; 24 min).*

STOP 6 MILL CREEK ANTICLINE AT HANGING ROCKS

At Hanging Rock Campground, near Wapocomo, you get a cross-sectional view of the Mill Creek Anticline where the South Branch of the Potomac River cuts across it. This north-northeast-trending anticline formed above a thrust ramp at depth, what is known as a fault-bend fold. The Lower Devonian Oriskany Sandstone forms the crest of the outcrop. Below the crest, the Brallier is complexly deformed. This fold extends over 50 km (30 mi) and has subsidiary folds on both limbs. A pervasive axial plane cleavage was caused by pressure solution (dissolving of minerals in areas of high stress) related to west-directed compression and shortening.

View north from Hanging Rock campground, where the South Branch of the Potomac River cuts through the Mill Creek Anticline.

The Oriskany Sandstone forms an important gas reservoir throughout the central Appalachians. In the Valley and Ridge Province, gas fields are found in anticlines such as this. For example, the fields around Augusta, 10 km (6 mi) to the southeast, produce from fractured Oriskany in a thrust-cored anticline.

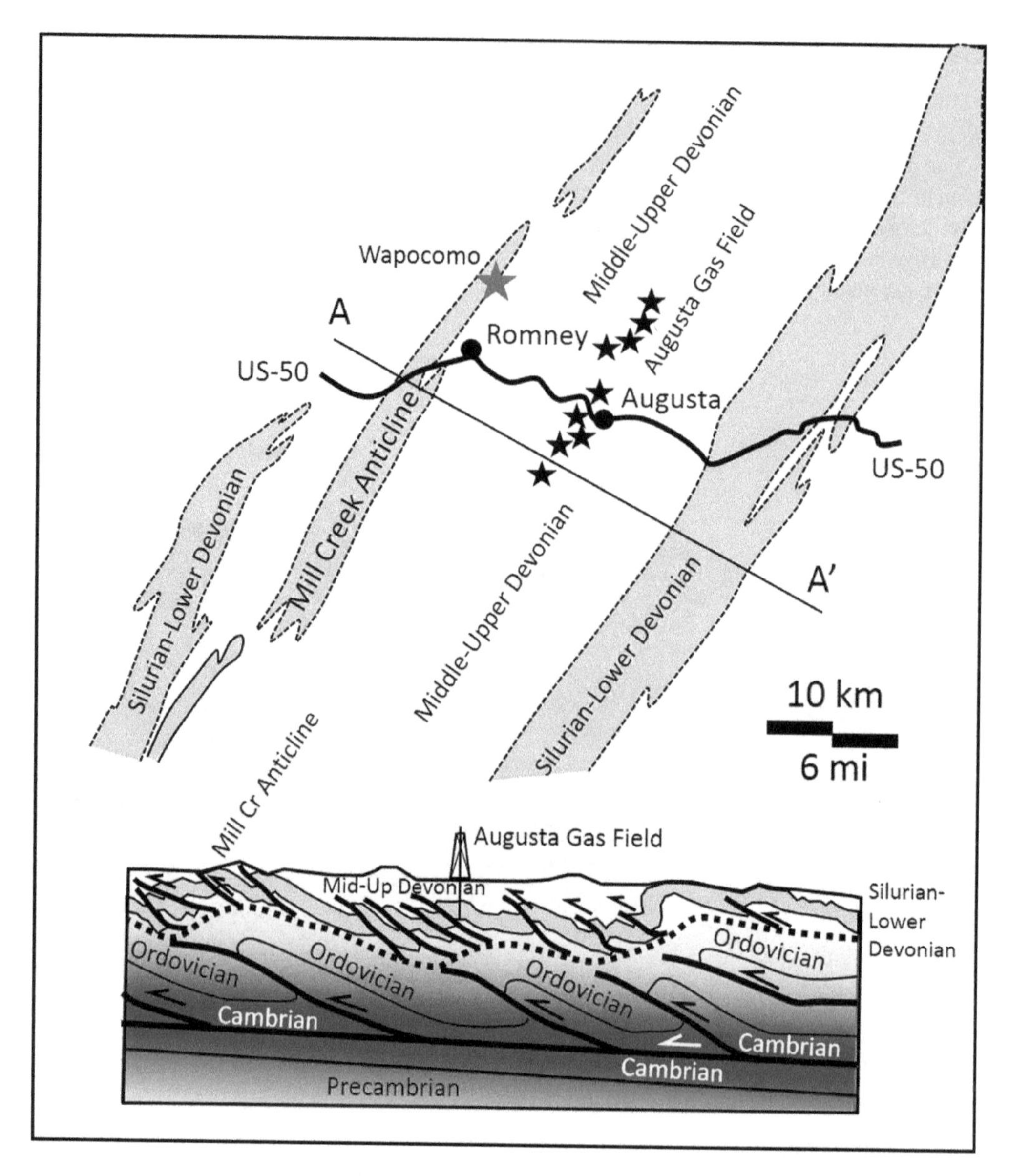

Simplified geologic map and cross section through the Mill Creek Anticline. Later thrusting re-folded earlier structures. (Modified after Pohn et al., 1985.)

Mill Creek Anticline to Pinto: *Return north on WV-28 to Springfield; turn left (northwest) and continue on WV-28 to Short Gap; turn left (northwest) onto WV-956 and drive to US-220/McMullen Hwy; turn right (northeast) onto US-220 and drive to Pinto Road on the right; turn right (east) onto Pinto Road and take it to the end by the RR trestle. Climb the left (east) side to the tracks. This is* ***Stop 7, Pinto Railroad Cut*** *(39.567924, −78.839921), for a total of 33.3 km (20.7 mi; 29 min).*

STOP 7 PINTO RAILROAD CUT

You leave the Valley and Ridge Province, and enter the Allegheny Plateau Province near Cresaptown and Pinto. A railroad cut along the North Branch of the Potomac River exposes deformation along the Allegheny Structural Front.

At the Pinto railroad cut, you can see "Z folding" exceptionally well in the Silurian McKenzie Formation. This folding indicates the beds are slightly overturned and that an anticline exists to the east. That would be the Wills Mountain Anticline, the westernmost fold of the Valley and Ridge Province. This outcrop is on the overturned west limb of the structure.

A couple of mines in the outcrop were used to extract limestone to make cement at nearby lime kilns.

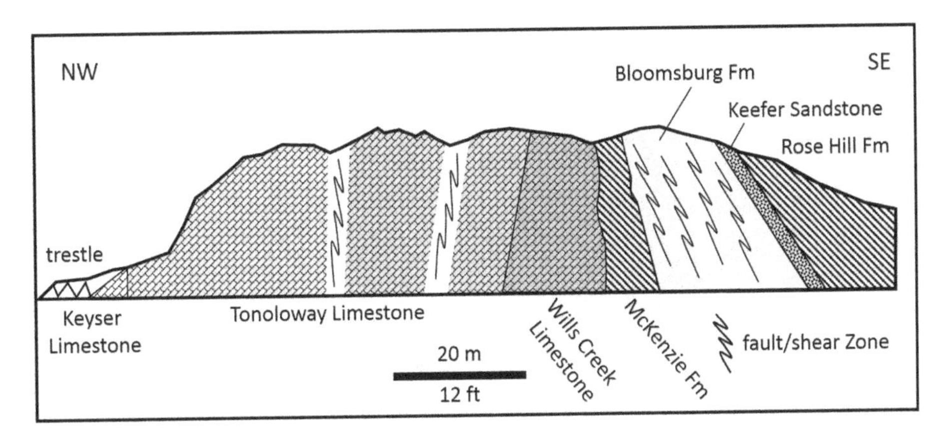

Northwest-southeast cross section through the Pinto railroad cut. (Modified after Pohn et al., 1985.)

"Z folds" in the Silurian McKenzie Formation, Pinto railroad cut.

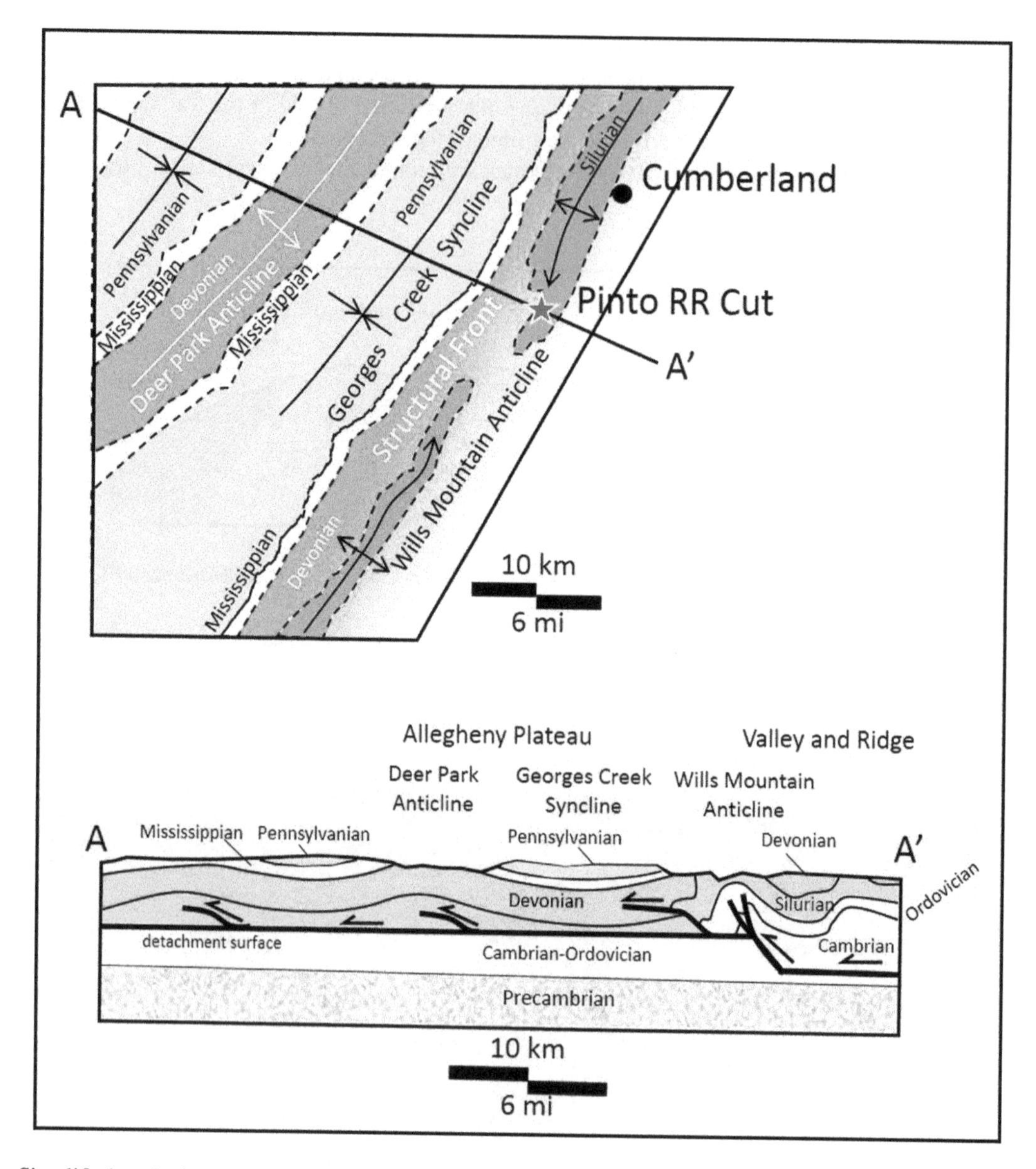

Simplified geologic map of the Pinto railroad cut area. (Modified after Pohn et al., 1985.)

Pinto to Keyser Cut: *Return to US-220 and turn left (southwest) and drive to WV-46 E/Armstrong Street; turn left (east) onto WV-46 and drive to Howell Automotive on the right, eastern outskirts of Keyser. This is* ***Stop 8, Keyser Roadcut*** *(39.434411, −78.954399), for a total of 23.3 km (14.5 mi; 20 min). There is no shoulder, and lots of traffic, so be careful!*

STOP 8 KEYSER CUT

Both sides of the roadcut are highly folded and faulted. On the eastern exposure, the folds vary from open to slightly overturned and then strongly overturned as you move west. Most folds verge to the west or northwest, the direction of movement of the main regional thrust.

Keyser roadcut, looking east. Folds verge to the northwest. McKenzie Formation and Rochester Shale.

Keyser roadcut looking west. Folding and thrusts are exposed. Units on this side of the road include the Rochester Shale, Keefer Sandstone, and the upper Rose Hill Formation.

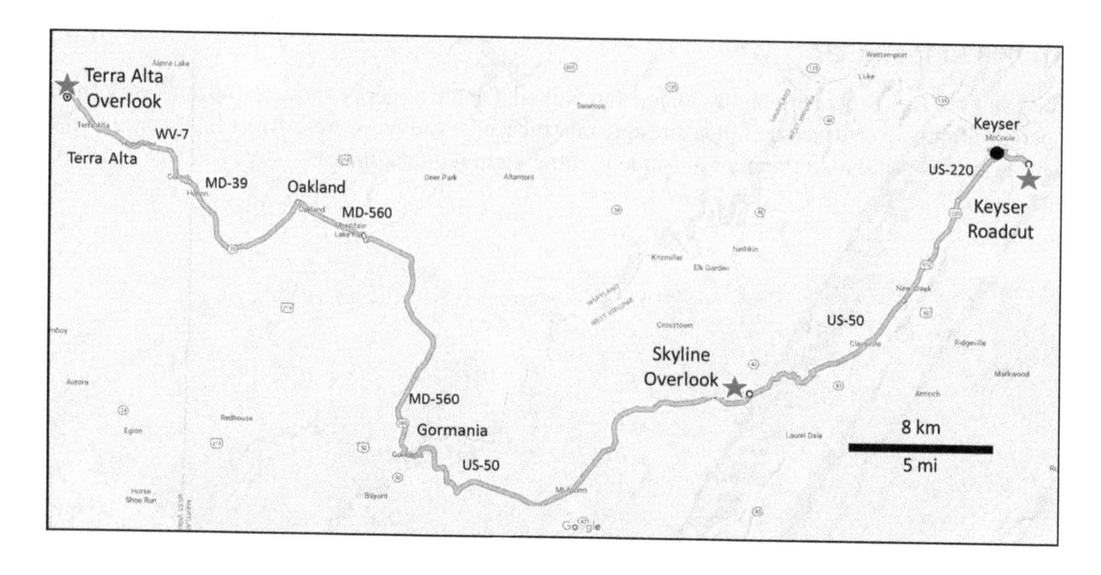

Keyser to Skyline Scenic Overlook: *Drive west on WV-46 to Keyser; turn left (south) on Water Street and drive to US-220/5; continue south on US-220 to WV-972 S (New Creek Hwy); continue south on US-50 (Northwestern Pike) to WV-42 N at Skyline; turn right (north) onto WV-42 and pull over in parking lot on the right just past the intersection. This is* ***Stop 9, Skyline Scenic Overlook*** *(39.323818, −79.129833), for a total of 23.2 km (15.7 mi; 23 min).*

STOP 9 SKYLINE SCENIC OVERLOOK, ALLEGHENY STRUCTURAL FRONT

You are now at the Allegheny structural Front, the boundary between the Allegheny Plateau Province and the Valley and Ridge Province. This boundary defines not only a change of topography, but it is also the farthest west surface expression of thrust faulting. West of here the folds are gentle (low amplitude, broad wavelength) rather than tight, and few thrust faults reach the surface.

Skyline Scenic Overlook, looking east toward Wills Mountain Anticline across the Allegheny Structural Front.

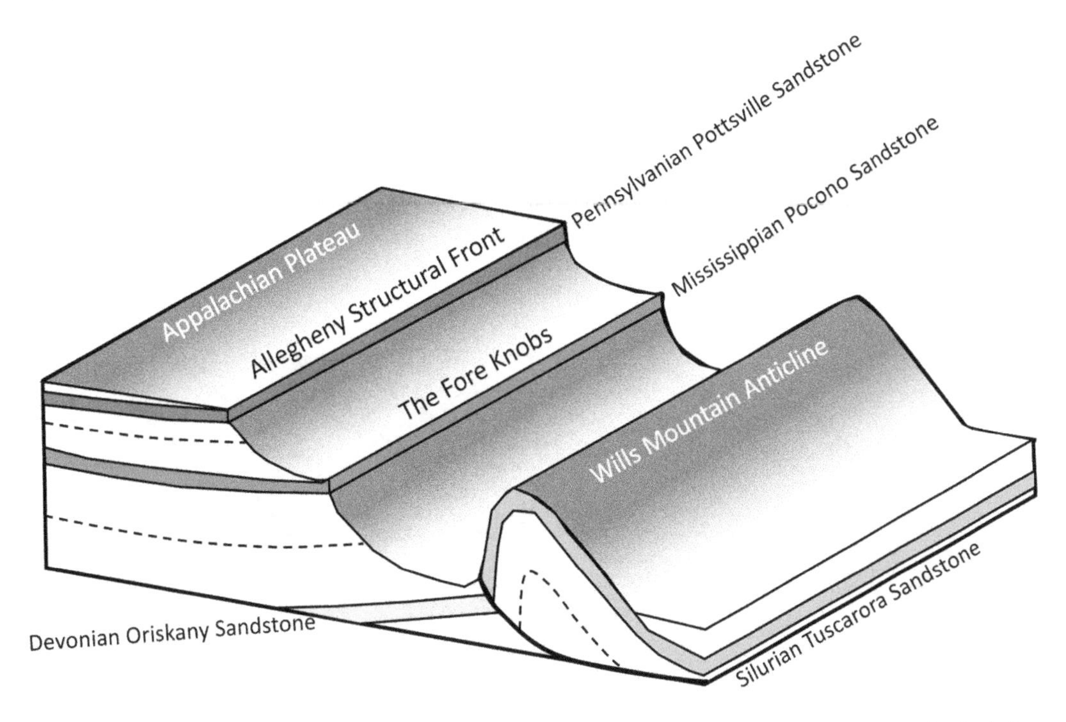

Schematic diagram of the Allegheny Structural Front and leading-edge thrusted Wills Mountain Anticline. Approximate location is Skyline Overlook. (Modified after Renton (online).)

From this stop, you have a view east toward the Appalachian Mountains. The first ridge to the east is the Wills Mountain Anticline, developed in the Lower Silurian Tuscarora Sandstone (resistant unit) and older rocks. It is the leading edge structure of the Valley and Ridge Province. Notches in the ridge are "wind gaps." These are similar to "water gaps," but the stream that had originally cut across the ridge could not erode fast enough and was captured and diverted by streams flowing down the valleys.

On the Allegheny Plateau to the west, the Middle Pennsylvanian Pottsville Sandstone forms the major ridges. The structures are gentler, and thrusts (believed to be in the subsurface) are not exposed.

Skyline Overlook to Terra Alta: *Return to US-50 W and turn right (southwest); take US-50 W to Gormania; turn right (north) on MD-560N to Oakland; turn left (west) onto MD-135 W; continue straight on MD-39 W; continue straight on WV-7 W to pullout on right alongside a hay field about 940m (3,000ft) west of White Church Road, or 2.8km (1.7 mi) west of Terra Alta. This is* ***Stop 10, Terra Alta View*** *(39.458715, –79.563253), for a total of 62.4km (38.8 mi; 53min).*

STOP 10 TERRA ALTA OVERLOOK OF THE ALLEGHENY PLATEAU

The topography becomes more rolling as you leave Skyline and approach Mt. Storm. You are entering the Allegheny Plateau. East of Gormania, you drive through the Blackwater Anticline. Between Gormania and Oakland, you are in the North Potomac Syncline. The large ridge north and west of Gormania is Backbone Mountain, east-dipping Lower to Middle Pennsylvanian Pottsville Sandstone that marks the east side of the Deer Park Anticline. The valley south of Oakland consists of shale and sandstone of the Upper Devonian Foreknobs Formation (formerly the Chemung Formation) in the core of the anticline.

Between Oakland and Hopemont, we cross the Mt. Carmel Syncline. The valley here contains easily eroded Mississippian Greenbrier Limestone. Crossing Pocono Ridge at Hopemont, we ascend the Briery Mountain Anticline. Whereas the core is breached and contains Foreknobs Formation, the flanking ridges consist of erosion-resistant Pocono and Pottsville formations.

The Briery Mountain Anticline lies beneath Terra Alta and contains the Terra Alta gas field. This once important field is now depleted and serves as a gas storage site; the local utility injects gas in the summer and withdraws it when needed in the winter.

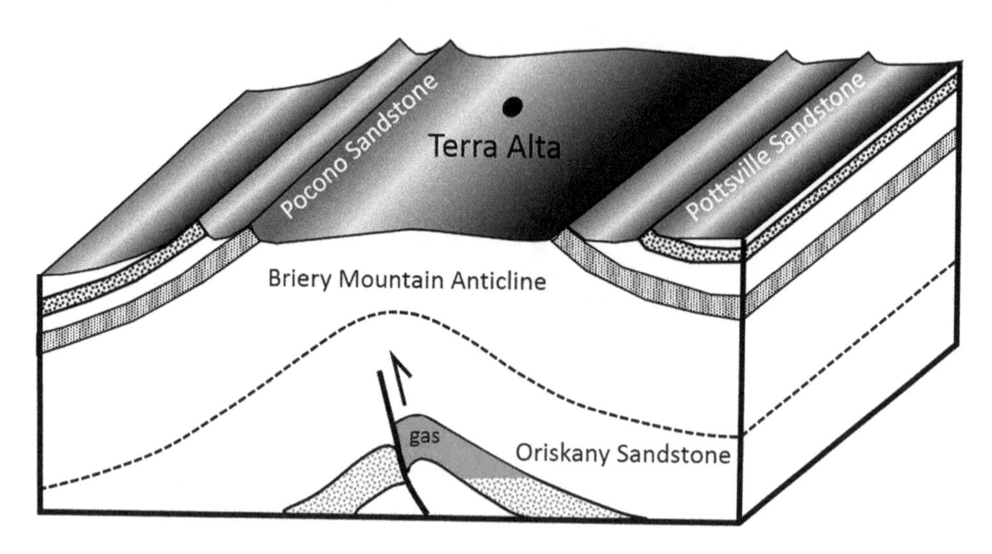

Briery Mountain Anticline at Terra Alta. (Modified after Renton, online.)

We end this transect in the gentle folds of the Allegheny Plateau. From this stop, we look west across the plateau to Preston Anticline between us and the skyline, and Chestnut Ridge Anticline on the far western horizon. These gentle folds are developed in Pennsylvanian age strata, perhaps above buried thrusts. The plateau covers much of West Virginia. It is bordered on the west by undeformed to gently deformed Devonian to Permian sediments of the Appalachian Basin. The Allegheny Plateau was glaciated in its northern and western parts: the high plateau seen here is unglaciated. Glacial till plains of the central continent extend west from the plateau.

The next geo-tour will traverse the Appalachian provinces to the south, this time from west to east, providing yet more access to structures and strata of these amazing mountains.

View west over the gently rolling Allegheny Plateau as seen from the first ridge (Pocono Sandstone) west of Terra Alta.

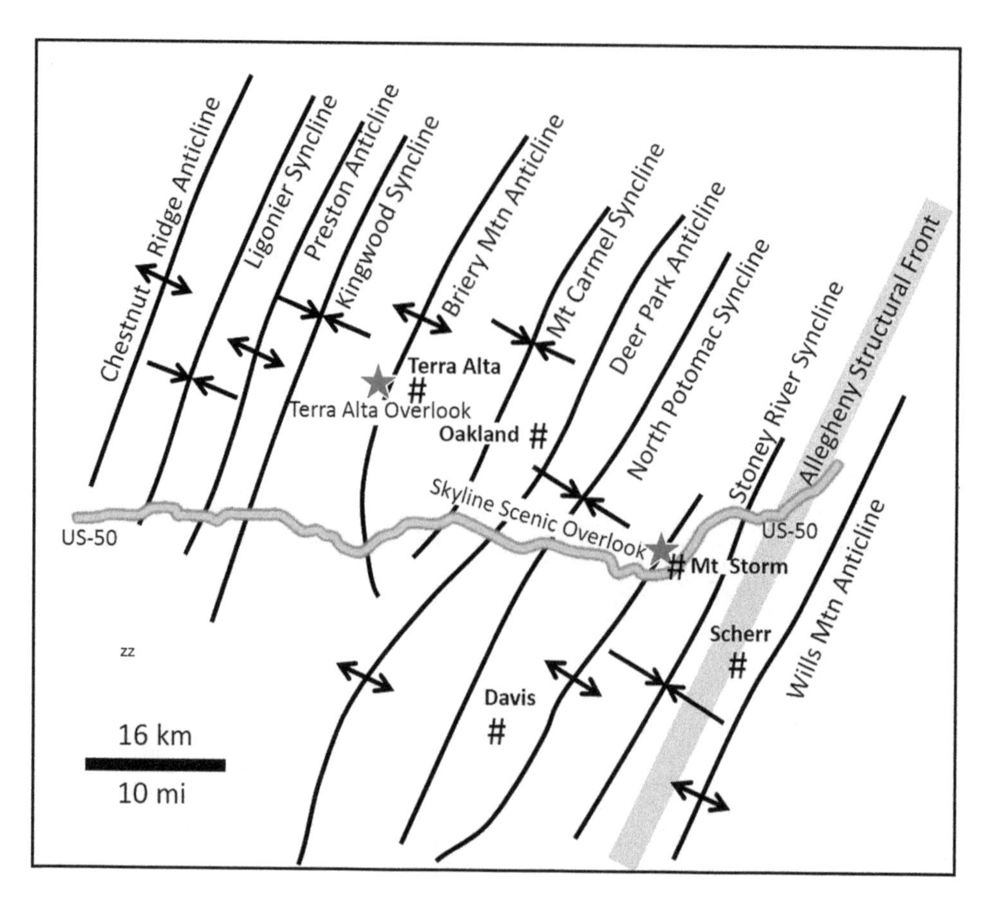

Geologic sketch map showing folds between Terra Alta and Scherr. Stops (including some from the following transect) are shown as red stars.

ACKNOWLEDGMENTS

This transect could not have been written without my co-leader Ben Prost. The assistance of Jim Coleman and a guidebook by Howard Pohn are gratefully acknowledged.

Howard A. Pohn, a U.S. Geological Survey geologist who worked to unravel the geology of the central Appalachian Mountains for over 20 years, passed away in December of 2015. He funded much of his own field expenses, because for many years during that time the USGS didn't have sufficient funding to cover all scheduled field work. He wrote and co-authored several papers and field guides on the Appalachians. His final publication, *Lateral Ramps in the Folded Appalachians and in Overthrust Belts Worldwide* (U.S. Geological Survey Bulletin 2163), came out in 2000 (https://pubs.usgs.gov/bul/b2163/). After retiring in 1996, he continued to go into the field to look at outcrops and sort out the geology. He is missed.

Howard Pohn at the Pinto railroad cut, October 1993.

REFERENCES

Brown, P.M., D.L. Brown, T.E. Shufflebarger Jr., and J.L. Sampair. 1977. *Wrench-Style Deformation in Rocks of Cretaceous and Paleocene Age, North Carolina Coastal Plain.* North Carolina Geological Survey Special Publication 5, Raleigh, NC, 47 p.

Coleman, J.L. Jr., R.T. Ryder, R.C. Milici, and S. Brown. 2014. Overview of the potential and identified petroleum source rocks of the Appalachian Basin, Eastern United States. U.S. Geological Survey Professional Paper 1708, 33 p.

Dana, J.D. 1895. *Manual of Geology.* American Book Company, New York, 974 p.

Derby, J.R., R.D. Fritz, S.A. Longacre, W.A. Morgan, and C.A. Sternbach (eds.) 2012. *The Great American Carbonate Bank: The Geology and Economic Resources of the Cambrian-Ordovician Sauk Megasequence of Laurentia. AAPG Memoir 98.* AAPG, Tulsa, OK, 528 p.

de Witt, W. Jr. 1993. Principal oil and gas plays in the Appalachian Basin (Province 131). U.S. Geological Survey Bulletin 1839-I, p. I1–37.

de Witt, W., Jr., and R.C. Milici. 1989. Energy resources of the Appalachian orogeny. *In* Hatcher, R.D. Jr., Thomas, W.A., and G.W. Viele (eds.), *The Appalachian–Ouachita orogen in the United States, the geology of North America*, Geological Society of America, Boulder, CO, v. F-2, pp. 495–510.

Evans, M. 1989. Day 3 – Great valley in northern Virginia. In Woodward, N.B., *Geometry and Deformation Fabrics in the Central and Southern Appalachian Valley and Ridge and Blue Ridge, Field Trip Guidebook T357.* American Geophysical Union, Washington, DC, 6 p.

Fichter, L.S., S.J. Whitmeyer, C.M. Bailey, and W. Burton. 2010. Stratigraphy, structure, and tectonics: An east-to-west transect of the Blue Ridge and Valley and Ridge provinces of northern Virginia and West Virginia. *Geological Society of America Field Guide*, v. 16, 23 p.

Frisch, W. 2010. *Plate Tectonics – Continental Drift and Mountain Building.* Springer, New York City, 212 p.

Hatcher, R.D. Jr., B.R. Bream, and A.J. Merschat. 2007. Tectonic map of the southern and central Appalachians: A tale of three orogens and a complete Wilson cycle. In Hatcher, R.D. Jr, M.P. Carlson, J.H. McBride, and J.R. Martinez Catalan (eds.), *The 4D Framework of Continental Crust: Geological Society of America Memoirs.* Geological Society of America, Boulder, CO, v. 200, pp. 595–632.

McLaurin, B.T., and W.B. Harris. 2001. Paleocene faulting within the Beaufort Group, Atlantic Coastal Plain, North Carolina. *Geological Society of America Bulletin*, v. 113, pp. 591–603.

Milici, R.C., and M.L. Upchurch. 2014. Oil and gas. Virginia Division of Geology and Mineral Resources, 4 p. www.dmme.virginia.gov/dgmr/pdf/oilandgas.pdf.

Mitchell, B. 2017. A look at the valley and ridge. www.thoughtco.com/a-look-at-the-valley-and-ridge-1441241. Accessed 18 October 2018.

Mixon, R.B., and W.L. Newell. 1977. Stafford fault system: Structures documenting Cretaceous and Tertiary deformation along the Fall Line in northeastern Virginia. *Geology*, v. 5, pp. 437–440.

Molnar, P. 2015. *Plate Tectonics: A Very Short Introduction.* OUP, Oxford, 160 p.

Núñez González, B. 2016. Geosyncline. https://commons.wikimedia.org/wiki/File:Geosinclinal,_2016.svg. Accessed 28 October 2018.

Oreskes, N. 2018. *Plate Tectonics – An Insider's History of the Modern Theory of the Earth.* CRC Press, Boca Raton, FL, 448 p.

Pohn, H.A., W. de Witt Jr., and A.P. Schultz. 1985. Disturbed zones, lateral ramps, and their relationships to the structural framework of the central Appalachians. Eastern section – AAPG November 1985 Meeting Field Trip Guidebook 3, 40 p.

Prost, G.L., and B.P. Prost. 2018. *The Geology Companion – Essentials for Understanding the Earth.* CRC Press, Boca Raton, FL, 486 p.

Prowell, D.C. 1988. Chapter 29: Cretaceous and Cenozoic tectonism on the Atlantic coastal margin. In Sheridan, R.E., and J.A. Grow, Jr. (eds.), *The Geology of North America*, v. 1–2, The Atlantic Continental Margin. The Geological Society of America, Boulder, CO, pp. 557–564.

Rader, E.K., R.C. McDowell, T.M. Gathright II, and R.C. Orndorff. 1996. Geologic map of Clarke, Frederick, Page, Shenandoah, and Warren Counties, Virginia: Lord Fairfax planning district. Virginia Department of Conservation and Economic Development, Division of Mineral Resources Publication 143, 1:100,000.

Redfern, R. 2001. *Origins: The Evolution of Continents, Oceans, and Life.* University of Oklahoma Press, Norman, OK, 360 p.

Renton, J.J. 2018. A geology field trip. 19 p. www.wvgs.wvnet.edu/www/geoeduc/geoeduc.htm and www.wvgs.wvnet.edu/www/geoeduc/FieldTrip/GeologyFieldTripGuide.pdf.

Ruppert, L.F., and R.T. Ryder. 2014. Executive summary, Chapter A.1 of coal and petroleum resources in the Appalachian Basin: Distribution, geologic framework, and geochemical character. US Geological Survey Professional Paper 1708, 8 p.

Sak, P.B., N. McQuarrie, B.P. Oliver, N. Lavdovsky, and M.S. Jackson. 2012. Unraveling the central Appalachian fold-thrust belt, Pennsylvania: The power of sequentially restored balanced cross sections for a blind fold-thrust belt. *Geosphere*, v. 8, no. 3, pp. 685–702.

Saunders, C. 2011. *What is the Theory of Plate Tectonics?* Crabtree Publishing, St. Catharines, ON, 64 p.

Sykes, L.R. 2019. *Plate Tectonics and Great Earthquakes: 50 years of Earth-shaking Events.* Columbia University Press, New York, 272 p.

Tollo, R.P., M.J. Bartholemew, J.P. Hibbard, and P.M. Karabinos. 2010. From Rodinia to Pangea: The lithotectonic record of the Appalachian region. *Geological Society of America Memoirs*, v. 206, pp. 1–19.

US Energy Information Administration. 2017. West Virginia state profile and energy estimates. www.eia.gov/state/seds/seds-data-fuel.php?sid=WV.

US Geological Survey. Origin of the Appalachian orogen. https://commons.wikimedia.org/wiki/File:Appalachian_orogeny.jpg.

Virginiaplaces.org. Coal in Virginia. www.virginiaplaces.org/geology/coal.html. Accessed 28 October 2018.

Wilson, J.T. 1972. *Continents Adrift - Readings from Scientific American.* W.H. Freeman and Company, San Francisco, CA, 172 p.

2 Central Appalachians
Blackwater Falls, West Virginia, to Shenandoah National Park, Virginia

View south along crest of the Blue Ridge, Shenandoah National Park. (Courtesy of National Park Service, www.nps.gov/shen/index.htm.)

OVERVIEW

Our traverse of the Appalachians starts at Blackwater Falls State Park near the contact between the Allegheny Plateau and the Valley and Ridge provinces in West Virginia. We traverse the Central Appalachian Mountains from west to east, from the fold-thrust belt to the hinterland of the Pennsylvanian-Permian Alleghanian Orogeny. We cross West Virginia and Virginia, along the way examining the rocks and structures that make this area unique. The traverse ends in Shenandoah National Park along the majestic Blue Ridge.

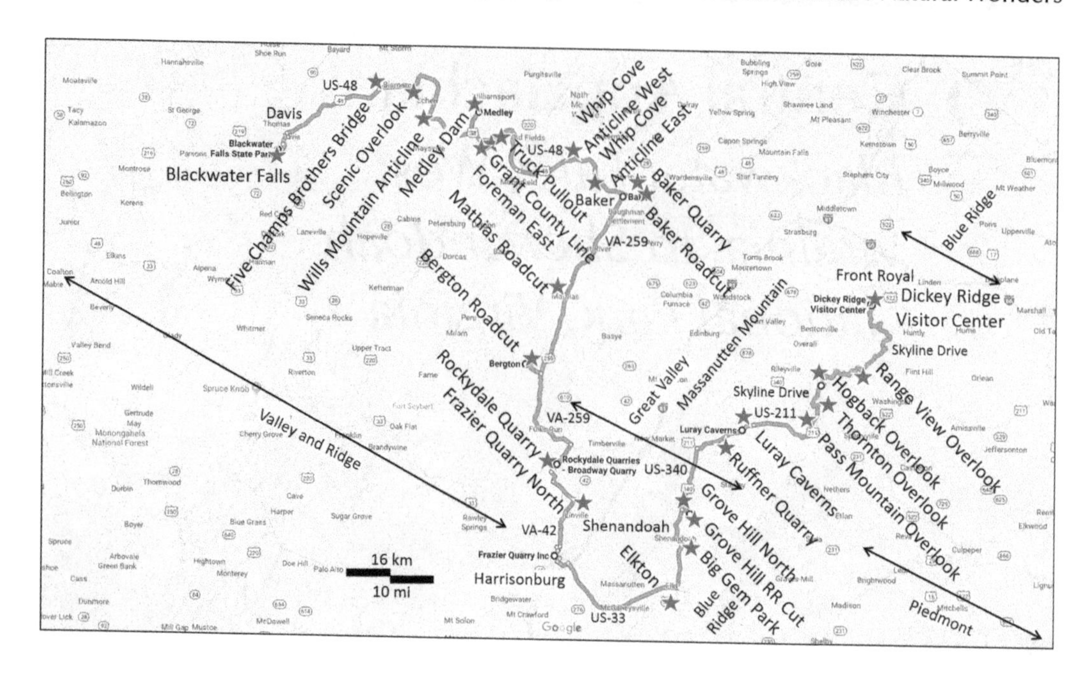

ITINERARY

Stop 1 Blackwater Falls
Stop 2 Five Champs Brothers Bridge: swamps and slump blocks (Mile 77.8)
Stop 3 Scenic Overlook, Allegheny Structural Front (Mile 81.8)
Stop 4 Wills Mountain Anticline
Stop 5 Medley Dam Cut
Stop 6 Foreman East (Mile 96)
Stop 7 Grant County Line Structure (Mile 97)
Stop 8 Truck Pullout (Mile 98)
Stop 9 Whip Cove Anticline West (Mile 112.8)
Stop 10 Whip Cove Anticline East (Mile 117)
11 Baker Area
  Stop 11.1 Baker Quarry
  Stop 11.2 Baker Roadcut (Mile 121.4)
Stop 12 Mathias Cut
Stop 13 Bergton Roadcut
Stop 14 Rockydale Quarry
Stop 15 Frazier North Quarry
Stop 16 Elkton
Stop 17 Big Gem Park
18 Grove Hill Area
  Stop 18.1 Grove Hill Landing
  Stop 18.2 Grove Hill North
Stop 19 Ruffner Quarry
Stop 20 Luray Caverns
21 Blue Ridge, Skyline Drive, Shenandoah National Park
  Stop 21.1 Pass Mountain Overlook
  Stop 21.2 Thornton Hollow Overlook
  Stop 21.3 Hogback Overlook

Stop 21.4 Range View Overlook
Stop 21.5 Dickey Ridge Visitor Center (end)

This geo-tour begins in the relatively young, gently deformed strata of the Allegheny Plateau and proceeds east through progressively older layers in the Valley and Ridge to the Great Valley, ending in extensively deformed and metamorphosed truly ancient units of the Blue Ridge. This is equivalent to crossing from the lightly deformed continental margin (tectonic foreland) to the core of the deformed Appalachian Fold-Thrust belt.

The geologic history and resources of the Appalachians are recounted in the preceding chapter. Suffice it to say that the central Appalachians are divided into geologic provinces, or geographic regions, based on their topography and geologic structural style. From west to east, there are the foreland Appalachian Basin, the mildly deformed Allegheny Plateau, the folded and thrusted Valley and Ridge, Great Valley, and Blue Ridge, and the deeply buried hinterland of the Piedmont. The Allegheny Plateau is a high plateau, partially glaciated, that has gentle folding at the surface and probably has blind or buried thrusts at depth. The youngest rocks in the region (Devonian to Permian) are exposed at the surface. The Valley and Ridge consists, as its name implies, of alternating resistant ridges and easily eroded valleys that are the expression of weathering and erosion of fairly tight thrust-cored folds. Overall, the thrusting was directed from east to west and the folds for the most part verge (are inclined) to the west. Rocks in the Valley and Ridge consist mainly of folded and faulted Cambrian to Devonian sandstone, shale, and carbonates (limestone and dolomite). The Great Valley Province contains thick Cambrian and Ordovician carbonate platform deposits similar to those seen in the Bahamas today. The "Great American Carbonate Bank" extended around Laurentia from present-day Virginia to Nevada. These rocks were folded and thrust from east to west during the Alleghanian Orogeny, with the intensity of folding and faulting decreasing westward. The Great Valley Province is characterized by karst topography, a result of rainfall and snowmelt dissolving the limestone. These limestones contain abundant cave systems. A key landmark of the Great Valley is Massanutten Mountain, an 80 km (50 mi) ridge within the Shenandoah Valley. It is folded into a U-shaped syncline and supported by resistant sandstone. The valley floor consists of easily eroded limestone and shale. East of the Great Valley lies the Blue Ridge. Metamorphosed Cambrian sedimentary rocks that overlie crystalline (igneous) basement are continuous with unmetamorphosed Cambrian rocks found farther west. Metamorphism increases eastward across this province. The Blue Ridge contains the oldest rocks in the region, a 1.2 billion-year-old granodiorite that has been pushed up and to the west by the Blue Ridge Thrust. Rocks above the granitic basement include metamorphosed sediments of the Swift Run Formation and Catoctin Formation greenstone that originated as basaltic lava flows. The Blue Ridge is a large anticlinorium, a thrusted and folded uplift that exposes the oldest and most deformed rocks in the region. Farther east, the Piedmont consists of rock pasted to the eastern margin of ancient North America during multiple mountain-building episodes. Most of the Piedmont basement rock is underlain by metamorphosed Late Precambrian and Paleozoic sedimentary and igneous rocks. These rocks were overprinted by Mesozoic pull-apart, or rift basins that contain continental, lake, and marine sediments reflecting early stages of opening of the Atlantic.

Not only are the rocks in the area old, but the mountain range itself is ancient. Compared to the relatively young Rocky Mountains (75–40 million years old), the Appalachians were formed by the collision of Laurentia (proto-America-Europe) with Gondwana (the southern continent) to form the supercontinent Pangea between 325 and 260 Ma. The collision caused the Alleghanian Orogeny, crumpling and thrusting that formed the Appalachian Mountains here, and the Ouachita-Marathon Orogeny from Oklahoma to west Texas. During their heyday, these mountains were every bit the equal of the Rockies and Alps. This mountain-building episode ended around 250 million years ago, leaving plenty of time for the mountains to erode and wear away to the gentle ridges and valleys seen during this traverse.

Begin—Blackwater Falls *Trading Post, Canyon Point Road (39.112777, −79.483965)*

STOP 1 BLACKWATER FALLS

Almost heaven, West Virginia. You are at the eastern margin of the Allegheny Plateau Province of the Central Appalachian Thrust Belt. The Blackwater River flows southwest through Davis and then enters Blackwater Canyon. The chasm is carved 148 m (485 ft) into the Pennsylvanian Pottsville Sandstone, starting at Blackwater Falls. The Pottsville Formation forms the west flank of the Blackwater Anticline, a northeast-trending fold whose axis lies 8 km (5 mi) east of Davis.

A short walk takes you to the falls, a 19 m (62 ft) cascade where the Blackwater River leaves the gentle Canaan Valley and enters steep Blackwater Canyon. The falls are at the level of the Connoquenessing Sandstone of the Pottsville Formation. These fluvial (river-derived) sandstones produce oil and gas elsewhere in West Virginia. Connoquenessing is from a Delaware Indian word meaning "for a long way straight."

Blackwater Falls and the Pottsville Formation.

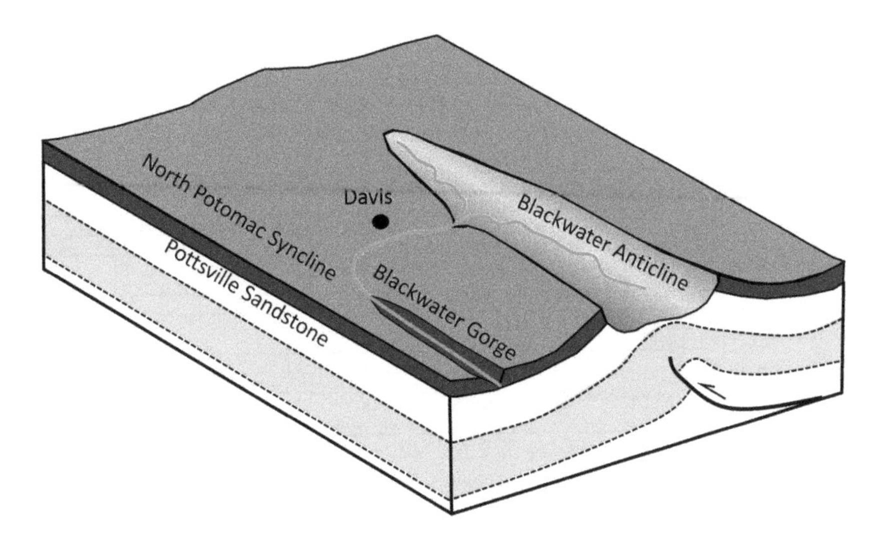

Structure underlying the Blackwater Gorge area. (Modified after Renton (online)).

During the early 1800s, this area was a hub for the tanning of hides. Tannic acid, used in tanning, is derived from leaves and bark of the local hemlock trees. It is the tannic acid that turns the Blackwater River dark.

A state park was formally established in 1937 and protects 954 ha (2,358 ac) of red spruce and eastern hemlock upland forest.

Visit

Blackwater Falls State Park has a lodge, cabins, a campground, and nature center. Hiking, biking, skiing, and fishing are popular pastimes. There are a number of trails, some of which are wheelchair accessible.

Address: 1584 Blackwater Lodge Rd, Davis, WV 26260
Hours: The park closes 10:00 pm
Phone: (304) 259-5216
Email: blackwaterfallssp@wv.gov
Web site: https://wvstateparks.com/park/blackwater-falls-state-park/

Age	Orogeny	Group	Formation	Thickness (m)	Lithology
Pennsylvanian	Alleghenian	Conemaugh Gp	Glenshaw	91-111	red and light to dark gray sandstone, siltstone, shale, coal, and lenticular non-marine limestones
			Allegheny	46-61	Cyclothemic sequences of coal, sandstone, shale, limestone
					Unconformity
		Pottsville Gp	Kanawha	110	Gray conglomerate; fine to coarse-grained sandstone; minor limestone, siltstone, shale, and coal
			New River		
			Pocahontas		
Miss.			Mauch Chunk	330	Sandstone, siltstone, shale, coal; plant fossils
	Calm		Greenbrier	0-120	Limestone
	Acadian		Pocono	90-520	Medium to light gray sandstone, conglomerate, thin black carbonaceous shales
Devonian			Hampshire	600	Red sandstone, siltstone, shale
		Greenland Gap	Foreknobs	600	Red and green sandstone, siltstone
			Scheer		
			Brallier	457-518	Shale and thin siltstones
		Millboro	Harrel	360-420	Dark gray fine sandstone and siltstone
			Mahantango		
			Marcellus	100-150	Black shale
			Needmore	30-160	Olive gray fine sandstone, siltstone, shale; fossils abundant in places
	Calm				Unconformity
			Oriskany	3-40	Resistant white, gray, or tan sandstone; abundant fossils
		Helderberg	Licking Creek	15-27	Mainly limestone, often chert-bearing, interbedded with shale and sandstone; abundant fossils
			Mandata	0-33	
			New Scotland	20-45	
			New Creek	5-15	
			Keyser	20-180	
Silurian			Tonoloway	15-75	Medium-dark gray limestone
		Bloomsburg	Wills Creek	0-120	Red, fine-grained sandstone, siltstone, and shale
			Williamsport		
			McKenzie	0-23	Buff to gray calcareous shale with fossils
			Rochester	12-15	Shale
	Taconic	Clinton	Keefer	21	Sandstone
			Rose Hill	200	Red fine to coarse sandstone and shale
			Tuscarora	15-76	Resistant sandstone
Ordovician			Juniata/Oswego	0-114	Red crossbedded sandstone/gray-white coarse crossbedded sandstone; red shale
			Martinsburg	915	Olive-green shale; brown sandstone; fossiliferous
			Edinburg	130-182	Massive black limestone and interbedded thin shales
	Divergent Continental Margin		Lincolnshire	8-52	Light gray to black limestone
			New Market	12-76	Medium to dark gray, shaly and partly dolomitic limestone; abundant fossils
					Unconformity
			Beekmantown	760	Thick-bedded dolomite; black chert
			Stonehenge	152	Blue-gray thick-bedded limestone
Cambrian			Conococheague	760	Limestone, dolomite, sandstone
			Elbrook	610	Blue-gray limestone, dolomite
			Rome	610	Red-green shale, dolomite, limestone
			Shady	488	Mainly dolomite; limestone at top and base
		Chilhowee	Antietam	150-457	Buff crossbedded sandstone
			Harpers	610	Thin-bedded gray phyllite and quartzite
			Weverton	244	Crossbedded coarse sandstone, conglomeratic sandstone, green-gray phyllite
Precambrian					Unconformity
			Catoctin	0-610	Greenstone (altered basalt flows)
			Swift Run	0-61	Dark green-brown to light gray conglomeratic quartzite and phyllite
					Unconformity
			Pedlar		Granodiorite, granite gneiss

Stratigraphy of the central Appalachian Mountains, northern Virginia, and West Virginia.

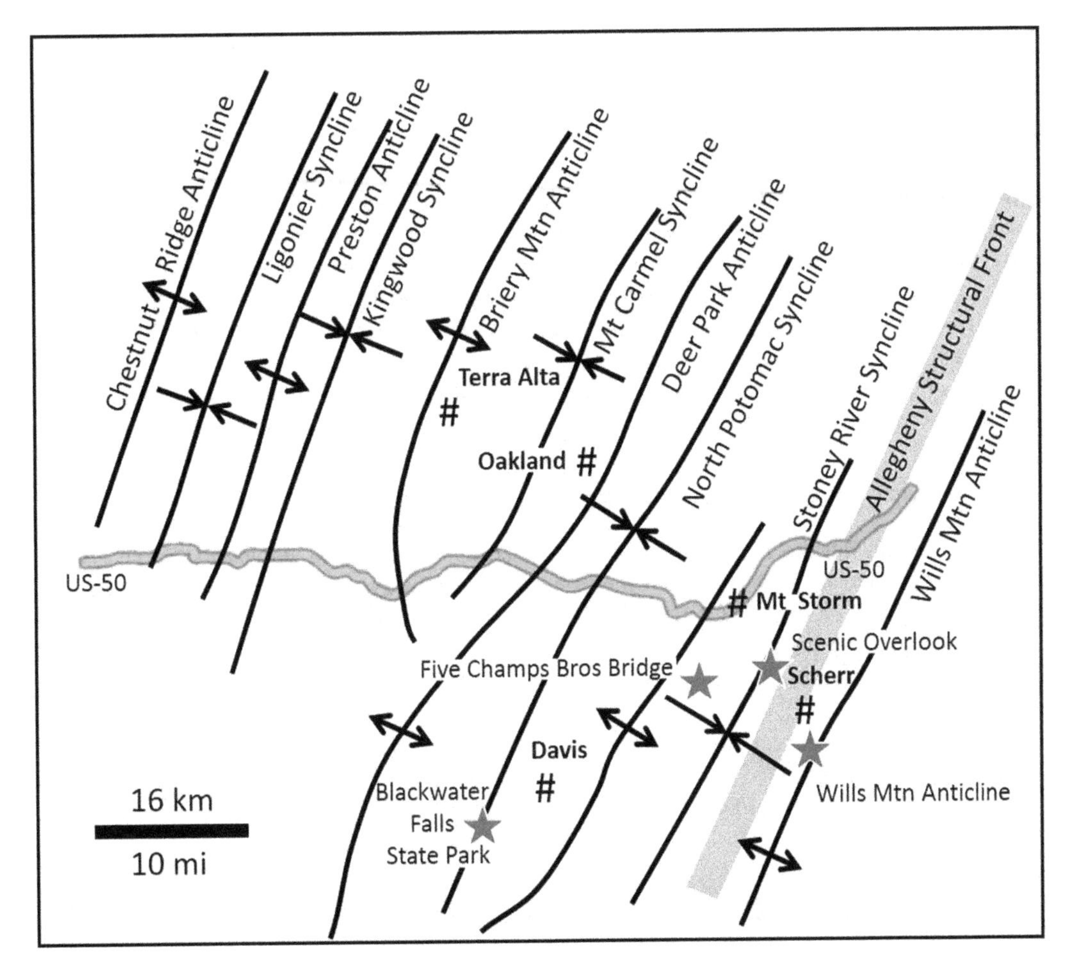

Geologic sketch map showing folds between Blackwater Falls and the Wills Mountain Anticline. Stops are shown as red stars.

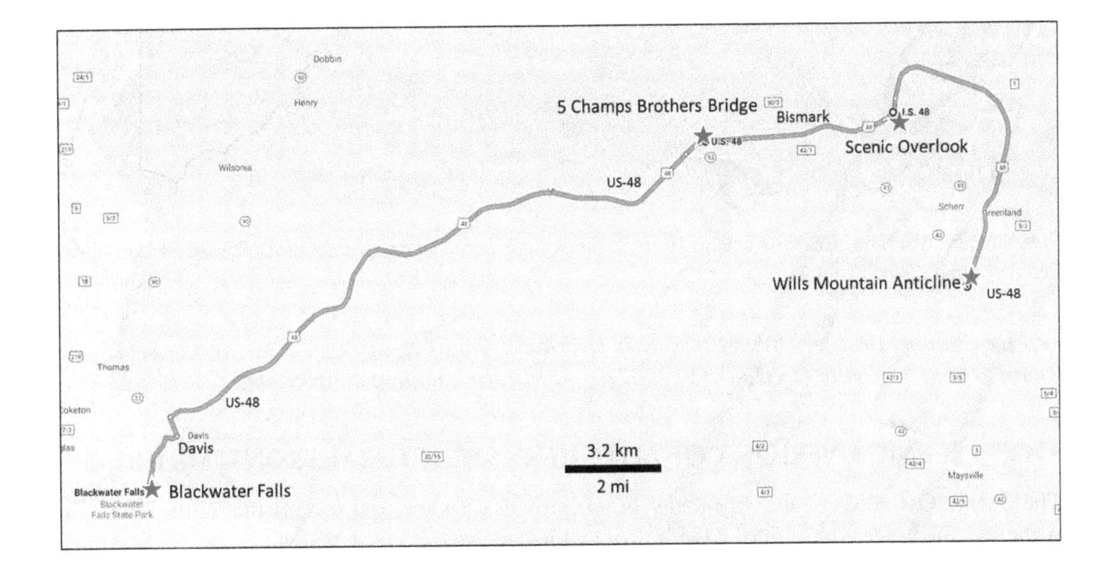

Blackwater Falls to Five Champs Brothers Bridge: *From Blackwater Falls Trading Post return to Davis; turn left (north) on WV-32; turn right (east) on WV-93/US-48 (Corridor H); drive to pullout on right, east side of Five Champs Brothers Bridge, approximately Mile 77.8. This is* ***Stop 2, Five Champs Brothers Bridge*** *(39.212801, −79.265714), for a total of 24.8 km (15.4 mi; 17 min).*

As you approach Bismark, you pass through the Devonian Hampshire Formation redbedded sandstone, Foreknobs Formation shale and sandstone, rusty shale and sandstone of the Brallier Formation, and, at Scherr, black shale of the Marcellus Formation. The Devonian Marcellus Shale is a major target of oil exploration in the region, although it requires hydraulic fracturing to produce. Oil in the low-permeability shale occurs in isolated pores and requires fracture stimulation to flow to a well.

Immediately east of Scherr, the Devonian Oriskany Sandstone outlines a small, subsidiary anticline on the west limb of the larger, regional Wills Mountain Anticline. The Oriskany was the main oil reservoir in this region when the industry was just getting started here in the mid-1800s.

STOP 2 FIVE CHAMPS BROTHERS BRIDGE ROADCUT (MILE 77.8)

This stop has an intra-formational unconformity and apparent rotated slump blocks. The Upper Pennsylvanian Glenshaw Formation of the Conemaugh Group contains coastal swamp sediments (sandstone, shale) and thin, discontinuous coal beds (Bakerstown Coal). Slumping may have been a result of channel down-cutting or small-scale growth faulting. The angular unconformity within the Glenshaw represents a time gap resulting from a period of localized erosion during deposition of the formation.

This location lies between the Blackwater Anticline to the west and Stoney River Syncline to the east. These structures formed during the Alleghany Orogeny.

The coal is uncommercial and is no longer mined.

Five Champs Brothers Bridge roadcut. Rotated blocks are seen below an unconformity, all in the Upper Pennsylvanian Conemaugh Group.

Five Champs Brothers Bridge to Scenic Overlook: *Continue east on US-48 for 6.6 km (4.1 mi; 4 min) to* ***Stop 3, Scenic Overlook*** *(39.220779, −79.192190) pullout on the right.*

STOP 3 SCENIC OVERLOOK OF ALLEGHENY STRUCTURAL FRONT (MILE 81.8)

The Scenic Overlook, at the Allegheny Front, provides a view east toward the Wills Mountain Anticline, the westernmost fold of the Valley and Ridge Province in this area.

View southeast toward Wills Mountain Anticline (middle distance) from the Scenic Overlook.

The roadcut north of US-48 exposes the Mississippian Mauch Chunk Formation. These redbedded sandstone, siltstone, and shale layers, with a limestone at the base, were deposited in delta, floodplain, and coastal plain environments at the beginning of the Alleghanian Orogeny. Most of the sediments were derived from growing mountains to the north. The Mauch Chunk appears similar to redbeds in the Devonian Hampshire Formation.

The Mauch Chunk contains gas and coal, but is uneconomic in that coals are thin and discontinuous, and gas accumulations are small.

Mauch Chunk Formation redbeds exposed in the roadcut at the Scenic Overlook.

Scenic Overlook to Wills Mountain Anticline: *Continue east on US-48 for 11.2 km (7.0 mi; 7 min) to* ***Stop 4, Wills Mountain Anticline*** *(39.170110, −79.164065) and pull over to the left at the center of the freeway.*

STOP 4 WILLS MOUNTAIN ANTICLINE

The Wills Mountain Anticline is outlined by the Lower Silurian Tuscarora Sandstone, the main ridge-forming sandstone in the Valley and Ridge Province. The fold is asymmetric, verging (inclined) to the west, as shown by the steep west flank and gentle east flank. This is one indication that the "push" that caused the folding came from the east.

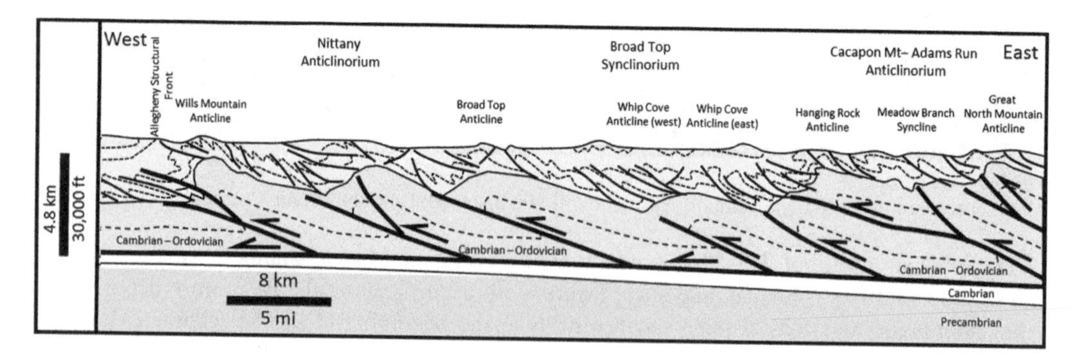

Structural cross section near US-48 from Scherr to Baker. (Modified after Kulander and Dean, 1986.)

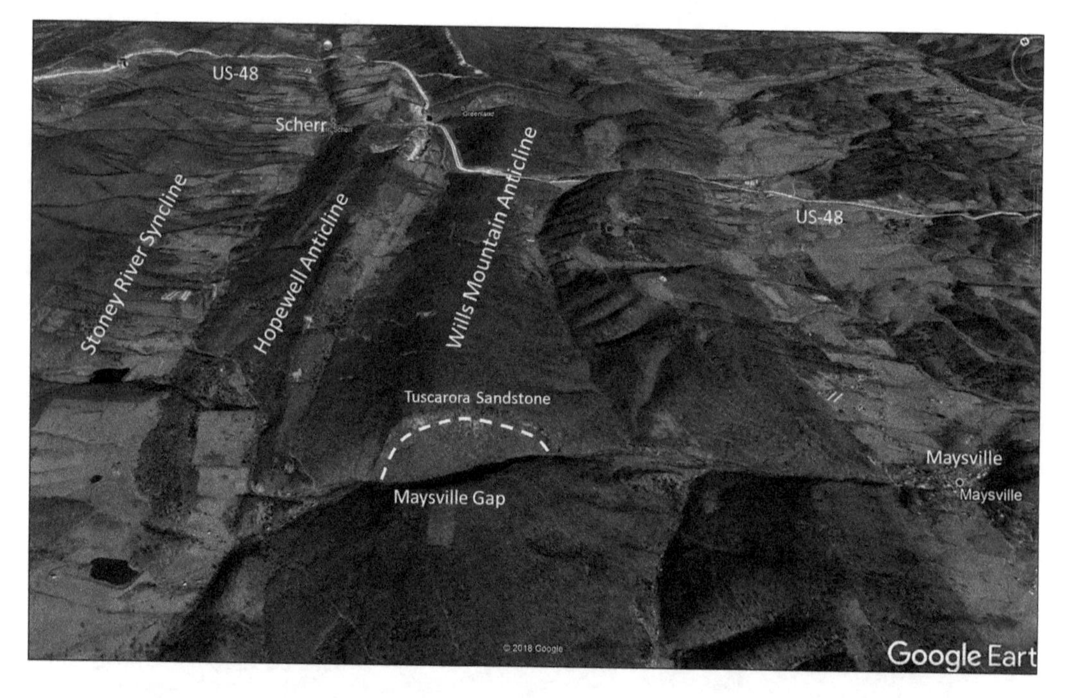

Wills Mountain Anticline, oblique view north, from Maysville Gap to the US-48 roadcut. Google Earth image, image © Landsat/Copernicus.

US-48 roadcut into west-flank Wills Creek Anticline. Tuscarora Sandstone caps the structure; red crossbedded sandstone of the Ordovician Juanita Formation lies in the core of the fold.

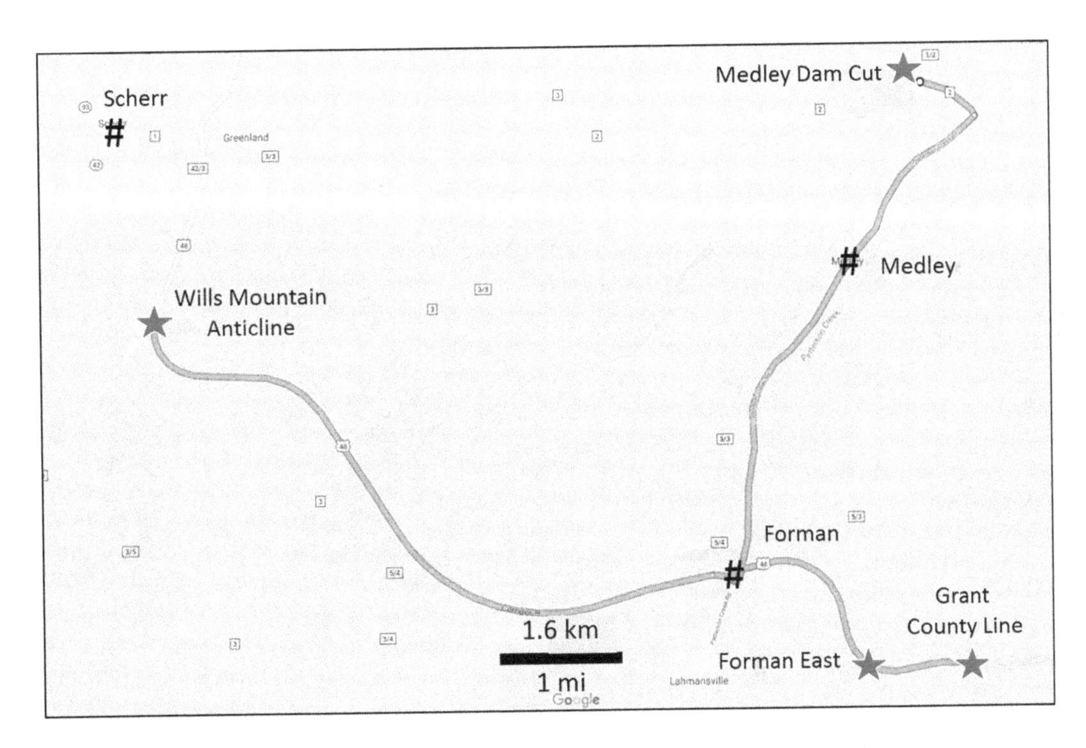

Wills Mountain Anticline to Medley Dam: *Continue east on US-48 to Patterson Creek Road exit; turn left (north) onto Patterson Creek Road and drive past Medley to Belle Babb Lane; turn left (west) and drive 940m (3,000ft) to dam cut on the right. This is* ***Stop 5, Medley Dam Cut*** *(39.196790, −79.046272), for a total of 17.5km (10.9 mi; 14min).*

STOP 5 MEDLEY DAM CUT

This somewhat weathered outcrop reveals intensely folded Brallier Formation with a number of small west-directed thrusts and west-verging folds. These structures may be small-scale analogs to the larger structures seen in the Valley and Ridge Province.

A simple anticline at the west end of the outcrop contains small parasitic folds. The central part consists of east-dipping layers with small west-directed thrusts. The east end shows small, tight, west-verging folds. The Brallier Formation shale and thin siltstone layers have been folded, jointed, faulted, and refolded by multi-stage thrusting on the east flank of the Bedford Syncline (Pohn et al., 1985; Hunt et al., 2017).

Medley dam cut in Brallier Shale. View north. The area in the box is shown in the next figure.

Detail from east end of cut, above, showing a minor west-directed thrust.

Medley Dam to Forman East: *Return to Patterson Creek Road and drive south to US-48; turn right to merge onto US-48 E and drive east to Mile 96. This is* ***Stop 6, Forman East*** *(39.129524, −79.054629); pull off on right shoulder for a total of 11.3 km (7 mi; 9 min).*

STOP 6 FORMAN EAST (MILE 96)

The Foreman East roadcut exposes folding in the Devonian Helderberg Group on the west flank of the regional Patterson Creek Mountain Anticline. A small displacement east-directed thrust cuts the core of a syncline.

Syncline in roadcut at Forman East stop. View north across US-48.

Forman East to Grant County Line: *Continue east on US-48 to Mile 97. This is* ***Stop 7, Grant County Line*** *(39.129774, −79.038875). Pull over on the right shoulder for a total of 0.9 km (0.6 mi; 1 min).*

STOP 7 GRANT COUNTY LINE (MILE 97)

This roadcut shows small-scale subsidiary folding on the east limb of the Patterson Creek Mountain Anticline. The anticline-syncline pair shows east-vergence (gentle west limb and steeper east limb), suggesting that there are minor east-directed backthrusts or bedding plane detachments nearby.

These folds are developed in marine limestones of the Devonian Helderberg Group.

Anticline-syncline pair developed in the Helderberg Group at the Grant County line stop. View north across US-48.

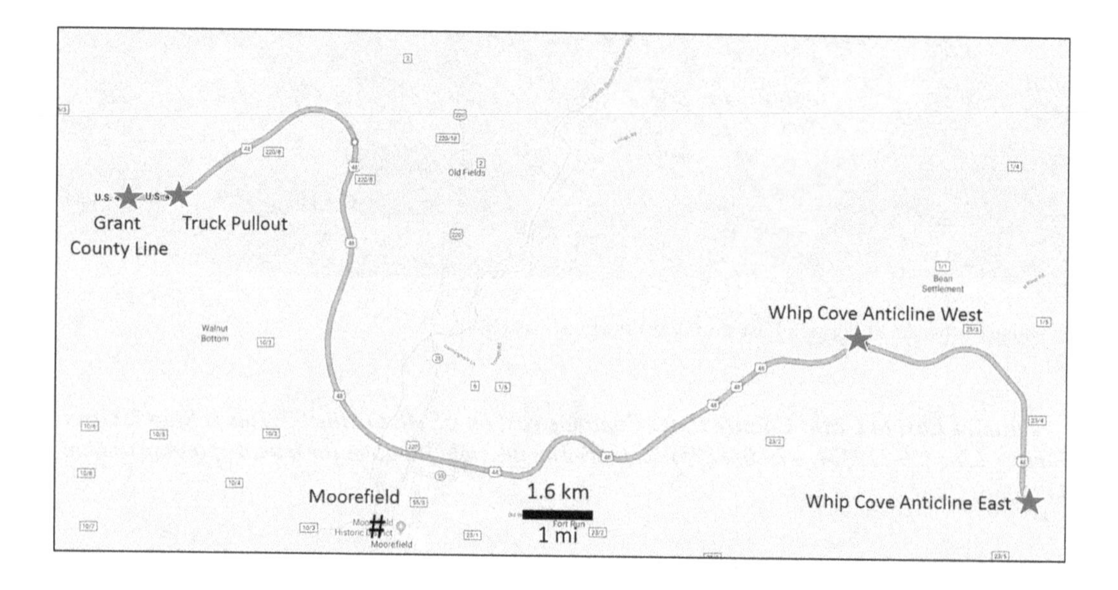

***Grant County Line to Truck Pullout:** Continue east on US-48 to Mile 98 and Truck Pullout on right. This is **Stop 8, Truck Pullout** (39.130080, −79.025873), for a total of 1.1 km (0.7 mi; 1 min).*

STOP 8 TRUCK PULLOUT (MILE 98)

The Truck Pullout stop offers a section of the Devonian Helderberg Group from just above the base to just below the contact with the overlying Oriskany Sandstone. The gently east-dipping layers of mostly marine limestones were deposited on a carbonate shelf during the lull between the Taconic and Acadian orogenies. Crinoid fragments are common in the New Creek Formation, whereas algal mounds are evident in the Keyser Formation.

The roadcut lies near the axis of a local syncline between the regional Patterson Creek Mountain Anticline and Clearville Syncline. A small syncline at the west end of the outcrop lies below a bedding-parallel detachment surface (thrust fault). Abundant calcite veins indicate both shear and extension during folding.

Low porosity restricts the occurrence of hydrocarbons in these units. Secondary porosity and fracture porosity create reservoirs for small gas accumulations, mainly in the New Creek Formation.

Truck Pullout roadcut, US-48. Google Street View looking southeast. The area inside the box is enlarged in the next figure.

Detail from the west end of the roadcut seen in the previous figure. This west-verging anticline-syncline pair is nicely exposed at the truck pullout.

Truck Pullout to Whip Cove Anticline West: *Continue east on US-48 to Mile 112.75. This is* ***Stop 9, Whip Cove Anticline West*** *(39.102600, −78.846713). Pull over on right shoulder for a total of 24.2 km (15.1 mi; 13 min).*

STOP 9 WHIP COVE ANTICLINE WEST (MILE 112.8)

A west-verging secondary fold developed on the east flank of the Whip Cove Anticline West is exposed in this roadcut. This type of fold is usually associated with small west-directed thrusts: it is unclear whether a fault actually cuts this feature.

The outcrop consists of Late Devonian Hampshire Formation redbeds. The Hampshire Formation contains shales and crossbedded sandstones deposited in river and alluvial environments west of the Middle to Late Devonian Acadian Mountains. Mudcracks and paleo-soils are common; plant fossils are rare.

Part of Whip Cove West Anticline developed in the Greenland Gap Group (formerly Chemung Formation). Fault is speculative. Google Street View north.

Whip Cove Anticline West to Whip Cove Anticline East: *Continue east on US-48 to Mile 117. This is* ***Stop 10, Whip Cove Anticline East*** *(39.071480, −78.803224). Pull over on right shoulder for a total of 6.5 km (4.1 mi; 4 min).*

STOP 10 WHIP COVE ANTICLINE EAST (MILE 117)

A west-verging asymmetric fold in the Foreknobs Formation is exposed in this roadcut. The secondary fold lies on the west flank of the Whip Cove East Anticline and may be associated with a small west-directed thrust.

The Foreknobs Formation is part of a westward-advancing clastic (sandstone, siltstone, shale) sediment package deposited during the Acadian Orogeny. The unit contains both non-marine and marine layers. Marine fossils are abundant in some sandstones and include brachiopods and crinoids. The Foreknobs Formation is equivalent to the Warren, Speechley, Balltown, Bradford, Riley, Benson, and Alexander gas-bearing sandstones of western West Virginia (Hunt et al., 2017).

West-verging subsidiary fold developed on the east flank, Whip Cove Anticline. Street View looking north.

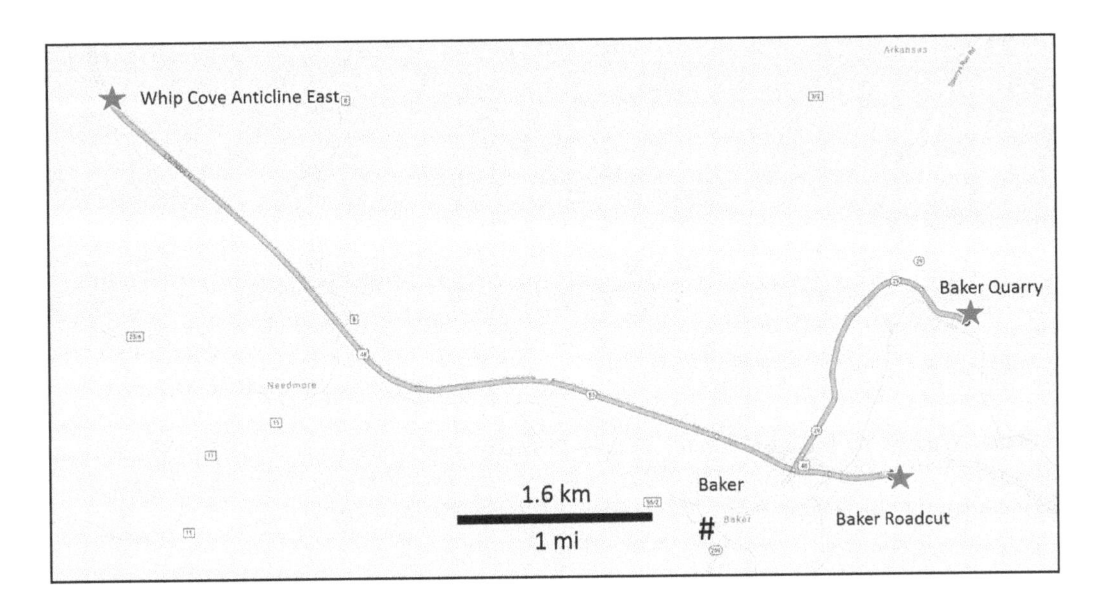

Whip Cove Anticline East to Baker Quarry: *Continue east on US-48 to the WV-259 S exit; turn left (north) on WV-29 and drive to Old WV-55; continue straight on Old WV-55 for 480 m (1,580 ft) to an abandoned quarry on the left. This is* ***Stop 11.1, Baker Quarry*** *(39.055493, −78.722900), for a total of 9 km (5.6 mi; 6 min).*

11 BAKER AREA

Baker lies in the eastern Valley and Ridge Province near the boundary with the Great Valley Province. The area from Baker south to Mathias and Bergton, where we go next, is a zone of tight anticlines and synclines and multiple small thrusts just west of the Little North Mountain Thrust. The formations are largely thin-bedded Silurian and Devonian strata.

STOP 11.1 BAKER QUARRY

The abandoned Baker Quarry outcrop provides an example of layer-parallel shortening expressed as bedding-plane thrusts. The Silurian Wills Creek Formation limestone is exposed here. A detachment in the upper part of the quarry cuts up-section. This is known as a "hanging wall ramp." Below the fault is a "footwall flat"; that is, the detachment is parallel to bedding below the fault. The detachment surface itself has been folded after thrusting, probably a result of a later thrust below this outcrop.

The west end of the outcrop contains Devonian Keyser Limestone to Oriskany Sandstone. Faulting has cut out some of the Wills Creek Formation, all of the Tonoloway Limestone, and part of the Keyser Limestone.

Caution: the vines on the outcrop are poison ivy. Don't touch them.

Hanging wall ramp on a footwall flat in the Wills Creek Formation, central part of the Baker Quarry.

Baker Quarry to Baker Roadcut: *Return to US-48 and drive east to the first roadcut, at Mile 121.4; pull off onto WV-55 on the right at the east end of the roadcut. This is* ***Stop 11.2, Baker Roadcut*** *(39.044766, −78.730081), for a total of 3.3 km (2.1 mi; 3 min).*

STOP 11.2 BAKER ROADCUT (MILE 121.4)

Generally west-dipping beds of the Silurian Wills Creek Formation have been deformed into a west-verging anticline at this roadcut. The roadcut lies between the Sideling Hill Syncline to the west and the Hanging Rock Anticline to the east. The Wills Creek Formation contains thin limestones, calcareous shales, and shaley limestones. It has slightly more shale than the overlying Tonoloway Limestone.

Deformed Wills Creek Formation in the Baker roadcut.

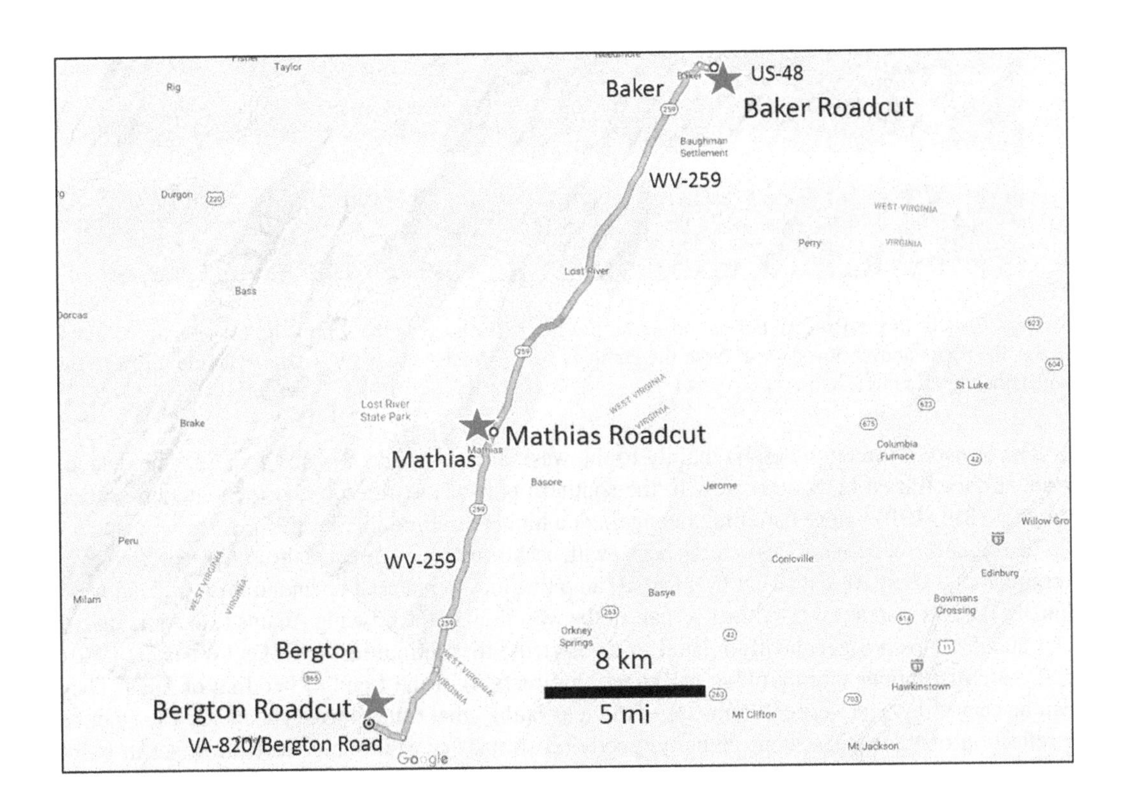

Baker Roadcut to Mathias: *Cross US-48 to the westbound lane and drive west to the WV-259 S exit at Baker; turn left (south) on WV-259 and drive to Clearview Drive (beside the sign "Mathias Unincorporated") on the right, northern outskirts of Mathias. This is* ***Stop 12, Mathias Roadcut*** *(38.886089, −78.861303), for a total of 22.4 km (13.9 mi; 17 min).*

STOP 12 MATHIAS ROADCUT

Thrust faults often take advantage of bedding planes to slip. These detachment surfaces can cut up-section in the direction of thrusting (a "thrust ramp") or perpendicular to the direction of transport (a "lateral ramp"). This outcrop shows a lateral ramp in the Brallier Formation (Pohn et al., 1985).

Mathias roadcut in the Brallier Formation looking west. A thrust near the top of the cut is exposed along strike: the rocks above moved west. Note the complex folding and a small, out-of-the-syncline thrust to the north (the lateral ramp).

The transport direction here is mainly to the west, and the roadcut is north-south. The detachment surface (thrust fault) starts low in the southern part of the outcrop and then cuts up-section roughly 12 m (40 ft) before flattening out again at a higher stratigraphic level.

The sequence of events at this outcrop began with backthrusts (east-directed thrusts) that created east-verging folds. These were then cut by a lateral ramp that moved material up and to the north and west. Finally, the entire area was tilted west as part of the west limb of the growing Adams Run Anticline.

Lateral ramps are genetically related to cross-strike discontinuities or CSDs (Wheeler, 1980). CSDs are disruptions (abrupt offset or change in direction) of the trend of bedding or folds. They can be caused by near-vertical faults, as with "tear faults" that offset thrust sheets. Or they can be a reflection of buried basement faults on overlying strata. For example, a vertical offset in basement rocks can cause a lateral ramp in an overlying thrust sheet. That may be what happened at the Mathias roadcut.

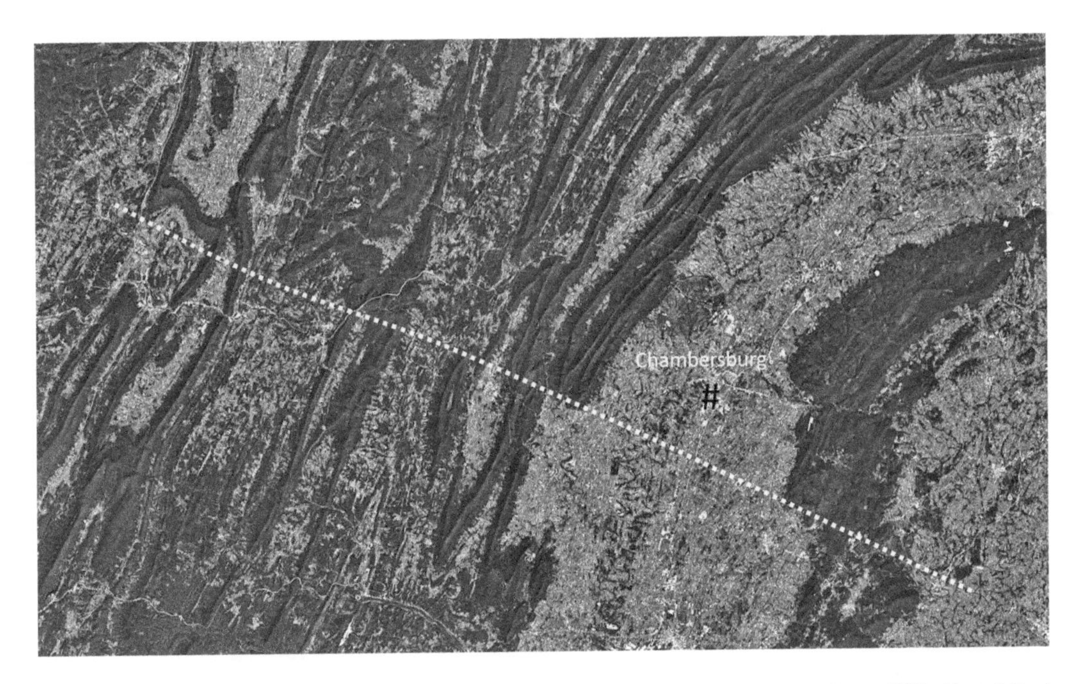

Cross-strike structural discontinuities in the Appalachian Thrust Belt. An east-southeast CSD (dotted line) is perpendicular to folding near Chambersburg, Pennsylvania, as indicated by fold terminations, aligned water gaps, and offset ridges. True color satellite image courtesy of TerraMetrics and Google Earth, image © Landsat/Copernicus.

Mathias to Bergton Roadcut: *Continue south on WV-259 to VA-820/Bergton Road to Bergton; turn right (west) on VA-820 and drive 1.8 km (1.1 mi) to a gravel pullout on the left beyond the guardrails. Walk back east along the roadcut. This is* ***Stop 13, Bergton roadcut*** *(38.758216, −78.935302), for a total of 17.9 km (11.1 mi; 12 min).*

STOP 13 BERGTON ROADCUT

The Bergton roadcut exposes at least three west-directed thrusts and numerous west-verging folds developed in the Brallier Formation shale. The normal sequence is for a thrust to fold the layers and then cut through the fold. Subsequent thrusts below the original will fold the earlier-formed structures. Sometimes later thrusts or subsidiary thrusts cut up through the earlier folds; sometimes they form at a higher stratigraphic level.

Folded and thrusted Brallier Formation at the Bergton roadcut. This photo shows mostly subthrust deformation. Looking north near the west end of the exposure just before the curve in the road. Farther east along this roadcut you can see deformation below a nearly horizontal thrust fault.

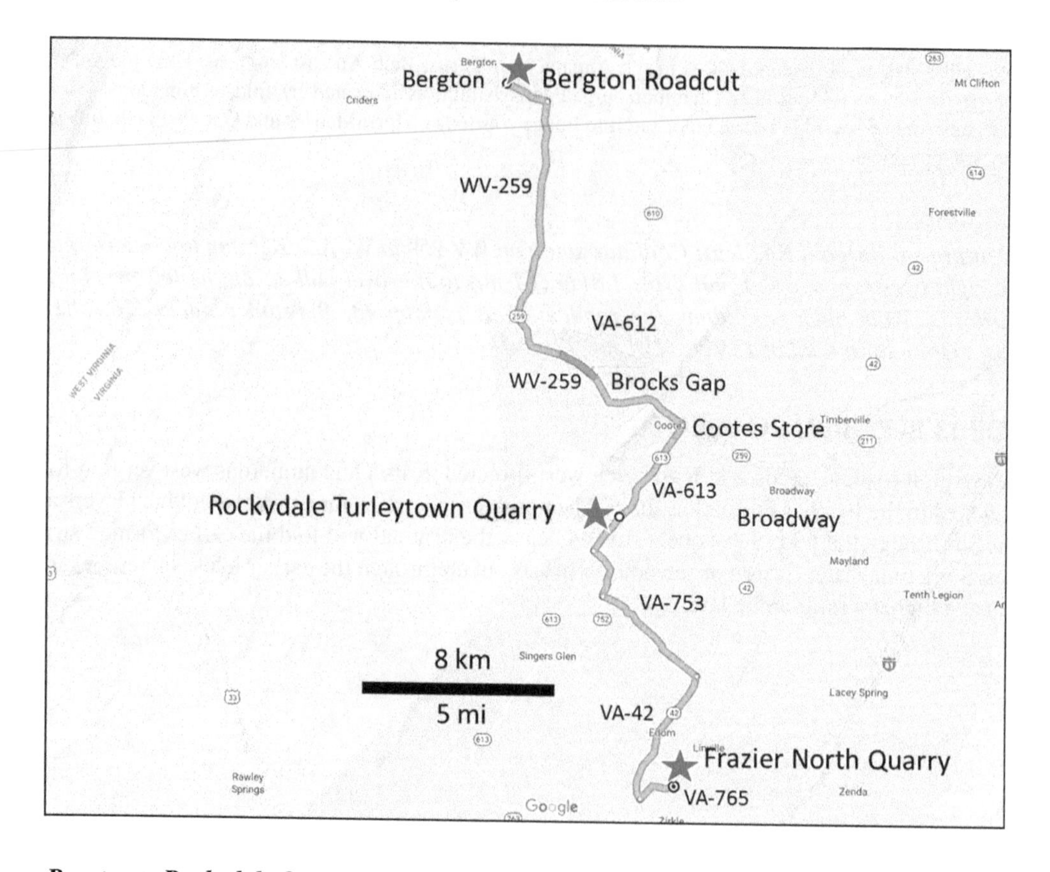

Bergton to Rockydale Quarry: *Backtrack east on VA-820 to VA-259; turn right (south) and drive to the intersection with VA-612 North (observe the vertical sandstone north of the road); continue east through Brocks Gap on VA-613 to Cootes Store; turn right (south) on VA-613 (Turleytown Road) and drive 4.7 km (2.9 mi) to the quarry on the right. This is* ***Stop 14, Rockydale Turleytown Quarry*** *(38.603175, −78.880516), for a total of 24.4 km (15.1 mi; 18 min).*

As you pass the intersection of VA-259 with VA-612 (38.645867, −78.867885), there is an impressive wall of rock on the north side of the road just west of Brocks Gap. Brocks Gap is a water gap of the North Fork of the Shenandoah River through Little North Mountain. This imposing outcrop is Chimney Rock, consisting of vertical to overturned Oriskany Sandstone (formerly called Ridgeley Sandstone here) below the Little North Mountain Thrust. The Little North Mountain Thrust carries intensely folded and faulted lower Paleozoic carbonates of the Great Valley over mainly folded Paleozoic rocks of the Valley and Ridge (Rader and Perry, 1976). Chimney Rock is on the slightly overturned east limb of the Supin Lick Syncline. This is the same thrust fault zone exposed in and near our next stop at Rockydale Quarry (Brent, 1960).

Chimney Rock, vertical to overturned Oriskany Sandstone at the intersection of VA-259 and VA-612. Google Street View north.

STOP 14 ROCKYDALE TURLEYTOWN QUARRY

Rockydale does not provide on-demand tours, but if you contact them in advance, you may be able to arrange a tour. Note that, due to ongoing excavation, the quarry walls are constantly changing. Thanks to John DePasquale and Rockydale Quarries Corp. for providing a tour of the quarry.

The east side of this quarry contains exposures of Cambrian Elbrook Formation dolomite carried by the Little North Mountain Thrust. The west side contains shattered Ordovician Beekmantown Formation dolomite. They are separated by several fault planes. Whereas the Elbrook Formation is clearly layered, the Beekmantown Formation is so intensely fractured and brecciated that bedding is not evident.

North face of Pit 1, Rockydale Turleytown quarry in the Ordovician Beekmantown Formation. West-verging folds are clearly exposed.

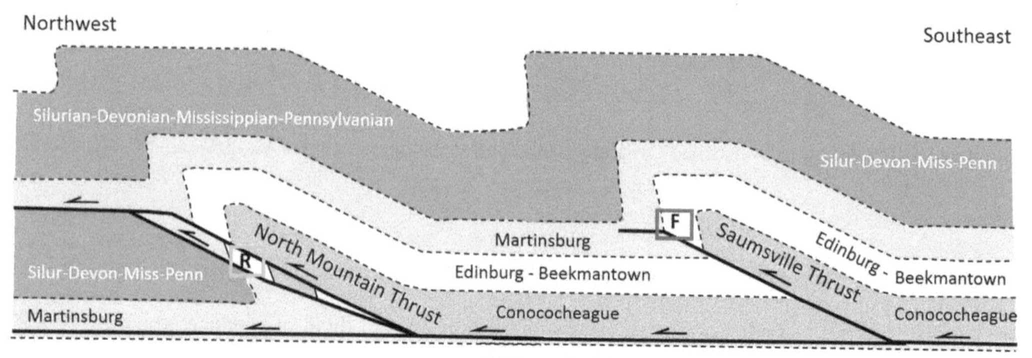

Schematic diagram of the Little North Mountain Thrust. Rockydale Quarry is at approximate position of the "R"; Frazier Quarry is at the "F." (Modified after Pohn et al., 1985.)

Visit

John DePasquale, P.E., Regional Manager
Rockydale Quarries Corporation
2343 Highland Farm Road, NW, Roanoke, VA 24017
Phone: 540-491-9021
Email: jdepasquale@rockydalequarries.com
Web site: rockydalequarries.com

Rockydale Quarry to Frazier North Quarry: *Continue south on VA-613 to VA-753; turn left (southeast) on VA-753 and drive to where it makes a sharp left; continue on VA-753/Wengers Mill Road to VA-42/Harpine Hwy; turn right on VA-42 and drive to VA-765; turn left on VA-765 and drive to the quarry entrance on the left. This is* ***Stop 15, Frazier North Quarry*** *(38.506383, −78.854037), for a total of 16.5 km (10.2 mi; 14 min).*

Northeast face, Frazier North quarry. Quarry is developed in the Ordovician Edinburg Formation.

STOP 15 FRAZIER NORTH QUARRY

Frazier Quarry offers tours to the public. Contact them in advance of your trip to make arrangements. Again, note that due to ongoing excavation, the outcrops along the quarry walls are constantly changing. Thanks are provided to David Zirk and the Frazier Quarry, Inc. for the tour of the quarry pit.

The Frazier North quarry contains a small, unnamed anticline carried on a thrust just west of the Mayland Anticline. A well-exposed ramp anticline is visible on the north face of the quarry. The fold formed on a northwest-dipping backlimb thrust to the main thrust. Shear zone veins and slickensides, evidence for bed-parallel slip, are also observable.

Visit

David Zirk, Safety Coordinator
Frazier Quarry, Inc.
75 Waterman Dr., Harrisonburg, VA 22802
Mobile: 540-820-5655
Email: David.Zirk@frazierquarry.com
Website: www.FrazierQuarry.com

Frazier North Quarry to Elkton: *Continue east on VA-765 to US-11; turn right (south) on US-11 and get on ramp to I-81 S; follow signs to Harrisonburg; drive south on I-81 to Exit 247A; merge onto US-33E/Market Street to Elkton; take US-33 east to VA-649; turn right (southeast) on VA-649 and drive to US-340N; turn left (northeast) onto US-340 and drive 4.5 mi (7.2 km), past Hawksbill Creek to the driveway on the right or pullout on the left just beyond the outcrop. This is* ***Stop 16, Elkton*** *(38.386607, −78.625321), for a total of 39.0 km (24.2 mi; 33 min).*

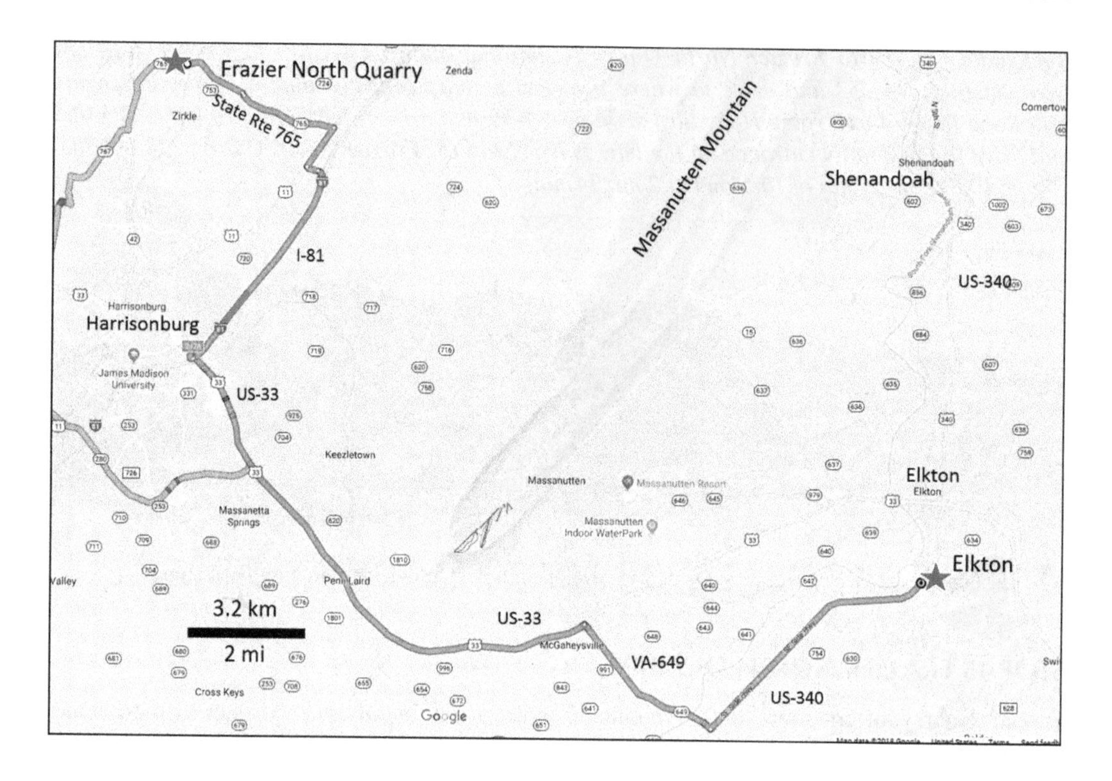

STOP 16 ELKTON

This outcrop contains Cambrian Antietam Formation sandstone and conglomeratic sandstone carried above the Blue Ridge Thrust. The rock is a highly fractured quartzite (quartz-rich metamorphosed sandstone).

As you drive north from here in the Shenandoah Valley, the high ridge to the west is Massanutten Mountain. Massanutten Mountain consists of resistant parallel ridges of Silurian Massanutten Sandstone (equivalent to the Tuscarora Sandstone farther west) overlying Ordovician Martinsburg Formation shale. Massanutten Sandstone is folded into a syncline (U-shaped fold) and outcrops along the ridge tops.

Cambrian Antietam Sandstone in the Elkton roadcut.

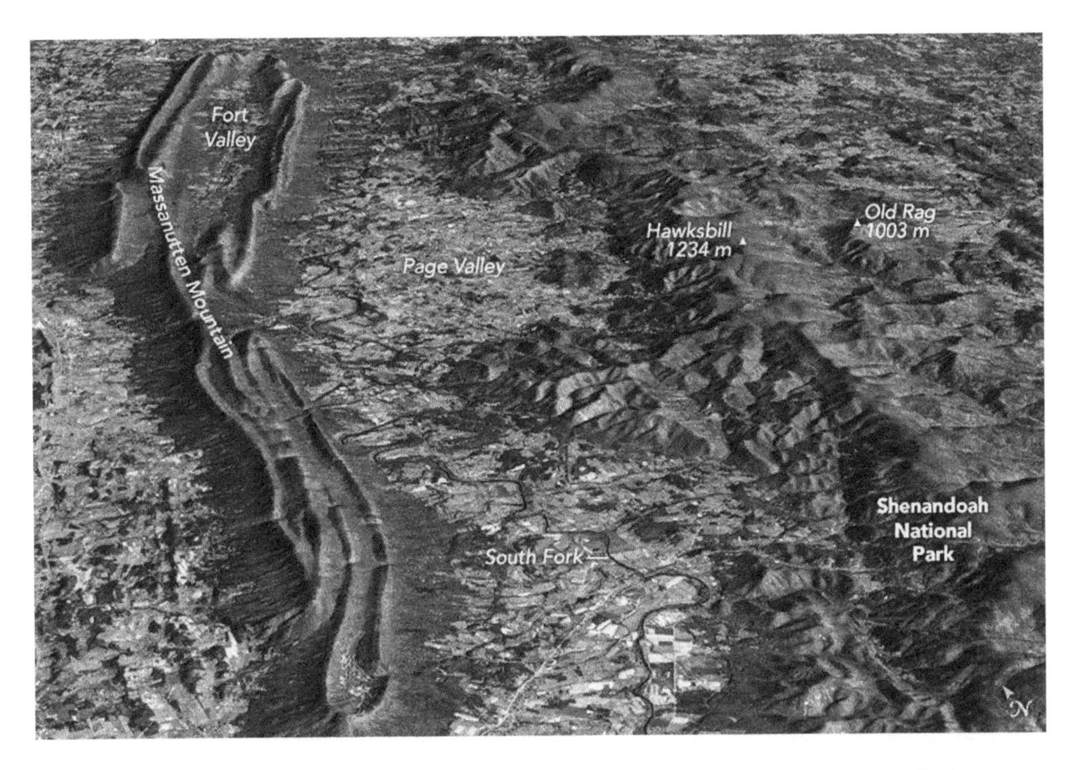

Massanutten Mountain is an 80 km (almost 50 mi) long synclinal fold held up by ridges of Massanutten Sandstone. This is an oblique view to the north courtesy of NASA and USGS. ASTER image and terrain data. https://commons.wikimedia.org/wiki/File:Contrasting_Ridges_in_Virginia_-_Labeled.jpg.

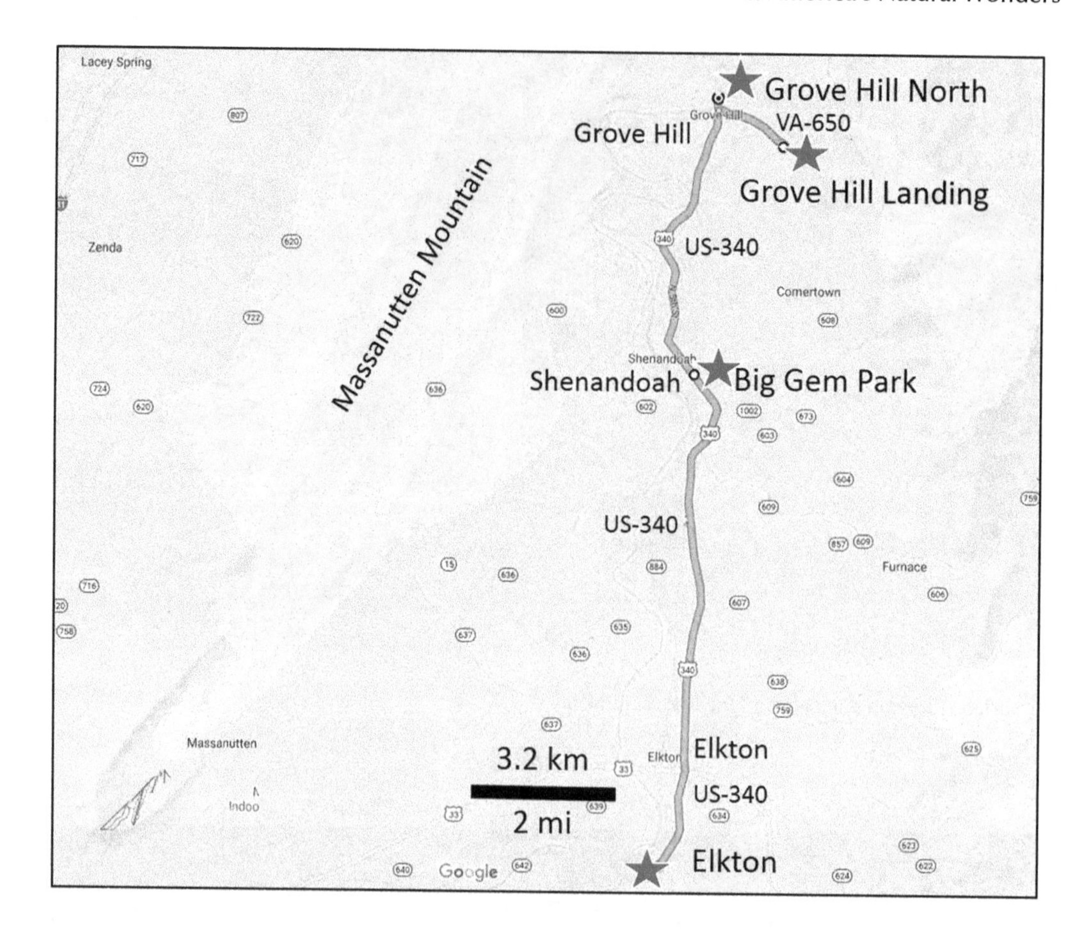

Elkton to Big Gem Park: *Continue north on US-340 to* ***Stop 17, Big Gem Park*** *(38.482077, −78.616868) at the south end of Shenandoah; turn right to park. A total of 11.2 km (7.0 mi; 10 min).*

STOP 17 BIG GEM PARK

Southwest-dipping Cambrian Conococheague Formation limestone at Big Gen Park has been carried west on the Blue Ridge Thrust. The limestone is inclined about 30o to the east and is lightly metamorphosed such that it has developed a pervasive cleavage.

Outcrops of Conococheague Formation limestone around Big Gem Pond.

Big Gem Park to Grove Hill Landing Railroad Cut: *Continue north on US-340 to Grove Hill; turn right (east) on VA-650 and drive to Crooked Run Road just after the boat ramp; turn right (south) on Crooked Run Road and pull over under the railroad trestle. This is* ***Stop 18.1, Grove Hill Landing Railroad Cut*** *(38.526505, −78.594654) for a total of 8.2 km (5.1 mi; 8 min). Climb up to the railroad cut on the east side of the road.*

18 GROVE HILL AREA

Half a kilometer (0.3 mi) south of the intersection of VA-650 and US-340 is a cherty fault breccia between the houses. This breccia is a result of brittle deformation along the Blue Ridge Thrust. The breccia can be traced at least 27 km (17 mi) north and is at least 50 m (164 ft) thick. Above this breccia is a ductile deformation zone in the Cambrian Conococheague Formation that can be seen in the Grove Hill Landing railroad cut. This tells us that late brittle (shallow) faulting was superimposed on an early ductile (deeply buried) deformation zone.

As we turn right onto VA-650 in Grove Hill, we are in Ordovician Beekmantown Formation dolomite and limestone. At the railroad cut, we are in Cambrian Conococheague Formation limestone and dolomite.

STOP 18.1 GROVE HILL LANDING RAILROAD CUT

This outcrop illustrates a ductile deformation zone at the leading edge of the Blue Ridge Thrust system. Ductile deformation is a result of increased pressure from burial and of rock type. Carbonate rocks are especially susceptible to ductile deformation, which is the ability to deform extensively without breaking. Rocks bend and flow in slow motion. Ductile rock responds to stress by folding rather than faulting or jointing. Look for mylonitic limestone (recrystallized limestone caused by crushing and grinding along fault zones) and flow structures such as boudinaged (stretched into sausage shapes) dolomite layers and flow cleavage within the limestone.

Footwall (below the thrust) deformation in the Cambrian Conococheague Formation, Grove Hill Landing railroad cut.

West-dipping outcrops of Cambrian Conococheague Formation, Grove Hill Landing railroad cut.

Grove Hill Landing to Grove Hill North: *Return to VA-650 and drive west to US-340; turn right (north) on US-340, cross South Fork Shenandoah River and immediately take first road on the right. Walk north along US-340. This is* ***Stop 18.2, Grove Hill North*** *(38.535724, −78.611486), for a total of 2 km (1.3 mi; 2 min).*

STOP 18.2 GROVE HILL NORTH

Just north of the bridge across the South Fork Shenandoah River are outcrops of Ordovician Beekmantown dolomite. The rocks are dipping west into the Massanutten Mountain Syncline. This is subthrust deformation below the Blue Ridge Thrust.

Beekmantown Formation dipping west, just north of Grove Hill. Google Street View looking north.

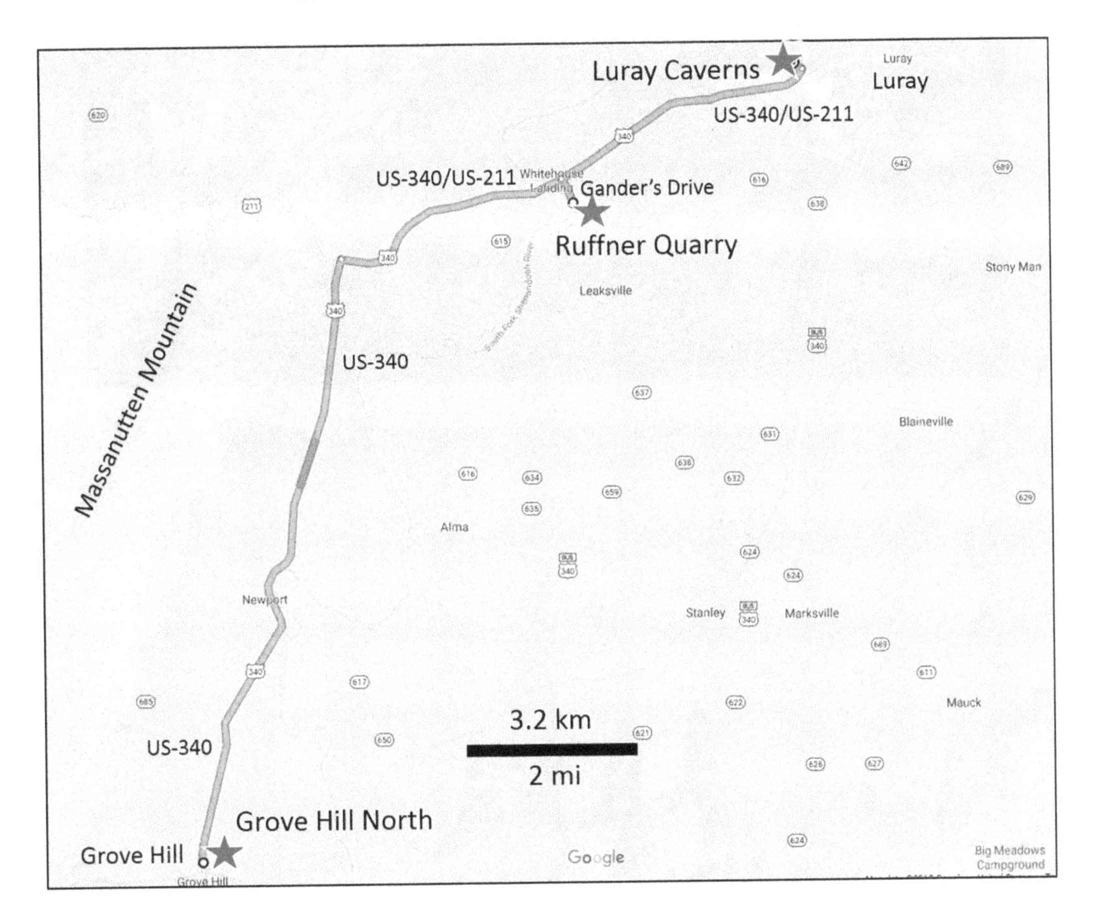

Grove Hill North to Ruffner Quarry: *Continue north on US-340 to US-211; turn right (east) onto US-211/340 and drive to Gander's Drive. Turn right (south) onto Gander's Drive and drive 500m (1,600ft) to abandoned Quarry on left. This is* ***Stop 19, Ruffner Quarry*** *(38.642581, −78.53020), for a total of 16.8km (10.5 mi; 13min).*

STOP 19 RUFFNER QUARRY

The abandoned and largely overgrown Ruffner quarry is in the Ordovician Edinburg Formation on the west flank of the leading edge anticline carried on the Blue Ridge Thrust. A recumbent fold (a fold lying on its side) is inclined to the northwest in the quarry walls. A small folded thrust fault and fault-propagation fold (a fold formed at the termination of a thrust fault) can be seen in the northeast face of the quarry.

Northeast quarry wall showing recumbent fold, folded fault, and fault-propagation fold.

The northeast quarry wall contains a northwest-inclined recumbent fold developed in the Edinburg Formation. The quarry wall approximately 7 m (22 ft) high.

Ruffner Quarry to Luray Caverns: *Return north to US-211; turn right (east) on US-211 and drive to Cave Hill Road; turn left (northwest) and drive to Luray Caverns parking lot. This is* ***Stop 20, Luray Caverns*** *(38.665079, −78.482259), for a total of 5.8 km (3.6 mi; 6 min).*

STOP 20 LURAY CAVERNS

Discovered in 1878, Luray Caverns was designated a World Heritage Site in 1973. It is the largest cavern system in the eastern United States. The caverns are developed in the Lower Ordovician Beekmantown Dolomite by slightly acidic groundwater that dissolves the dolomite slowly over many years. The dissolved dolomite later precipitates as groundwater drips from the cave ceiling to form stalactites, hanging like icicles from the top of the cave, and stalagmites growing up from the cave floor. Whereas the rock is hundreds of millions of years old, the cave system itself probably isn't more than a few million years old at most. The rooms range from 9 to 43 m (30 to 140 ft) high and are at a constant temperature of 12°C (54°F). Colors come from impurities in the rock: reds from iron, black from manganese, blues and greens from copper and cave algae. The dripstone formations grow at about 16 cm^3 (1 in^3) per 120 years.

Dripstone formations (stalactites and stalagmites) in Luray Caverns.

Visit

Luray Caverns are open daily from 9:00 am to 7:00 pm

Tours follow lighted, paved pathways and depart every 20 min starting at 9:00 am.

Admission: See website for current entry fees.
Website: https://luraycaverns.com/
Address: Luray Caverns, 101 Cave Hill Road, Luray, VA 22835
Phone: 540-743-6551

Luray Caverns to Pass Mountain Overlook: *Return to US-211; turn left (east) on US-211 and drive to Skyline Drive exit; turn left (north) onto Skyline Drive; enter Shenandoah National Park; drive north to* ***Stop 21.1, Pass Mountain Overlook*** *(38.675135, −78.333908), for a total of 19.5 km (12.1 mi; 20 min).*

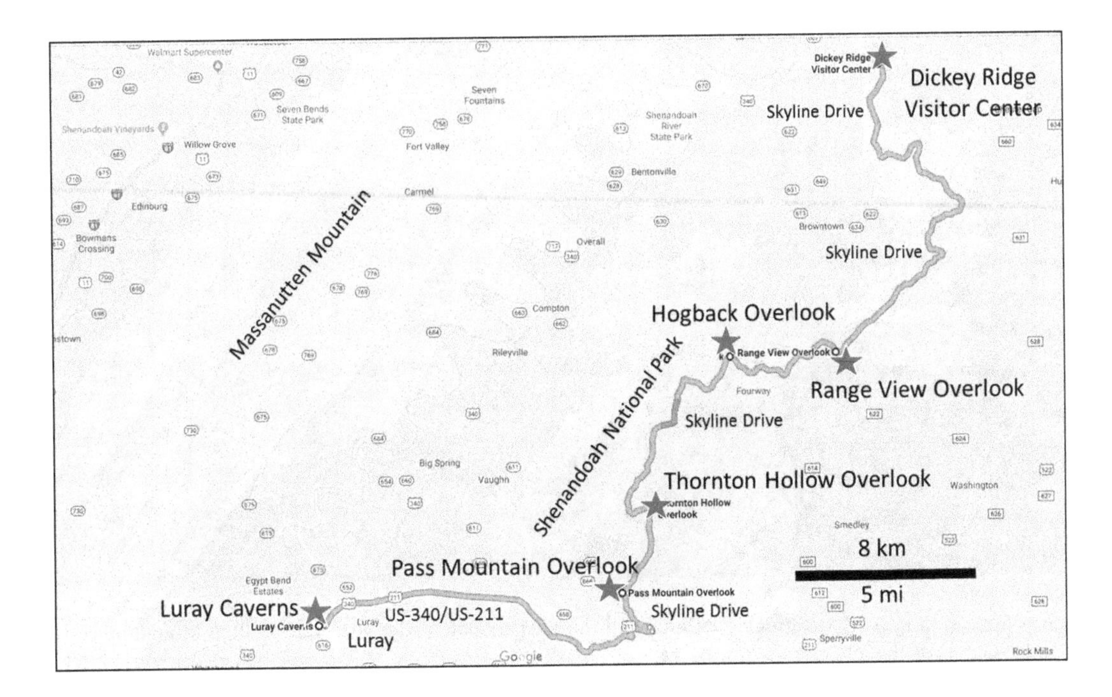

21 SKYLINE DRIVE, BLUE RIDGE, AND SHENANDOAH NATIONAL PARK

A national park for the Appalachians in Virginia was first proposed in 1901, but the land was largely private and many owners were against the idea. Virginia slowly acquired land through eminent domain and then gave it to the federal government. Shenandoah National Park was eventually established in 1935 and formally opened by President Roosevelt in 1936. The park was originally segregated, but in December 1945, the National Park Service mandated that all concessions in national parks be desegregated. By 1950, they were.

The park extends 170km (105 mi) from near Front Royal in the north to Waynesboro in the south. Its roughly 32,000ha (80,000 ac) include most of the Blue Ridge, from the Shenandoah Valley on the west to the Piedmont on the east. The highest peak, at 1,234m (4,048ft), is Hawksbill Mountain. Much of the area has been designated wilderness, and the park is a favorite of hikers and day trippers.

STOP 21.1 PASS MOUNTAIN OVERLOOK

You are now atop the Blue Ridge Mountains. Below you to the west the ridges of the Blue Ridge are underlain by steeply dipping quartzite beds of the Chilhowee Group (Weverton, Hampton, and Erwin formations; Gathright, 1976). The Luray area of the Shenandoah Valley can be seen looking west. Farther west is Massanutten Mountain and New Market Gap.

View west from Pass Mountain Overlook toward the Shenandoah Valley and Massanutten Mountain. New Market Gap is on the left.

Pass Mountain Overlook to Thornton Hollow Overlook: *Continue north on Skyline Drive to* ***Stop 21.2, Thornton Hollow Overlook*** *(38.706240, −78.319200), for a total of 4 km (2.5 mi; 4 min).*

STOP 21.2 THORNTON HOLLOW OVERLOOK

This stop provides a view east to the Blue Ridge foothills and the Piedmont.

Greenstone (epidote-rich metabasalt) of the 600 Ma Catoctin Formation outcrops on the north side of the road. The Catoctin Formation is cut by fault surfaces containing fibrous anthophyllite, a variety of asbestos (Gathright, 1976).

View southeast from the Blue Ridge over the Piedmont Province.

Catoctin greenstone exposed at Thornton Hollow Overlook.

Thornton Hollow Overlook to Hogback Overlook: *Continue north on Skyline Drive to* ***Stop 21.3, Hogback Overlook*** *(38.762732, −78.279997), for a total of 10.7 km (6.7 mi; 11 min).*

STOP 21.3 HOGBACK OVERLOOK

Several ridges (hogbacks) of the Valley and Ridge Province can be seen looking west from Hogback Overlook. The view west from here is one of the best in the park. You can see most of the Shenandoah Valley and Blue Ridge. The valley is underlain by Precambrian and Cambrian rocks that were thrust several miles west over younger Cambrian and Ordovician rocks. Browntown Valley is in the foreground, with the Blue Ridge escarpment and Mount Marshall to the northeast (Gathright, 1976).

Erosion of the Catoctin Formation has exposed 1.2 billion-year-old Pedlar Formation granodiorite. Iron staining of the granodiorite along fractures indicates ancient weathering of the rocks before they were covered by Catoctin lavas.

View west from Hogback Overlook. At least four separate ridges in the Valley and Ridge Province can be seen. The nearest ridges are part of the Massanutten Mountain Syncline.

Precambrian Pedlar granodiorite and granite gneiss at Hogback Overlook.

Hogback Overlook to Range View Overlook: *Continue north on Skyline Drive to* ***Stop 21.4, Range View Overlook*** *(38.764480, −78.227391) pullout on the right for a total of 6.1 km (3.8 mi; 6 min).*

STOP 21.4 RANGE VIEW OVERLOOK

This overlook provides a panoramic view of the Blue Ridge. The geography of the Piedmont Province to the east is clearly different from that of the Valley and Ridge Province to the west. The Blue Ridge is underlain by massive granitic rocks and, locally, by Catoctin greenstone. The ridge and geologic structure generally trends northeast. The Massanutten Mountain Syncline to the west, and other ridges of the Valley and Ridge Province, are long, uniform ridges that maintain their trend for many kilometers. These ridges consist of resistant layers of sandstone and limestone that form long, parallel ranges, as opposed to the more irregular crest of the Blue Ridge formed by massive igneous rocks (Gathright, 1976).

Outcrops at Range View Overlook are greenstones of the Catoctin Formation.

View south along the Blue Ridge from Range View Overlook. It is obvious why this was named the Blue Ridge.

Range View Overlook to Dickey Ridge Visitor Center: *Continue north on Skyline Drive to* ***Stop 21.5, Dickey Ridge Visitor Center*** *(38.871073, −78.204192), a total of 20.1 km (12.5 mi; 21 min).*

STOP 21.5 DICKEY RIDGE VISITOR CENTER

We end this transect at the Dickey Ridge Visitor Center for Shenandoah National Park at the crest of the Blue Ridge. The visitor center and picnic area are underlain by greenstones of the Catoctin Formation. Enjoy expansive views of the Blue Ridge to the south and Great Valley to the west.

Panoramic view west from the visitor center toward the Valley and Ridge Province.

This small visitor center has an information desk and nature exhibits. There is a bookstore with publications and maps, and backcountry permits are available. There are a first-aid station and restrooms as well.

Visit

The Visitor Center is open daily from 9:00 am to 5:00 pm from early April to late November (check with the park for exact dates).

Phone: 540-999-3500
Address: Shenandoah National Park, 3655 U.S. Highway 211 East, Luray, VA 22835.
Website: www.nps.gov/shen/index.htm
www.nps.gov/shen/planyourvisit/visitorcenters.htm

ACKNOWLEDGMENTS

This transect could not have been written without the field assistance and observations of Ben Prost. The support of Jim Coleman in offering stop ideas, reviewing the manuscript, and providing a guidebook by Howard Pohn is gratefully acknowledged.

REFERENCES AND FURTHER READING

Allen, R.M. Jr. 1967. *Geology and Mineral Resources of Page County.* Virginia Department of Conservation and Economic Development, Division of Mineral Resources Bulletin 81, Charlottesville, VA, 101 p.

Bently, C., D. Doctor, and A. Pitts. 2014. Geology of Corridor H. Geological Society of Washington fall field trip, 28 p. www.gswweb.org/GSW_Corridor_H_field_trip_draft-4.pdf.

Brent, W.B. 1960. *Geology and Mineral Resources of Rockingham County.* Virginia Department of Conservation and Economic Development, Division of Mineral Resources Bulletin 76, Charlottesville, VA, 205 p.

Butts, C., and R.S. Edmundson. 1966. *Geology and Mineral Resources of Frederick County.* Virginia Department of Conservation and Economic Development, Division of Mineral Resources Bulletin 80, Charlottesville, VA, 173 p.

de Witt, W. Jr., and R.C. Milici. 1989. Energy resources of the Appalachian Orogeny. In Hatcher, R.D. Jr., W.A. Thomas, and G.W. Viele, (eds.), *The Appalachian–Ouachita Orogen in the United States*, v. F-2, The Geology of North America. Geological Society of America, Boulder, CO, pp. 495–510.

Frye, K. 1986. *Roadside Geology of Virginia.* Mountain Press Publishing Company, Missoula, MT, 278 p.

Gathright, T.M. II. 1976. *Geology of the Shenandoah National Park, Virginia.* Virginia Department of Conservation and Economic Development, Division of Mineral Resources Bulletin 86, Charlottesville, VA, 146 p.

Gathright, T.M. II., and P.S. Frischmann. 1986. *Geology of the Harrisonburg and Bridgewater Quadrangles, Virginia.* Virginia Department of Conservation and Economic Development, Division of Mineral Resources Publication 60, Charlottesville, VA, 37 p.

Hunt, P.J., R.R. McDowell, B.M. Blake Jr., J. Toro, and P.A. Dinterman. 2017. What the H!? Paleozoic Stratigraphy Exposed, Pre-Meeting Field Trip Guide for the *46th Annual Meeting, Eastern Section of the American Association of Petroleum Geologists (ESAAPG)*, Morgantown, West Virginia, September 24 and 25, 2017. West Virginia Geological and Economic Survey, Field Trip Guide FTG-9, 87 p.

José F. Vigil from This Dynamic Planet -- a wall map produced jointly by the U.S. Geological Survey, the Smithsonian Institution, and the U.S. Naval Research Laboratory.) http://pubs.usgs.gov/gip/dynamic/Vigil.html.

Kulander, B.R., and S.L. Dean. 1986. Structure and tectonics of central and southern Appalachian Valley and Ridge and Plateau provinces, West Virginia and Virginia. *AAPG Bulletin*, v. 70, pp. 1674–1684.

Lampsire, L.D. 1992. Crustal structures and the eastern extent of the Lower Paleozoic shelf strata within the Central Appalachians: A seismic reflection interpretation. Virginia Polytechnic Institute and State University *MSc Thesis*, 84 p.

Orndorff, R.C. 2012. Fold-to-fault progression of a major thrust zone revealed in horses of the North Mountain Fault Zone, Virginia and West Virginia, USA. *Journal Geological Research*, v. 2012, 13 p.

Orndorff, R.C., and J.B. Epstein. 1994. A structural and stratigraphic excursion through the Shenandoah Valley, Virginia. US Geological Survey Open File Report 94-573, 26 p.

Pohn, H.A. 2000. Lateral ramps in the folded Appalachians and in overthrust belts worldwide – A fundamental element of thrust-belt architecture. US Geological Survey Bulletin 2163, 64 p.

Pohn, H.A., W. de Witt Jr., and A.P. Schultz. 1985. Disturbed zones, lateral ramps, and their relationships to the structural framework of the central Appalachians. Eastern Section – AAPG November 1985 Meeting Field Trip Guidebook 3, 40 p.

Rader, E.K., and W.J. Perry, Jr. 1976. Reinterpretation of the Geology of Brocks Gap, Rockingham County, Virginia. Virginia Minerals v. 22, no. 4. Virginia Department of Conservation and Economic Development, Division of Mineral Resources. 12 p.

Rader, E.K., and J.F. Conley. 1995. Geologic map of Warren County, Virginia. Virginia Department of Conservation and Economic Development, Division of Mineral Resources Publication 138, 1:50,000.

Rader, E.K., and T.M. Gathright II. 2001a. Geologic map of the front royal 30×60 minute quadrangle: Portions of Clarke, Page, Rockingham, Shenandoah, and Warren Counties, Virginia. Virginia Department of Conservation and Economic Development, Division of Mineral Resources Publication 162, 1:100,000.

Rader, E.K., and T.M. Gathright II. 2001b. Geologic map of the Augusta, Page, and Rockingham Counties portion of the Charlottesville 30×60 minute quadrangle. Virginia Department of Conservation and Economic Development, Division of Mineral Resources Publication 159, 1:100,000.

Rader, E.K., R.C. McDowell, T.M. Gathright II, and R.C. Orndorff. 2001. Geologic map of the Virginia portion of the Winchester 30×60 minute quadrangle. Virginia Department of Conservation and Economic Development, Division of Mineral Resources Publication 161, 1:100,000.

Rader, E.K., and G.P. Wilkes. 2001. Geological map of the Virginia portion of the Staunton 30×60 minute quadrangle. Virginia Department of Conservation and Economic Development, Division of Mineral Resources Publication 163, 1:100,000.

Renton, J.J. 2018. A geology field trip. 19 p. www.wvgs.wvnet.edu/www/geoeduc/geoeduc.htm and www.wvgs.wvnet.edu/www/geoeduc/FieldTrip/GeologyFieldTripGuide.pdf.

Ryder, R.T., R.C. Milici, C.S. Swezey, and M.H. Trippi. 2014. Appalachian Basin oil and natural gas: Stratigraphic framework, total petroleum systems, and estimated ultimate recovery. US Geological Survey Professional Paper 1708, 7 p.

Schultz, A.P., R.C. McDowell, and H. Pohn. 1989. *Structural Transect of the Central Apallachian Fold-and-Thrust Belt, Harpers Ferry, West Virginia to Cumberland, Maryland, July 15, 1989.* Field Trip Guidebook T227. American Geophysical Union, Washington, DC.

Southworth, S., A.A. Drake Jr., D.K. Brezinski, R.P. Wintsch, M.J. Kunk, J.N. Aleinikoff, C.N. Naeser, and N.D. Naeser. 2006. Central Appalachian Piedmont and Blue Ridge tectonic transect, Potomac River Corridor. Geological Society of America Field Guide 8, 34 p.

US Geological Survey. Figure 53. Origin of the Appalachian Orogen, a result of three separate continental collisions involving the North American Continent with the Taconic and Acadian terranes, and finally the collision of the African and North American continents during the Alleghanian Orogeny at the end of the Paleozoic. http://3dparks.wr.usgs.gov/nyc/images/fig53.jpg and http://3dparks.wr.usgs.gov/nyc/common/captions.htm.

West Virginia Geological and Economic Survey. 2006. Bedrock geology, Moorefield quadrangle, Hardy County, WV. www.wvgs.wvnet.edu/www/statemap/483/483_Bed.htm. Accessed 28 October 2018.

Wheeler, R.L. 1980. Cross-strike structural discontinuities: Possible exploration tool for natural gas in Appalachian Overthrust Belt. *American Association of Petroleum Geologists Bulletin*, v. 64, pp. 2166–2178.

3 Colorado Rocky Mountain High

Denver, Colorado, to Grand Junction, Colorado

Rocky Mountains as seen from Denver. (Photo courtesy of Adam Ginsburg, https://commons.wikimedia.org/wiki/File:Mountains_from_westlands.jpg.)

OVERVIEW

Tour 3 explores the geology of the Colorado Rockies between Denver and Grand Junction. Starting with an overview of the Denver Basin, we visit the world-famous Colorado School of Mines Geology Museum, Red Rocks Park, and Dinosaur Ridge, all on the outskirts of Denver. Following Interstate 70 west, we pass the Hogback Roadcut, the mining districts of Idaho Springs, Central City, and Georgetown, traverse the continental divide, review the geology of Vail and the Colorado River gorge at Glenwood Canyon, and cross the distinctive Grand Hogback, the border between the Rockies and the Colorado Plateau. We end at the matchless and historic Book Cliffs at Grand Junction. Along the way, we discuss petroleum resources of the Denver and Piceance basins and review the mineral deposits and mining history of the Colorado Mineral Belt.

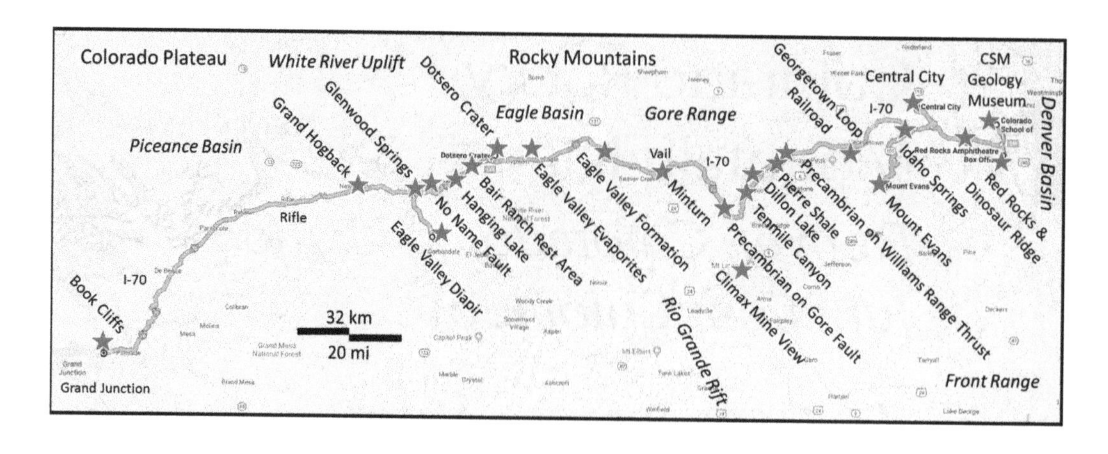

ITINERARY

Begin in Golden, Colorado
Stop 1 Colorado School of Mines Geology Museum
2 The Front Range and Red Rocks Park
  Stop 2.1 Red Rocks Visitor Center and Amphitheater
  Stop 2.2 Red Rocks Geologic Marker
3 Dinosaur Ridge
  Stop 3.1 Dinosaur Ridge Discovery Center
  Stop 3.2 Dinosaur Tracks
Stop 4 I-70 Roadcut and the Golden Fault
Stop 5 Precambrian at Evergreen Parkway (Exit 252)
Side Trip 1 Central City
  ST 1.1 Hidee Gold Mine Tour
  ST 1.2 Idaho Springs Formation, Central City Parkway
Stop 6 Idaho Springs, the Colorado Mineral Belt, and Argo Mine (Mile 241)
Side Trip 2 Mount Evans
Stop 7 Georgetown Loop Railroad and Mines (Mile 227)
Stop 8 Precambrian carried on Williams Range Thrust (Mile 208.1)
Stop 9 Pierre Shale (Mile 206.4)
Stop 10 Dillon Reservoir Scenic Overlook (Mile 203.5)
Stop 11 Tenmile Canyon and the Gore Range (Mile 200)
Side Trip 3 Climax Mine Overlook
Stop 12 Precambrian carried on the Gore Fault (Mile 193.5)
Stop 13 Maroon Formation, Vail Pass (Mile 191.7)
Stop 14 Minturn Formation, West Vail (Mile 171.9)
Stop 15 Eagle Valley Formation at Wilmore Lake, Edwards Rest Stop (Mile 160.3)
Stop 16 Deformed Eagle Valley Evaporites (Mile 141.5)
Stop 17 Dotsero Crater and Lava Flow (Exit 133, Mile 135–135.5)
Stop 18 Bair Ranch Rest Area—Glenwood Canyon and White River Uplift (Mile 129)
Stop 19 Hanging Lake Rest Area (Mile 125)
Stop 20 No Name Fault (Mile 118)
Stop 21 Glenwood Springs and Hot Springs Pool
Side Trip 4 Eagle Valley Evaporite Diapir
Stop 22 Grand Hogback and Piceance Basin
Stop 23 Book Cliffs at Grand Junction—End

THE DENVER BASIN

We begin at the boundary between the Rocky Mountains and the continental craton, the mostly stable, long-term core of North America. East of here, there hasn't been any major deformation since the Pennsylvanian Appalachian uplift that extended from New England south to Georgia and then west through the Ouachita uplift of Arkansas and Oklahoma to the Marathon uplift in west Texas. Deformation in the craton has been limited to broad, regional warping known as epeirogenic uplift or subsidence. Deformation in the Rockies to the west has been going on sporadically since Pennsylvanian time around 300 million years ago.

The Denver Basin, also known as the Denver–Julesburg Basin or DJ Basin, covers 180,000 km^2 (70,000 mi^2) along the east flank of the Rocky Mountains in eastern Colorado, southeast Wyoming, southwest Nebraska, and western Kansas. The roughly oval, north-south-elongated basin stretches from near Pueblo to near Cheyenne and is highly asymmetric, being deepest along the mountain front near Denver. It has a steep west flank and a gentle east flank. This geometry suggests that it is a foreland basin linked to uplift in the Rockies rather than an epeirogenic (regional) downwarp typical of the midcontinent.

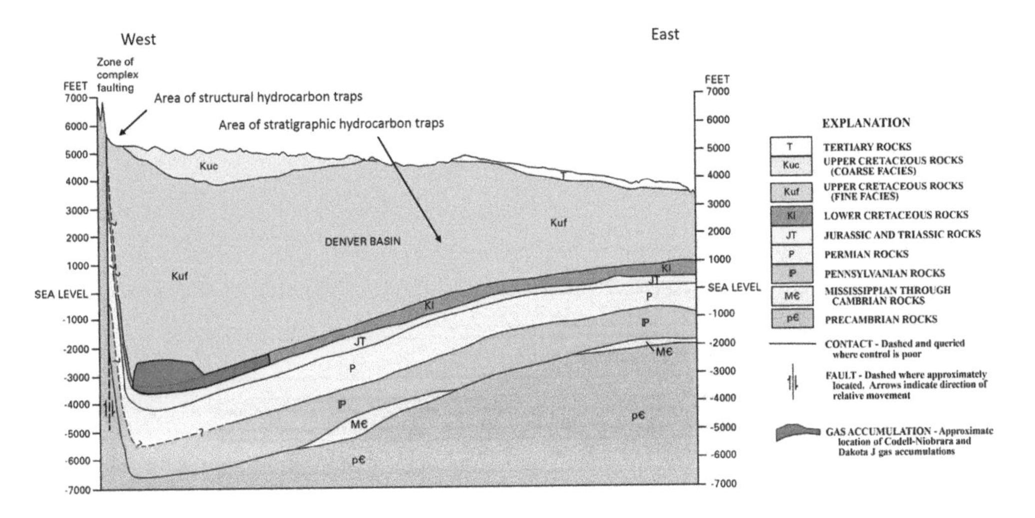

Cross section through the Denver Basin. (Courtesy of Philip Nelson and Stephen Santus, U.S. Geological Survey, https://commons.wikimedia.org/wiki/File:Denver_Basin_Cross_Section.png.)

In the deepest part of the basin, about 3,900 m (13,000 ft) of sedimentary rock lies over the 1.6 billion-year-old Precambrian basement. The oldest sedimentary rock in the basin is the Pennsylvanian Fountain Formation, roughly 300 million years old, deposited when the Ancestral Rocky Mountains were being uplifted and eroded and the basin was just starting to subside. Since then, the basin has mostly been just above or just below sea level. We know, for example, that the Jurassic Morrison Formation was deposited on a coastal plain, that the Cretaceous Dakota Group was deposited as beach and barrier bar deposits, and that the Pierre Shale was deposited as mud on the bottom of the Cretaceous Interior Seaway. The basin deepened further and acquired its current geometry during the Laramide Orogeny 70–50 million years ago. By about 68 million years ago, the area was again above sea level as the Arapaho Conglomerate was shed off the rising Rockies. By 64 million years ago, the area was covered in rainforest; 37 million years ago, the Wall Mountain Tuff erupted, dropping ash over the area. The mountains had eroded to the level of the plains, forming what came to be called the Late Eocene Erosion Surface. The Eocene Erosion Surface was gently rolling topography where remnants of the Laramide-age Rockies were buried beneath alluvial sediments and only a few isolated peaks rose above the plains. Regional uplift over the past 10 million years caused renewed erosion, mainly in the plains and Denver Basin, and unearthed the Rocky Mountains.

Age			Formation
Tertiary			Green Mountain gravels
Tertiary			
Tertiary/Cretaceous			Denver Formation/Dawson Formation
Cretaceous			Arapahoe Formation
Cretaceous			Laramie Formation
Cretaceous			Fox Hills Sandstone
Cretaceous	Pierre Shale		Richard Sandstone Member
Cretaceous	Pierre Shale		Terry/Sussex Sandstone Member
Cretaceous	Pierre Shale		Hygiene/Shannon Sandstone Member
Cretaceous	Niobrara Formation		Smoky Hill Shale Member
Cretaceous	Niobrara Formation		Fort Hays Limestone Member
Cretaceous			Codell Sandstone Member
Cretaceous			Carlile Shale
Cretaceous			Greenhorn Limestone
Cretaceous			Graneros Shale "D" Sandstone
Cretaceous			Mowry Shale
Cretaceous	Dakota Group	South Platte Fm	Muddy ("J") Sandstone
Cretaceous	Dakota Group	South Platte Fm	Skull Creek Shale
Cretaceous	Dakota Group	South Platte Fm	Plainview/Dakota Formation
Cretaceous	Dakota Group		Lytle/Lakota Formation
Jurassic			Morrison Formation
Jurassic			Ralston Creek Formation
Jurassic			Sundance Formation
Tri			Jelm Formation
Permian			Lykins Formation
Permian			Lyons Sandstone
Permian			Owl Canyon Formation
Permian			Ingleside Formation
Pennsylvanian			Fountain Formation
Miss			
Dev			
Sil			
Ord			
Cam			
Pre-cam			Idaho Springs Formation

Strata found in the Denver Basin. Gray indicates periods of erosion or non-deposition. (Modified after Higley and Cox, 2007.)

The Denver Basin is rich in mineral resources. As of 2007, more than 167 million m^3 (1.05 billion barrels) of oil and 106 billion m^3 (3.67 trillion ft^3) of gas have been produced from the basin since the first well was drilled near a surface seep in the Cañon City field in 1862. Today, there are over 52,000 wells in 1,500 oil and/or gas fields in the basin. Producing layers are Pennsylvanian to Cretaceous sandstones and limestones, although most production is from the Cretaceous. Depth of production ranges from 270 m (900 ft) to 2,700 m (9,000 ft).

Bituminous coal was mined underground along the western edge of the basin between the late 1850s and 1979. The coal is in the Cretaceous Laramie Formation and was deposited on coastal plains and in lagoons during and after withdrawal of the Cretaceous Interior Seaway. Many of the old workings are unmapped. Collapse of these mines causes surface subsidence and is a source of concern for subdivisions along the mountain front.

Placer gold was found in small amounts along Clear Creek near Golden and Denver starting in 1858. Some sand and gravel operations still find a little gold.

Small amounts of uranium were mined from the Dakota Sandstone near Morrison.

Start*—Colorado School of Mines Geology Museum is located in the General Research Lab and* ***Geology Museum Building****, 1310 Maple St., Golden (39.751586, −105.224615). Park in Lot V off Maple Street.*

GOLDEN, COLORADO

Golden was founded during the Pike's Peak Gold Rush of 1859. The mining camp on Clear Creek at the foot of the Rockies was originally named "Golden City" in honor of Thomas Golden, an early prospector in the region. Its location made it a center of trade between the gold fields in the mountains to the west and settlements on the plains to the east. Golden City became the capital of the Territory of Jefferson in 1860 and capital of the Territory of Colorado from 1862 to 1867. In the early 1860s, a fierce competition developed between Denver, 19 km (12 mi) to the east, and Golden, to see which could build a railroad spur to the transcontinental railroad in Cheyenne 160 km (100 mi) to the north. Denver won the race, and in 1867, the territorial capital was moved to Denver City. Today, Golden is famous as the home of Colorado School of Mines and Coors Brewery.

Golden sits in a valley between Table Mountain and the Front Range of the Colorado Rockies. North and South Table Mountains, once continuous but now dissected by Clear Creek, consist of Late Cretaceous to early Paleocene sediments of the Denver Formation capped by 64–63 million-year-old basaltic lava flows (technically a "shoshonite porphyry"). The Denver Formation contains sandstone and shale deposited as alluvial fans, in river channels, and in swamps, and is between 180 to 480 m (600 to 1,600 ft) thick. The lava has conspicuous columnar jointing along the flanking cliffs. The source of the flows is thought to be Ralston Dike, a volcanic plug about 5 km (3 mi) north of Golden.

Golden and South Table Mountain looking east from Mount Zion. (Courtesy of R. Kimpel, https://commons.wikimedia.org/wiki/File:South_Table_Mountain_from_Lookout_Mountain.jpg.)

STOP 1 COLORADO SCHOOL OF MINES GEOLOGY MUSEUM

We begin this transect with a visit to the Colorado School of Mines Geology Museum. CSM, known as "the world's foremost college of mineral engineering," was established in 1874 to train miners and engineers. That role has expanded steadily until today the institution is dedicated to engineering and applied science in general and to earth science in particular. It consistently ranks as one of the best universities in the country.

The Geology Museum is the second most visited geology museum in the United States with over 26,000 visitors in 2016. It has displays of minerals and mining artifacts, including some of the best specimens in the world of fossils, gemstones, and meteorites, and has 17 moon rocks. There is a gift shop, walk-through mine, videos about local geology and critical natural resources, and an outdoor geology trail and guided tours. Visitors can bring in rock and mineral samples for identification.

The Geology Museum gift shop features mineral specimens, lapidary equipment, fossils, books, fluorescent minerals, and starter kits. Specimens range from less than $1 to fine minerals for advanced collectors.

Guided tours: The Geology Museum now charges for guided tours. Make reservations by phone 1 week in advance. Call 303-273-3815 for rates, details, and other inquiries.

Visit

Location: Colorado School of Mines, General Research Lab (GRL) building, 1310 Maple St., Golden, CO 80401
Phone: 303-273-3815
Email: geomuseum@mines.edu

Admission: Free

Hours: Monday–Saturday, 9 am–4 pm; Sunday 1–4 pm

The Geology Museum is closed Christmas Day and New Year's Day. It may be closed during the university's winter break.

Identification of specimens is performed between 10 am and noon, Tuesdays and Thursdays.

Gift Shop: Monday–Saturday, 9 am–3:45 pm, Sunday 1–3:45 pm

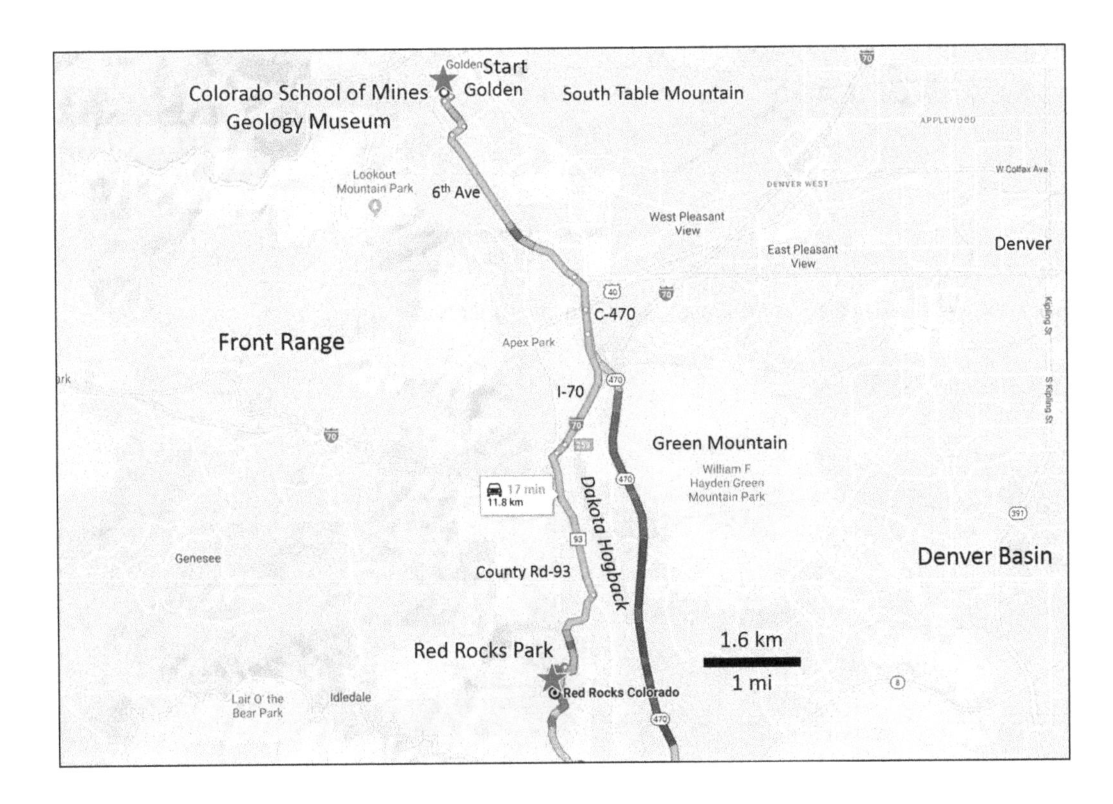

Geology Museum to Red Rocks Amphitheater: *Drive south on Maple to 19th Street; turn right (west) and drive to US-6. Turn left (south) and follow the signs to I-70 West (to Grand Junction); take Exit 259 to Morrison and Red Rocks; turn left (south) on County Road 93 and drive to Entrance 1, Red Rocks Park. Turn right (west) and follow signs to* ***Stop 2.1, Red Rocks Visitor Center and Amphitheater*** *parking lot (39.667214, −105.206464), for a total of 11.8 km (7.3 mi; 17 min).*

2 THE FRONT RANGE AND RED ROCKS PARK

The Front Range is the north-south range that emerges abruptly from the west edge of the Great Plains. It extends from near Pueblo in Colorado to near Caspar, Wyoming. The range climbs nearly 3,000 m (10,000 ft) above the Great Plains and has numerous peaks over 4,250 m (14,000 ft).

The Colorado Rockies are fundamentally different from the Appalachians we visited earlier, and the Idaho-Wyoming and Canadian Rockies we saw in Volume I. Whereas those ranges are a consequence of compression acting on thick sedimentary sections with resulting bedding plane detachments causing "sled runner thrusts," the Colorado Rockies are characterized by uplifted basement. Some have likened it to the kind of pop-up resulting from squeezing a watermelon seed between your fingers. The Idaho-Wyoming Fold-Thrust Belt and Canadian Rockies are considered Sevier-style deformation. Basement-cored uplifts are typical of Laramide-style deformation.

There are two competing ideas regarding uplift of the Colorado Rockies and Front Range. The older concept is that there has been mainly vertical uplift and that the mountain range is bounded by faults that steepen at depth. As the mountains rose, the flanks extended over the adjacent sedimentary basin such that the bounding faults appear to be low- to high-angle reverse faults at the surface. These faults would have inherited their orientation from older (Precambrian or Pennsylvanian) faults.

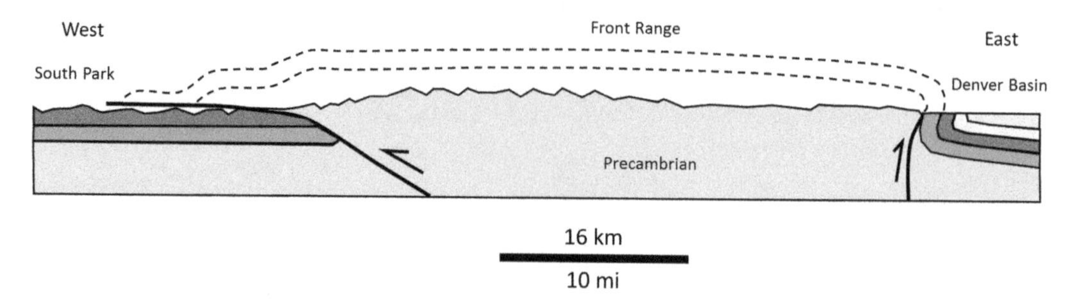

Schematic cross section through the Front Range. (Modified after Nesse, 2006.)

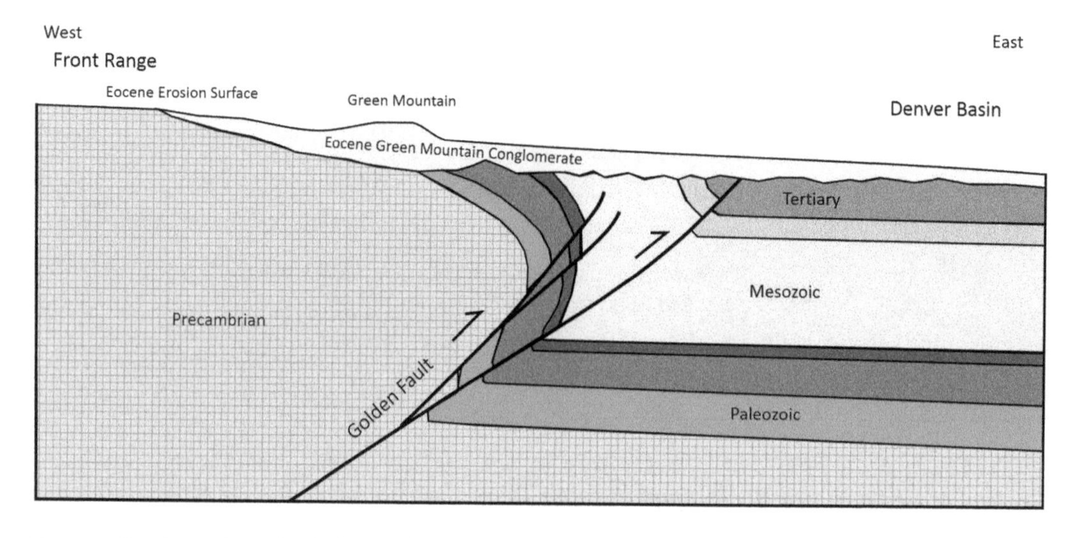

Cross section through the Rocky Mountain front near Denver. (Modified after Delvaux, 2016.)

The more recent concept is that Front Range uplift is the result of thrust faults originating near the crust-mantle boundary. In this model, the main thrusting is west-directed, with thrusts on the east flank being subordinate backthrusts. The west-directed crustal scale thrusting was a result of flat subduction of the Farallon Plate along the west coast of North America; that is, rather than plunging into the mantle near the west coast, the plate remained nearly flat beneath the crust and contributed to compression and shortening of the crust well into the mid-continent. The main west-flank thrusts are estimated to have over 10.5 km (6.3 mi) of lateral displacement.

By 10 million years ago, the core of the Front Range had eroded to the point where granite and gneiss were exposed. About 2 million years ago, Ice Age glaciers began carving U-shaped valleys and cirques. Miocene-to-Recent up-warping of the mid-continent led to rapid down-cutting of rivers and carving of the steep-sided canyons seen today.

The core of the Front Range west of Denver consists of Precambrian gneiss, granodiorite, and granite (Scott, 1972; Kellogg et al., 2008). The range is bounded by steep reverse faults on the east and shallow thrust faults on the west. The northwest to north-south faults that bound the Front Range are thought to have originated in Precambrian and Pennsylvanian time. *Frontrangia* was part of the Pennsylvanian Ancestral Rocky Mountains and shed sediments both to the east and west. Those deposited on the west side make up the Maroon Formation; those deposited on the eastern slopes became the Fountain Formation.

The Ancestral Rockies eroded to low-lying hills and plains near sea level by Early Cretaceous time. The Western Interior (or Cretaceous) Seaway covered most of the area from 75 to 85 million years ago. In Late Cretaceous time, the Laramide Orogeny began uplifting the modern Rocky Mountains. Pausing briefly during the Eocene for erosion to create the Late Eocene Erosion Surface, the mountains were nearly buried in their own eroded sediment. Uplift resumed in Oligocene time and continues to the present. Vertical uplift relative to the Denver Basin exceeds 6 km (20,000 ft).

Late Eocene Erosion Surface (arrows) and Mount Evans. View west from Denver. (Photo courtesy of Jeremy McCreary, www.cliffshade.com/colorado/mt_evans/me15.jpg.)

Green Mountain, the conspicuous hill between Red Rocks Park and Denver, consists of Green Mountain Conglomerate above Denver Formation sandstone. The Green Mountain Conglomerate was eroded from the Front Range between 64 and 55 million years ago and dumped near the mountain front. Petrified wood is common in these sediments on the slopes of Green Mountain.

STOP 2.1 RED ROCKS VISITOR CENTER AND AMPHITHEATER

Geology

The red sandstone that gives the park its name is the Pennsylvanian Fountain Formation, formed of the debris of the Ancestral Rocky Mountains. Equivalent to the Maroon Formation seen in central Colorado, this coarse-grained sandstone and conglomerate with abundant feldspar grains was eroded off the mountain chain known as Frontrange or *Frontrangia*. These Ancestral Rocky Mountains were formed during the collision and suturing of South America/Gondwana to proto-North America during the assembly of Pangea. The same scenic formation can be seen at Garden of the Gods west of Colorado Springs, in Roxborough State Park southwest of Denver, and at the Flatirons west of Boulder. Coarse sand and conglomerate are not carried far by rivers and streams, and feldspar quickly weathers to clay, indicating that this sandstone was deposited close to its source. Formed around 296–290 million years ago, around 300–1,350 m (1,000–4,430 ft) of material was deposited as alluvial fans and braided river channel sands along the ancient mountain front. The red color comes from pink feldspar grains and weathered iron minerals.

Above the Fountain Formation in the park is about 37 m (120 ft) of cream-colored Permian Lyons Sandstone, deposited as fine-grained quartz sand dunes roughly 250 million years ago.

About 245 million years ago, spanning the Permian-Triassic boundary, a shallow sea moved into this area. Algae found this an inviting environment and built algal mounds known as stromatolites. The easily weathered Lykins Formation consists of 122–330 m (400–1,080 ft) of red mudstone-gypsum-limestone deposited in an arid, low-relief tidal flat setting.

During the Laramide uplift of the Rocky Mountains between 75 and 40 million years ago, the layers were folded and tilted to between 40° and near-vertical. Erosion has since removed the softer rocks, leaving the hard, well-cemented monoliths seen at the park today.

Red Rocks looking south. The rocks are Pennsylvanian Fountain Formation shed east off the Ancestral Rockies.

History

Red Rocks Park and amphitheater, 32 km (20 mi) west of downtown Denver at the foot of the mountains, is one of the most popular music venues in the United States. First seen in 1820 by an army scouting expedition, it was named Garden of the Angels in 1870. In 1906, it was renamed Garden of the Titans by owner John Walker. After Walker sold the property to the city of Denver in 1928, it became known as Red Rocks. During the Great Depression, the Civilian Conservation Corps carved the amphitheater out of the native rock formations. With its spectacular scenery and nearly perfect acoustics, the amphitheater opened with an Easter Sunrise Service in 1947 and has since hosted concerts by musicians such as Nat King Cole, The Beatles, Joan Baez, Peter Paul and Mary, the Grateful Dead, Hall and Oats, Neil Young, and the Moody Blues, among others. The park was added to the list of National Historic Landmarks in 2015.

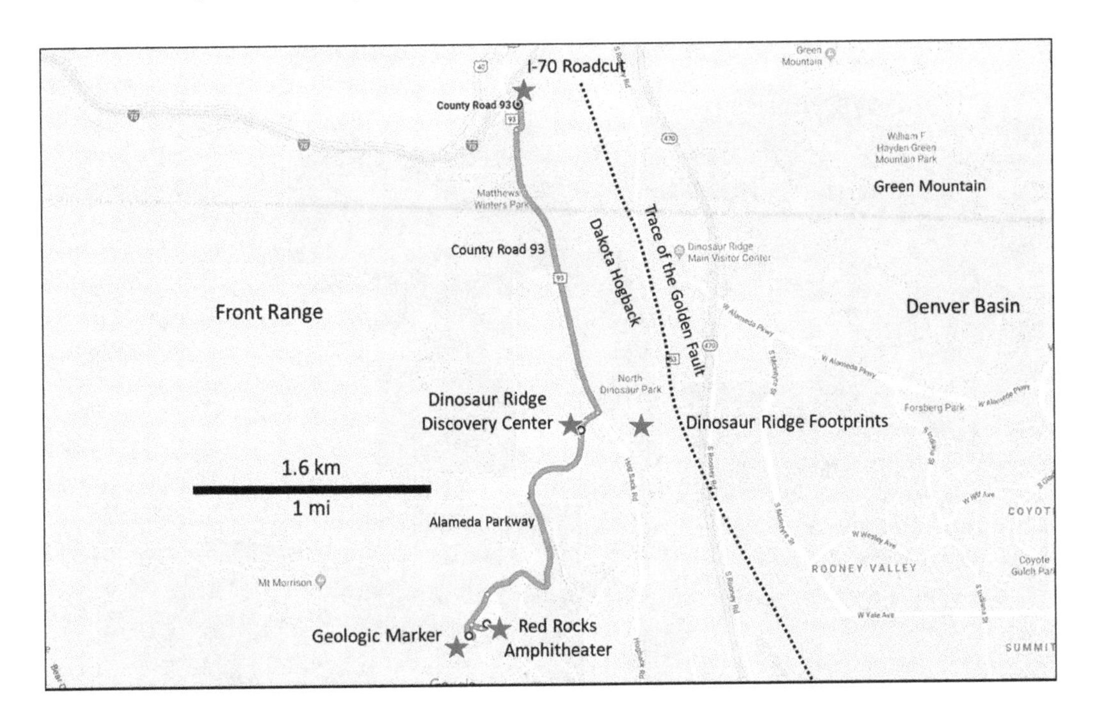

Red Rocks Amphitheater to Geologic Marker: *Exit the parking lot and turn left on Alameda Parkway; drive 270 m (0.2 mi; 1 min) to* ***Stop 2.2, Red Rocks Geologic Marker*** *(39.666530, −105.207977) on the right.*

STOP 2.2 RED ROCKS GEOLOGIC MARKER

The Fountain Formation was deposited directly on 1.7 billion-year-old gneiss. This means that you can touch a time gap of 1.4 billion years. This doesn't mean that no sediments were deposited over that time interval. It means that any sedimentary rocks that were deposited have since been eroded away. This is the same Great Unconformity we see in Death Valley, near Las Vegas, in the Grand Canyon, on the Uncompahgre Uplift, and in Glenwood Canyon.

The Great Unconformity at Red Rocks Park. A plaque marks the spot between Pennsylvanian Fountain Formation (300 million years old) and Precambrian gneiss (1.7 billion years old). There are 1,400 million years missing across the span of a hand.

Red Rocks Geologic Marker to Dinosaur Ridge: *Drive east on Alameda Pkwy 2.2 km (1.4 mi; 4 min) to* ***Stop 3.1, Dinosaur Ridge Discovery Center*** *(39.679861, −105.198521), for a total of 2.6 km (1.6 mi; 8 min).*

3 DINOSAUR RIDGE

In 1877, Arthur Lakes, geology professor at Colorado School of Mines, discovered bones in the Morrison Formation along the Dakota Hogback near Morrison. In an effort to gain support to continue excavating, he wrote to Othniel Marsh, a leading dinosaur expert at Yale, and sent him sketches and bone samples. "A few days ago, I discovered . . . some enormous bones apparently a vertebra and a humerus bone of some gigantic saurian in the Upper Jurassic or Lower Cretaceous at the base of Hayden's Cretaceous No. 1 Dakotah group." Marsh didn't respond. So he sent a letter and more bones to Marsh's rival Edward Cope. Once Marsh heard Cope was involved, he responded immediately. The Morrison quarries, along with a site at Como Bluff, Wyoming, became key locations in the "bone wars" between Marsh and Cope. The quarries provided the first fossils of Stegosaurus, Apatosaurus (Brontosaurus), Allosaurus, and Diplodocus. Lakes documented his work with detailed notes and drawings, and then abandoned the quarries in 1879. The sites were lost for over 120 years.

In 1937, workers extending Alameda Parkway uncovered dinosaur tracks in the Dakota Sandstone. New exploration during 1992–1993 discovered 335 tracks and 37 trackways. These are from dinosaurs that lived around 50 million years after those found in the underlying Morrison Formation. You can tell a lot from just tracks. Four-legged iguanodon-like herbivores, perhaps *Eolambia*, moved at about 2 mi/h, and parallel tracks indicate that they traveled in large groups. Carnivorous theropod tracks are about 22 cm (9 in) long and indicate animals that weighed 45 kg (100 lb), were solitary, and walked on two legs at around 8 km (5 mi) per hour. In 2002, researchers used Lakes' notes to rediscover the quarries. Further digging in 2003 uncovered the first Stegosaurus tracks, and in 2006, the world's first baby Stegosaurus prints. In 2016, they discovered two-toed raptor tracks, only the second time these have ever been found in North America.

STOP 3.1 DINOSAUR RIDGE DISCOVERY CENTER

The Dinosaur Ridge Discovery Center has touch-friendly exhibits, an Allosaurus, crocodile, and the Colorado State Dinosaur, the Stegosaurus. They are focused on the environments and animals that lived in Colorado oceans, beaches, and swamps during the Late Jurassic to Late Cretaceous (150–68 million years ago). Guided bus tours and adult field trips are also available.

VISIT

Address: 16831 W Alameda Pkwy., Morrison, CO 80465 (Red Rocks Park Entrance #1 off of Highway 93)
Phone: 303-697-3466
See the website for tour times and current fees.
info@dinoridge.org
Open 7 Days a week except New Years, Thanksgiving, and Christmas
Summer Hours—May 1–October 18
  Monday through Saturday 9 am–5 pm
  Sunday 12 pm–5 pm
October 19–October 31
  Monday through Saturday 10 am–5 pm
  Sunday 11:30 am–5 pm
Winter Hours—November 1–April 30
  Monday through Saturday 10 am–4 pm
  Sunday 11:30 pm–4 pm

Dinosaur Ridge Discovery Center to Dinosaur Tracks: *Take a short hike up the ridge to see tracks and bones at* ***Stop 3.2, Dinosaur Tracks*** *(39.681056, −105.192265). No driving is needed.*

STOP 3.2 DINOSAUR TRACKS

Dakota Ridge is part of the Dakota Hogback, so-called for the resemblance to the spine of a razorback hog. The hogback runs along the mountain front from Pueblo to the Wyoming border. At Dakota Ridge, it consists of Late Jurassic Morrison Formation mudstone and overlying Cretaceous Dakota Sandstone inclined 45°–60° to the east.

The Morrison Formation mudstone was deposited 156–145 million years ago in a nearly flat coastal plain similar to the Gulf Coast of the United States today. Meandering rivers crossed the plain and left thin channel sandstones. Easily eroded gray, green, and maroon shales between 80 and 111 m (270 and 365 ft) thick are typical of the Morrison Formation. Abundant dinosaur bones indicate a lush, supportive environment with lots of plants.

By around 115 million years ago, the Cretaceous Interior Seaway was in eastern Colorado. By around 110 million years ago, the Dakota Sandstone was being deposited along the flanks of the sea as beach, offshore bar, and barrier island sands, and lagoon muds and coals. The climate was temperate, with coastal forests and mangrove swamps. Dinosaurs migrating along the western coast of the seaway created a "dinosaur freeway" with abundant tracks of both two-legged carnivores and four-legged herbivores. The light gray to cream-colored sandstone is 90–113 m (300–370 ft) thick.

Dinosaur tracks in Cretaceous Dakota Group sandstone, Dinosaur Ridge near Morrison, CO. (Courtesy of Footwarrior, https://commons.wikimedia.org/wiki/File:Dinosaur_Ridge_tracks.JPG.)

Around 65 million years ago, the sea drained as the Laramide Orogeny lifted the modern Rocky Mountains. The originally horizontal layers were tilted to their present position by Laramide uplift of the Front Range. Intense erosion over the past 10 million years or so left us with the resistant ledges of the Dakota Hogback.

Above the Dakota Sandstone is about 150m (500ft) of dark, organic-rich shale and thin limestones of the Benton Formation. The Benton Shale lies beneath the valley east of the hogback.

Visit

Dinosaur Ridge was designated part of the Morrison Fossil Area National Natural Landmark in 1973. In 2001, Colorado declared it a State Natural Area, and in 2006, the Colorado Geological Survey made it a Point of Geological Interest. There are two trails with interpretive signs that explain the geology, bone sites and tracks, the ancient environments, and economic aspects of coal and oil found in these rocks. Visitors can see bones still in the rock in Arthur Lakes' Quarry #5 in the Morrison Formation on the west side of the ridge. The main trail climbs 61 m (200 ft) and is a 3.2 km (2 mi) round-trip from the museum to the top of the ridge where you can see fossils and tracks and get panoramic views of the mountain front. A shuttle bus tour is available for a small fee.

Dinosaur Ridge to I-70 Roadcut: *From Dinosaur Ridge trailhead drive north on County-93 for 2.4 km (1.5 mi; 3 min) to the Stegosaurus Parking Lot on the right at* ***Stop 4, I-70 Roadcut*** *(39.698750, –105.203608).*

STOP 4 I-70 ROADCUT AND THE GOLDEN FAULT

The Dakota Hogback at the eastern front of the Rockies was tilted up by the forces that created the modern mountains, the Laramide Orogeny. The I-70 Roadcut was excavated through the Dakota Hogback in 1971. These are the same rock layers that form Dinosaur Ridge a few miles to the south. The roadcut spans about 50 million years, from the youngest, 100 million-year-old rocks on the east to the oldest, 150 million-year-old rocks on the west side of the outcrop. Both sides of the roadcut have trails with interpretive signs that explain the ancient deposition environments of the layers, from marshes to tidal flats and beaches to the ocean floor.

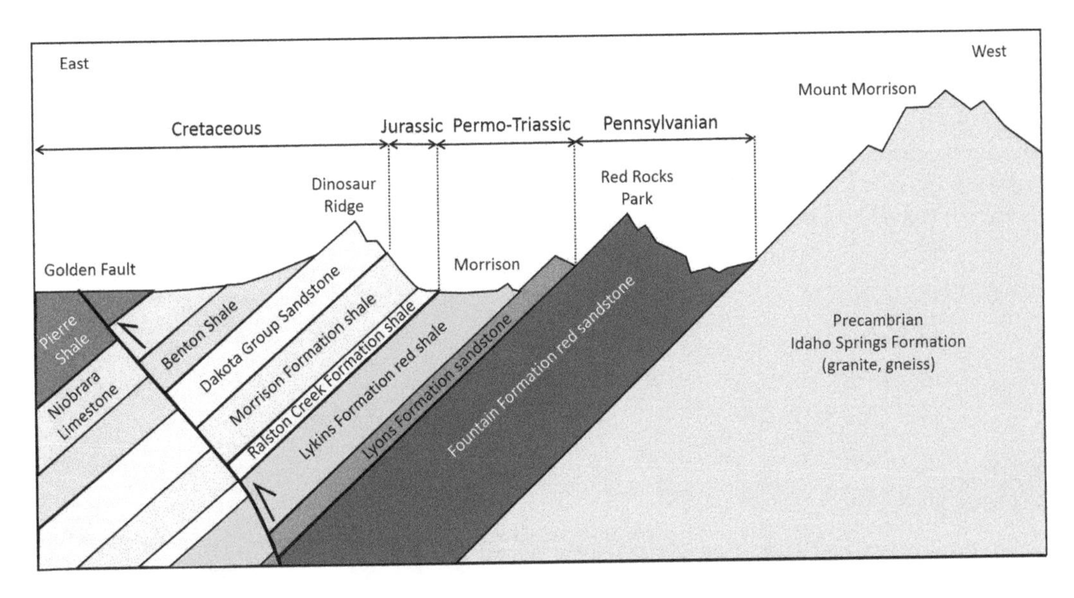

Layers seen in the I-70 roadcut. (Modified after public road sign.)

I-70 roadcut looking south.

The Golden Fault, the mountain-front fault here, lies in the valley east of the roadcut. The fault is inclined to the west, making it a reverse fault or thrust that puts older rock over younger rock. The fault cuts the Cretaceous Benton Formation at the surface. The main period of movement was during the Laramide Orogeny.

The valley to the east is underlain by, from youngest to oldest, the Cretaceous Pierre Shale, Niobrara Formation, and Benton Formation. The youngest layer in the roadcut belongs to the Cretaceous Dakota Group. The term "Dakota" has been applied to both a formation and a group. The Dakota Formation (or Dakota Sandstone) of central Colorado is primarily a sandstone that lies above the Morrison Formation and below the Benton Shale. The Dakota Group, mainly in northeast Colorado, comprises a lower Lytle Formation and upper South Platte Formation (Waagé, 1955). The Dakota Group at the I-70 roadcut consists of cream-colored sandstone and gray shale with thin coal beds. The top of the ridge is Lytle Sandstone. The oldest rocks, on the west side, are Jurassic Morrison Formation comprising gray, green, and maroon shales.

About 1.6 km (1 mi) west of this roadcut, at Mile 258, is the contact between the Pennsylvanian Fountain Formation and the Precambrian rocks that core the Rocky Mountains. Here, Precambrian rocks consist of a dark, 1.7–1.9 billion-year-old gneiss called the Idaho Springs Formation. The gneiss is intruded by 1.0–1.7 billion-year-old pink granitic plutons and dikes.

Driving up to this bridge from the east you get your first view of the Continental Divide. (Photo courtesy of Elizabeth Prost.)

As you pass Exit 254, you get your first view of the Continental Divide framed by the Mt. Vernon Country Club Road overpass. The Continental Divide sits as a series of peaks above the relatively flat Eocene erosion surface. The highway follows this surface for much of the way to the Divide. Later, during the Pleistocene, the peaks and valleys were sculpted by glaciers (Hughes et al., 2009).

If you chose to take Exit 254, you can stop at the Buffalo Herd Overlook to see the local bison.

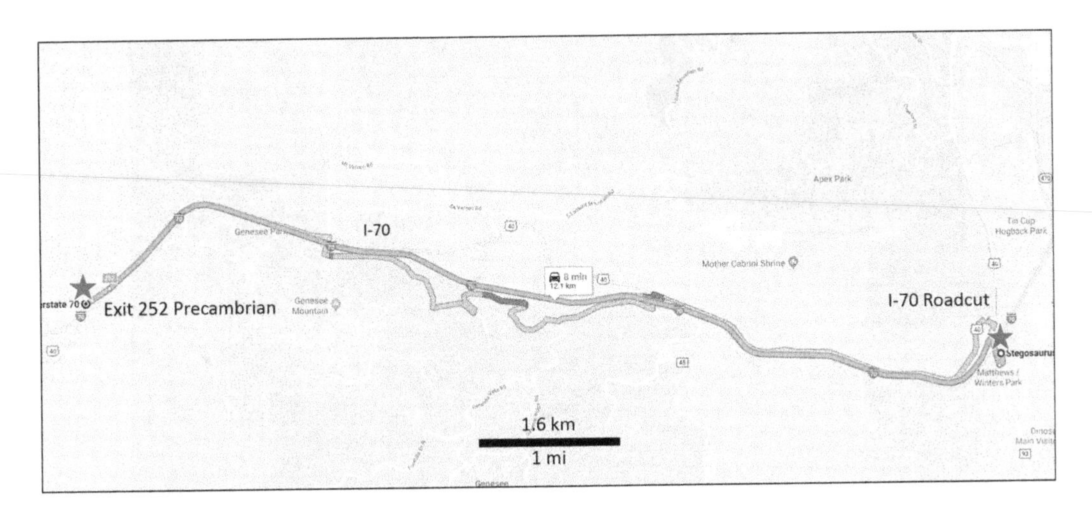

I-70 Roadcut to Exit 252 Precambrian: *Drive north on County Road 93 to the I-70 West onramp; turn left (west) onto I-70 and drive to Exit 252; pull over on the right after the barriers. This is* ***Stop 5, Precambrian*** *(39.704575, −105.327335), for a total of 12.1 km (7.5 mi; 8 min).*

STOP 5 PRECAMBRIAN AT EVERGREEN PARKWAY (EXIT 252)

The rocks here consist of granites and gneisses lumped into the Idaho Springs Formation. The granites, granitic dikes, and pegmatites seen between here and Dillon Reservoir represent three separate orogenies: The Boulder Creek Pluton (1.7 Ga), the Silver Plume Plutons (1.4 Ga), and the Pikes Peak Pluton (1.0 Ga). The dark amphibolite (hornblende-rich) gneiss was created by metamorphism of a volcanic arc terrane between 1.9 and 1.7 Ga. In addition to hornblende gneiss and felsic gneiss, there are mica schists, quartzites, marbles, and metamorphic cherts that record accretion of the Colorado Province to the older Wyoming craton during the Precambrian (Lytle, 2004; Stracher et al., 2007).

The roadcut at this exit exposes biotite-quartz-plagioclase gneiss, amphibolite gneiss, and pegmatite dikes (Hughes et al., 2009).

Gneiss in the Exit 252 roadcut indicates metamorphism and ductile deformation of an accreted volcanic arc and associated sediments. This package was later intruded by granitic dikes.

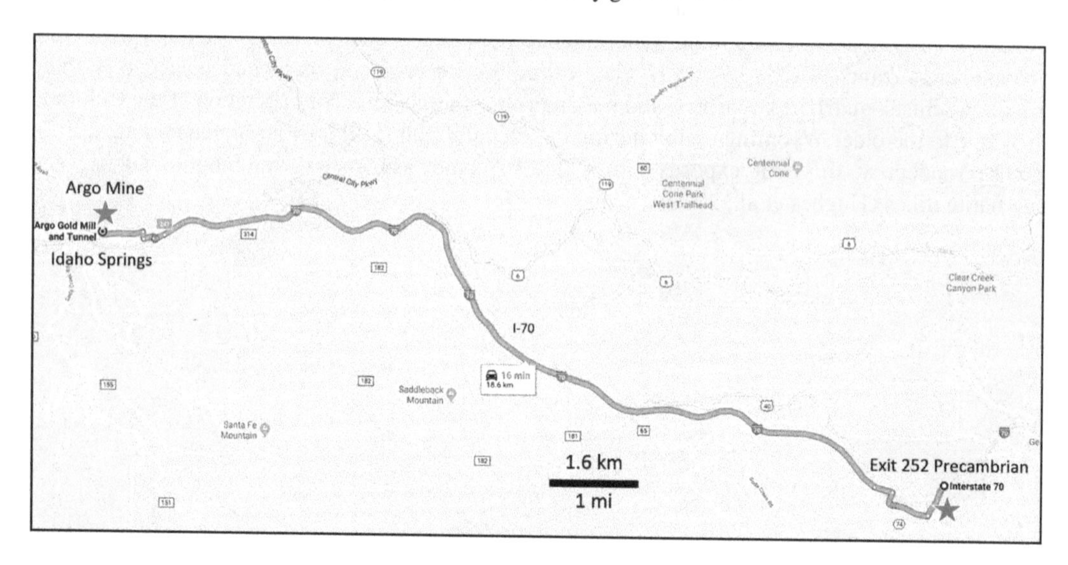

***Exit 252 Precambrian to Idaho Springs:** Continue west on I-70 to Exit 241; exit freeway to roundabout; take Colorado Blvd west to Gilson St; turn right (north) on Gilson St and drive to Riverside Drive; turn left (west) on Riverside Drive and drive to **Stop 7, Argo Mine** (39.742771, −105.506990), for a total of 18.6 km (11.6 mi; 16 min).*

***Side Trip 1, Exit 252 Precambrian to Central City:** Continue on I-70 west to Exit 243 to Central City. Turn right onto Central City Parkway going north to Central City and drive to Hidee Mine Road; turn left and drive to **Stop ST1, Hidee Mine, Central City** (39.787117, −105.500385), for a total of 28.7 km (17.8 mi; 25 min).*

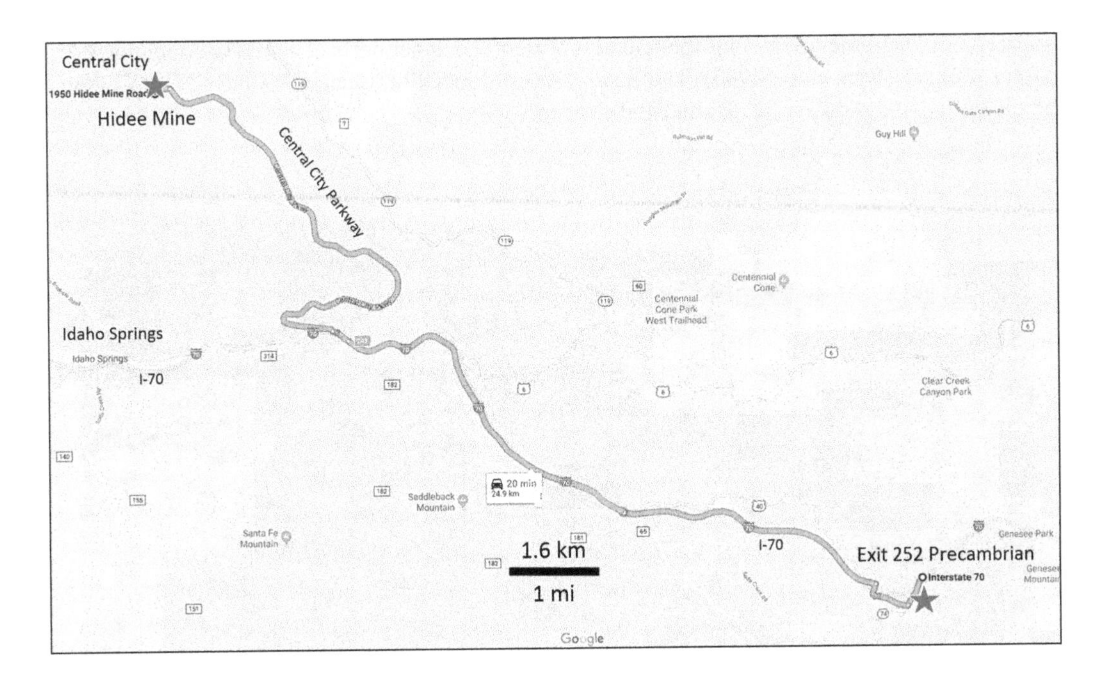

SIDE TRIP 1 CENTRAL CITY

History

In May 1859, while searching for the source of the placer gold in Clear Creek, John Gregory discovered the Gregory Lode, a gold-bearing vein between Black Hawk and Central City. Within a year, as many as 10,000 prospectors had come to Mountain City, as the area was then known. The most important mining district in the Front Range, the Central City-Blackhawk area produced gold, silver, lead, zinc, and copper, but declined in the early 1900s. Most of the veins were exhausted by 1920, but the increase in the price of gold from $20 to $35/oz in the early 1930s revived mining temporarily. The district was effectively dead by World War II. From 1950 to 1955, prospecting was stimulated by the search for uranium in the district. Uranium mining proved unsuccessful, and the population of Central City and Black Hawk fell to a few hundred.

The new gold rush is in casinos. Gambling was introduced in the early 1990s, and as of 2018, the thriving industry supports 6 casinos in Central City and 18 casinos in Black Hawk.

Geology

The road to Central City contains several excellent roadcuts exposing Precambrian gneiss and schist of the Idaho Springs Formation. Rocks that were originally volcanic are now dark gray amphibolite, a metamorphosed basalt. The Central City-Blackhawk gold mining district lies in the same Precambrian gneiss-granite-pegmatite terrane. Abundant faults cut these rocks. Derived from hot, mineral-rich fluids associated with small Tertiary-age (Laramide) intrusions, the ore occupies veins and stockworks that filled these faults. The veins have been classified into pyrite veins and galena-sphalerite veins. The pyrite veins contain pyrite and quartz and locally chalcopyrite (copper ore), tennantite (copper), enargite (copper), sphalerite (zinc), and galena (lead with minor silver). The galena-sphalerite veins comprise lead and zinc but also minor copper mineralization. A central ore zone about 5 km (3 mi) in diameter contains the pyrite veins and is surrounded by a zone of galena-sphalerite veins (Simms et al., 1963). The distribution of minerals represents a zoned hydrothermal system centered on a Laramide quartz-trachyte or rhyolite breccia pipe. The mineralization is thought to represent the upper parts of a molybdenum porphyry system at depth (Rice et al., 1985). The district has produced between $100 and $200 million of mainly gold, but also silver, lead, zinc, and copper.

Central City, Colorado circa 1862. (Wikipedia, Central City, https://commons.wikimedia.org/wiki/File:Central_City,_Colorado_(1862).png.)

Central City and old mines today.

The Coeur d'Alene Mine Shaft House is a relic of Central City's mining heritage. Developed in 1885, the Coeur d'Alene mine produced ore until the 1940s. Tours of the red building, on Academy Hill overlooking Central City, allow visitors to see antique mining equipment, the inner workings of a mine shaft house, and panoramic views of Central City/Black Hawk mining district.

ST1.1 HIDEE GOLD MINE TOUR

The Hidee Gold Mine in Central City offers tours and teaches gold panning. Tourists are allowed to keep samples of gold ore and any gold found during panning.

Visit

See website for current cost of Mine Tours and panning lessons:
- www.hideegoldmine.com/
- www.hideegoldmine.com/hidee.htm

Hours
- November to April 30
  - Open 10:30 am–2:30 pm
  - Tours: 11, 12, 1, and 2 pm
- May
  - Monday through Friday from 2 to 3:30 pm; Saturday and Sunday from 10:30 am to 3:30 pm, tours at 11, 12, 1, 2, 3 pm.
- June–July–August
  - Open 7 days a week, 9:30 am to 4:30 pm, tours at 10, 11, 12, 1, 2, 3, and 4 pm

September–October
Open 7 days a week, 9:30 am to 3:30 pm
Tours at 10, 11, 12, 1, 2, and 3 pm
Address: 1950 Hidee Mine Road, Central City, CO 80427
Phone: (720) 548-0343

Side Trip 1—Hidee Gold Mine to Central City Parkway: *Return south on the Central City Parkway and pull over on the right just before the final hairpin curve. This is* ***ST1.2, Idaho Springs Formation*** *(39.749665, −105.469572), for a total of 9.3 km (5.8 mi; 8 min).*

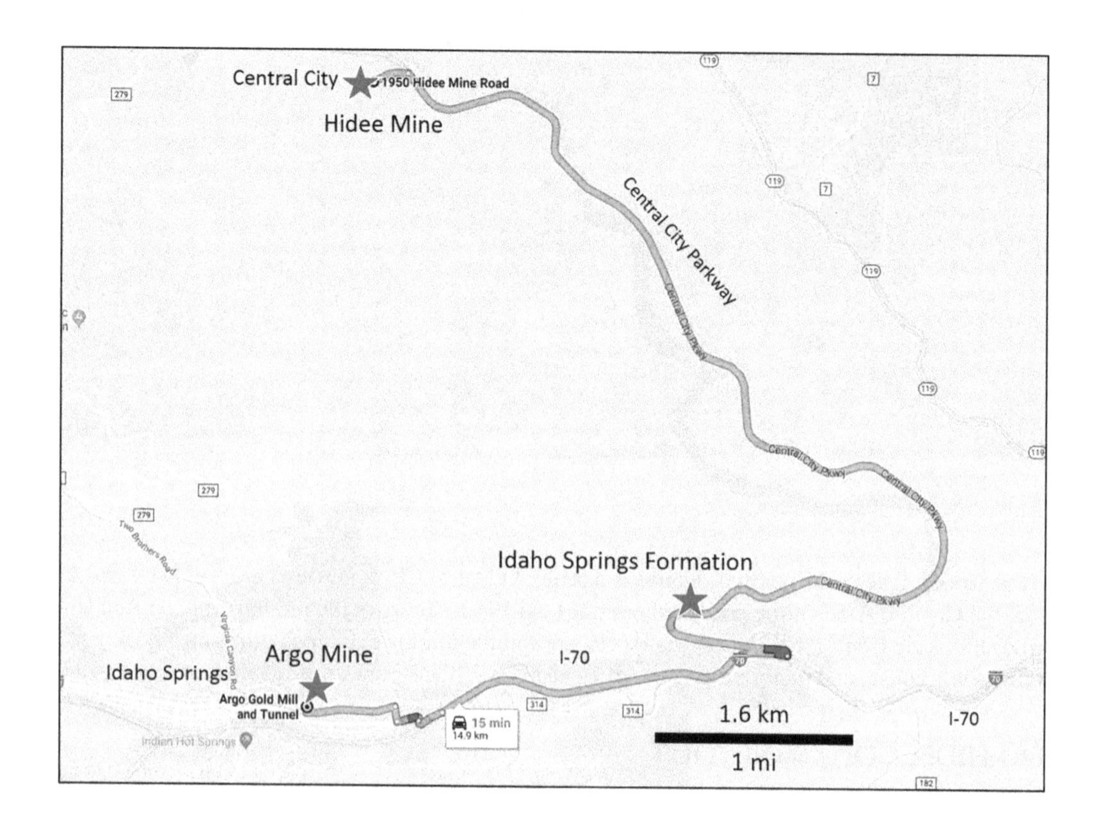

ST1.2 IDAHO SPRINGS FORMATION, CENTRAL CITY PARKWAY

The roadcut at this pullout exposes Idaho Springs Formation quartz-feldspar gneiss and hornblende gneiss intruded by granitic dikes.

Idaho Springs Formation gneiss and intrusive dikes exposed in roadcut, Central City Parkway. Google Street View looking northwest.

Side Trip 1—Central City Parkway to Idaho Springs: *Merge onto I-70 westbound and drive to Exit 241, Idaho Springs. Take the exit to Colorado Blvd and go west to Gilson Street; turn right (north) on Gilson to Riverside Drive; turn left (west) on Riverside to 23rd Ave.; turn right (north) on 23rd to Wall St.; turn right (east) on Wall St. to* ***Stop 6, Argo Mine,*** *for a total of 5.6 km (3.5 mi; 8 min).*

STOP 6 IDAHO SPRINGS, THE COLORADO MINERAL BELT, AND ARGO MINE

I-70 enters the Colorado Mineral Belt at Idaho Springs. What, exactly, is the Colorado Mineral Belt?

The Mineral Belt is a cluster of metallic mineral deposits that extends 400 km (250 mi) diagonally from the San Juan Mountains of southwest Colorado northeast to near Boulder in north-central Colorado. It includes the major mining districts of Ouray-Sneffels-Silverton-Telluride in the San Juan Mountains, the Leadville district on the Continental Divide, and the Central City-Idaho Springs district in the Front Range. The mineral deposits are thought to be controlled by zones of weakness in the continental crust inherited from a Precambrian northeast shear zone. The shear zone localized intrusion of Laramide (75–42 million years), mid-Tertiary (40–26 million years), and late Cenozoic (25 million years and younger) magmas and ore-bearing fluids. Intrusions and mineral deposits appear to be located at the intersection of the regional northeast shear zone with local northwest shear zones. Mining districts within the mineral belt have produced over 25 million troy oz (778 metric tons) of gold, along with silver and minor lead and zinc, between discovery in 1858 and 2000.

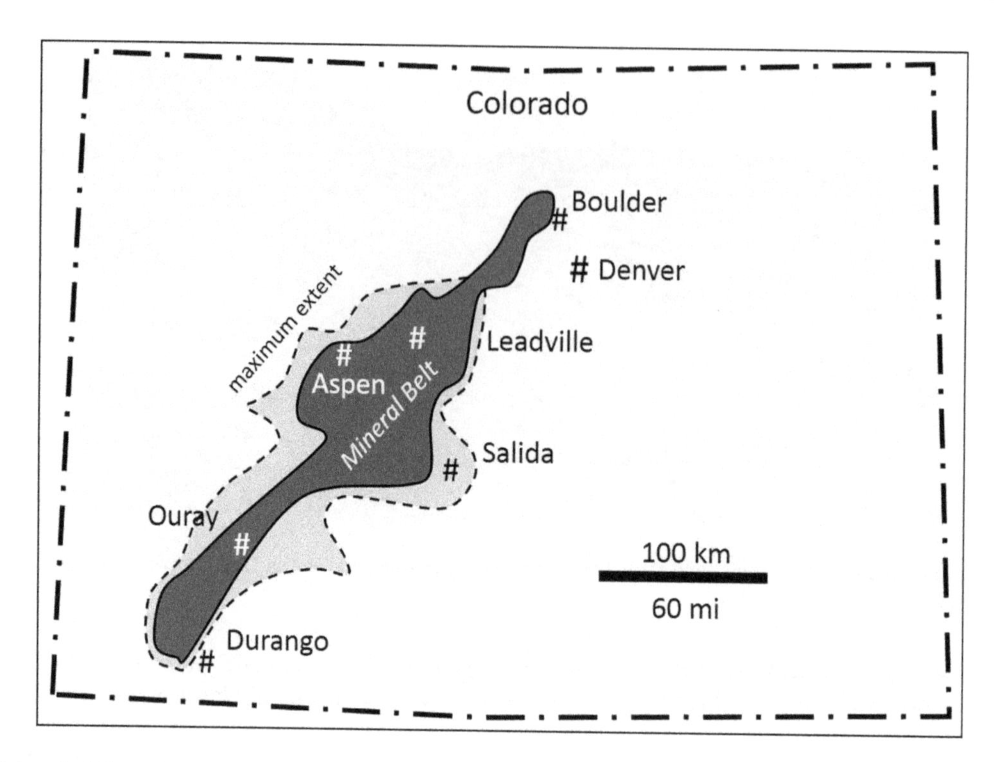

Colorado Mineral Belt. (Courtesy of Omphacite, https://commons.wikimedia.org/wiki/File:CO_Mineral_Belt.jpg.)

In 1859, George A. Jackson discovered placer gold at the present site of Idaho Springs. This was the first major gold discovery in Colorado. The placer deposits were soon played out, and prospectors then found gold veins in the valley walls.

Local legend has it that Native Americans from Idaho traveled here to bathe in the hot springs, hence the town's name. Be that as it may, the site was originally known as Jackson's Diggins. It was later called "Sacramento City," "Idahoe," "Idaho City," and finally "Idaho Springs." The town was the center of regional mining due to railroad access to Denver.

Idaho Spring's Argo Mine was a gold mine and gold milling operation. It sits at the entrance of the Argo Tunnel, bored between 1893 and 1910 to drain water from the gold mines in the Idaho Springs-Central City mining district. The mill was built at the tunnel entrance to process ore from the many mines drained by the tunnel. The property was closed from 1943 until 1976, when it was refurbished to preserve a sample of Colorado's mining history. The tour is popular with visitors.

The Argo Gold Mine and mill, Idaho Springs. (Photo courtesy of Elizabeth Prost.)

Visit

Argo Mine Tours: See website for current tour fees.
Website: http://argomilltour.com/
Hours
Open 7 days a week, 10 am to 6 pm
Tours Daily: 11 am, 12 pm, 1 pm, 2 pm, 3 pm and 4 pm
Address: Argo Gold Mill & Tunnel, 2350 Riverside Dr., Idaho Springs, CO 80452
Phone: (303) 567-2421

Argo Gold Mine to Georgetown Loop Railroad: *Return to I-70 and drive west to Exit 228, Georgetown; exit and turn right (south) on Argentine Street; bear right onto Loop Drive and continue to* ***Stop 7, Georgetown Loop Railroad*** *(39.701429, −105.706399), for a total of 21.7 km (13.5 mi; 18 min).*

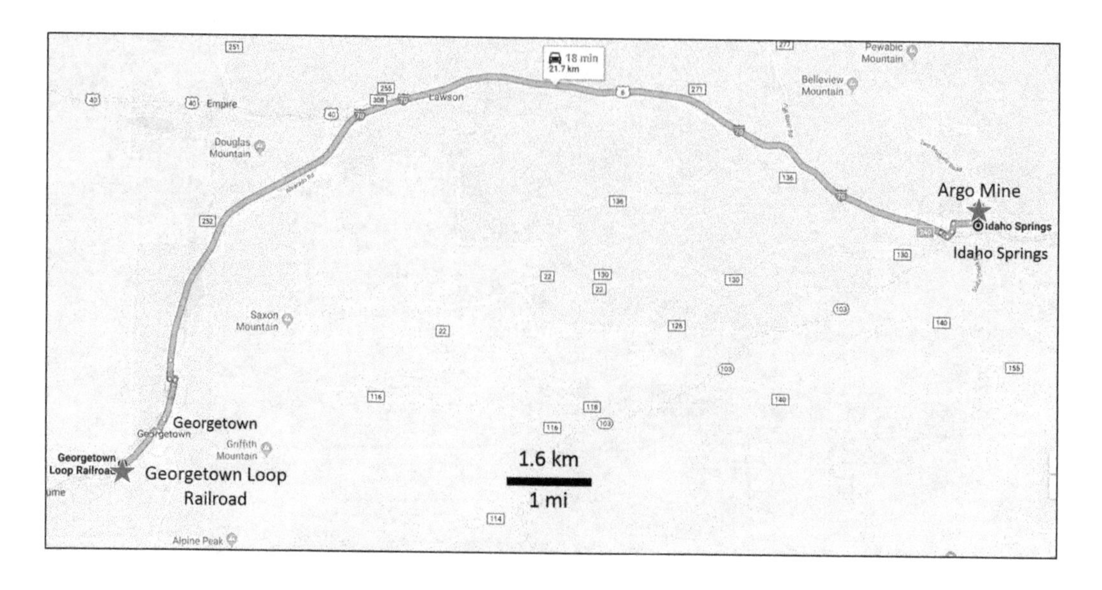

Side Trip 2—Argo Gold Mine to Mount Evans: *Return to Colorado Blvd. Take Colorado Blvd west to 13th Ave.; turn left (south) on 13th/CO-103 and take it to the summit of* ***Mount Evans*** *(39.587778, −105.642372), for a total of 45 km one way (28.3 mi; 1 h 2 min).*

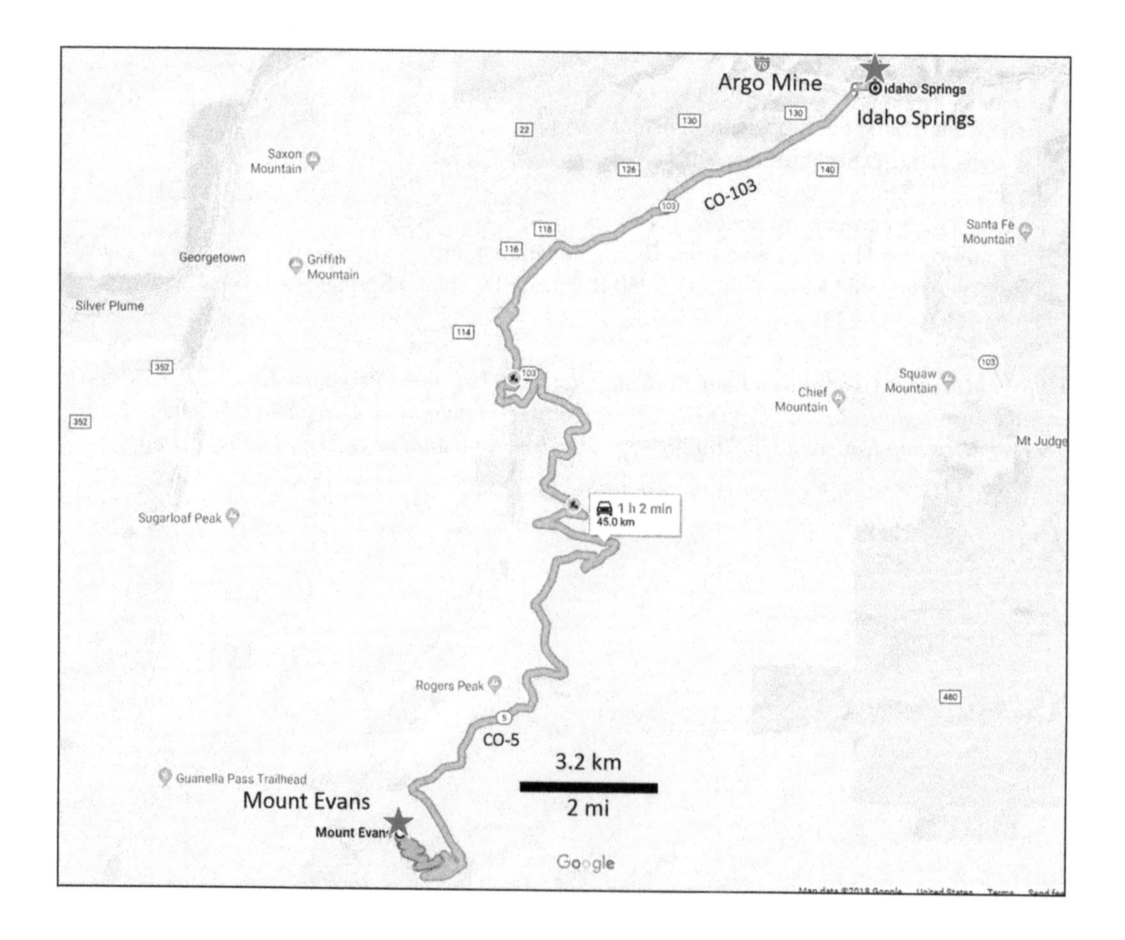

SIDE TRIP 2 MOUNT EVANS

You can drive to the 4,350 m (14,271 ft) summit of Mount Evans on "the highest paved road in North America." The road is generally open from Memorial Day to Labor Day: it is closed in the winter due to snow. The peak can be seen from 160 km (100 mi) away in the clear, dry air. Likewise, the view from the top is breathtaking.

The average annual temperature is −8°C (18°F); the highest recorded temperature is 18°C (65°F), and the lowest was −40°C (−40°F). The average wind speed is 51 km/h (31 mph). Because of the elevation, many suffer altitude sickness.

The rock underpinning Mt. Evans is the Mount Evans batholith, a 1.4 Ga granodiorite intrusion. It was part of *Frontrangia*, a range of the Ancestral Rocky Mountains. By Eocene time, it was one of a handful of resistant peaks rising above a flat erosion surface. Since then, uplift of the modern Rockies has led to down-cutting of canyons and glaciation. Most of the surrounding lakes are glacial tarns formed in cirques at the head of now-vanished glaciers.

Mount Evans and Summit Lake. (Courtesy of Boilerinbtown, https://commons.wikimedia.org/wiki/File:Mt_Evans.JPG.)

In 1972, the University of Denver built the 0.6 m (24 in) Ritchey-Chrétien telescope on the summit. In 1996, the university erected the Meyer–Womble Observatory nearby. At the time, it was the highest optical observatory in the world.

Side Trip 2—Mount Evans to Georgetown Loop Railroad: *Return north to I-70 and drive west to Exit 228, Georgetown; exit and turn right (south) on Argentine Street; bear right onto Loop Drive and continue to* ***Stop 7, Georgetown Loop Railroad,*** *for a total of 66.7 km (41.5 mi; 1 h 20 min).*

STOP 7 GEORGETOWN LOOP RAILROAD AND MINES (MILE 227)

History

Georgetown, the "Silver Queen of the Rockies," was established when George Griffith discovered gold at the head of Clear Creek in 1859. In 1864, a rich silver vein, the "Belmont lode," was discovered southwest of Silver Plume and marked the beginning of the silver boom. Opening of the Blackhawk smelter in 1867 greatly stimulated mining in the area. For a short time, this district was the world's largest silver producer. Production peaked in 1894. Zinc production became important during World Wars I and II.

More than 200 of the original buildings still stand in Georgetown's historic district, and mine tailings dot the canyon walls above the town.

Geology

Country rocks are Precambrian Silver Plume Granite and Idaho Springs biotite gneiss. Tertiary stocks and dikes of various compositions are abundant. Mineralization occurs either as silver-rich veins containing galena-sphalerite-pyrite with little gold, or as pyritic gold veins containing pyrite-chalcopyrite and gold with some galena, sphalerite, and silver minerals. Ore near the surface was rich in gold; silver ore occurred at depth.

Railroad and Mine Tours

The "Georgetown Loop" is a narrow gage railroad operated by the Colorado State Historical Society that passes over its own tracks to gain 195 m (640 ft) between Georgetown and Silver Plume. The railroad runs 4.8 km (3 mi) long between the two towns and crosses the canyon on a 29 m (95 ft) high trestle.

The original railroad was completed in 1884 and was considered an engineering marvel. The line was dismantled in 1939, but in 1959, the Georgetown Loop Historic Mining & Railroad Park was formed by the Colorado Historical Society to restore it. Construction began in 1973 with track and ties donated by the Union Pacific Railroad, and a new high bridge was built. The restored railroad opened in 1984.

The train ride includes an optional walking tour of the Lebanon Silver Mine, located near the halfway point on the railroad.

Historic Georgetown Loop Railroad and high trestle. (Courtesy of D. O'Brien, https://commons.wikimedia.org/wiki/File:Georgetown_Loop,_c._1885.jpg.)

Visit

Lebanon Silver Mine Tour:

The tour takes you 150 m (500 ft) into a mine tunnel bored in the 1870s. Guides point out rich silver veins and describe mining history. The temperature inside the mine is a constant 44°F, so bring a jacket.

Lebanon Extension Mine Tour:

The Lebanon Extension tour takes you over 275 m (900 ft) into Leavenworth Mountain following an ore body that extends from the mines of Silver Plume to the Silver Queen mine in Georgetown.

The Everett Gold Panning and Silver Mine Tour:

The Everett Mine gives visitors a sense of a mine as it was in the late 1880s. Guests also pan for gold and can take home everything they find!

Georgetown Loop Railroad Tour

Call For Reservations: 1-888-456-6777
Email: info@historicrailadventures.com
Address: P.O. Box 249, Georgetown, CO 80444
Devil's Gate Depot: 646 Loop Drive, Georgetown, CO 80444
Silver Plume Depot: 825 Railroad Avenue, Silver Plume, CO 80476
Tickets can be purchased online or reserved by calling 1-888-456-6777. Prices vary based on seating, whether you include a mine tour, time of year, and age. All prices subject to change. See website for current pricing.
Website: georgetownlooprr.com

Just east of Silver Plume on I-70, the Georgetown Loop Overlook (eastbound lanes only) provides scenic views of the Georgetown-Silver Plume mining district. The cliffs along the highway between Georgetown and Silver Plume are known for rock falls, so be aware.

Continuing west on I-70, the highway goes through the Eisenhower Tunnel, but you can drive to the Continental Divide following US-6, the scenic route. As you approach the Eisenhower Tunnel, there are avalanche chutes on both sides of the highway. The rocks at the tunnel and at Loveland Pass consist of Silver Plume Granite and Idaho Springs Formation gneiss. A glaciated landscape extends in all directions from the pass.

Mile 216–Avalanche Chutes south of I-70 just west of the Loveland Ski Area turnoff (Exit 216).

Georgetown Loop Railroad to Williams Range Thrust: *Return through Georgetown to I-70 west. Continue west to* ***Stop 8, Precambrian on Williams Range Thrust*** *(39.650338, −106.018414), at Mile 208.1 (just west of the runaway truck ramp). Pull over on the right shoulder, for a total of 33.4 km (20.7 mi; 25 min).*

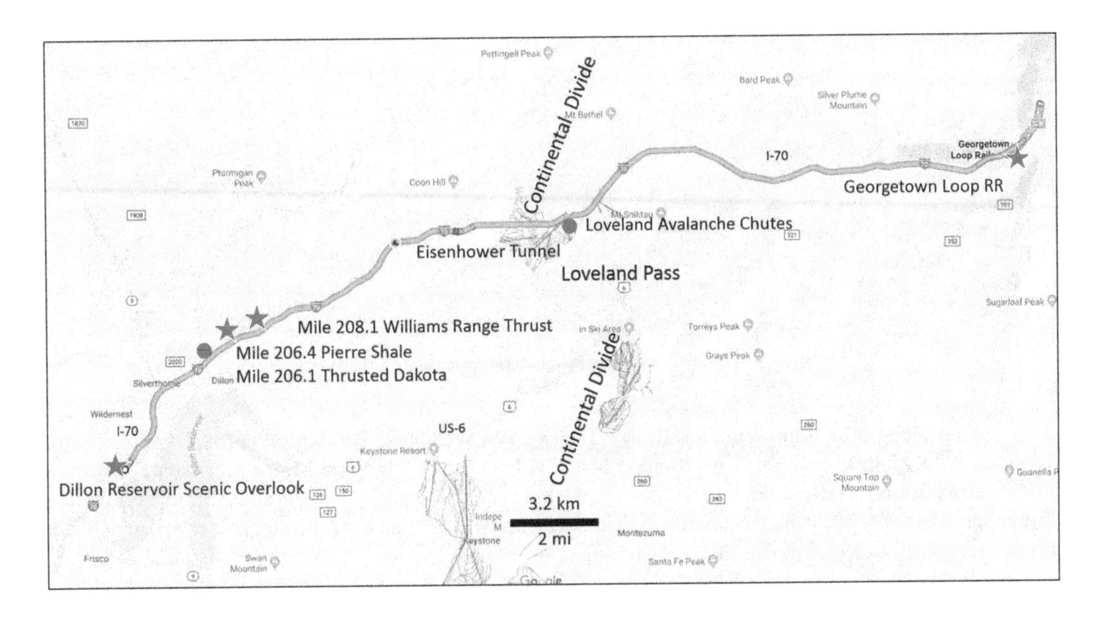

STOP 8 PRECAMBRIAN CARRIED ON WILLIAMS RANGE THRUST (MILE 208.1)

The Williams Range Thrust forms the western edge of the Laramide Front Range Uplift. Although it is covered here, roadcuts show Precambrian granitic rocks adjacent to Late Cretaceous Pierre Shale. A cross section shows Precambrian thrust over Cretaceous rocks. The thrust has at least 9 km of west-directed displacement. The fault was confirmed during the construction of the Roberts Tunnel, built to carry water from Dillon Reservoir to Denver. The thrust is mapped at the break in topography.

Precambrian granite exposed between Miles 208 and 213 was carried west by the Williams Range Thrust.

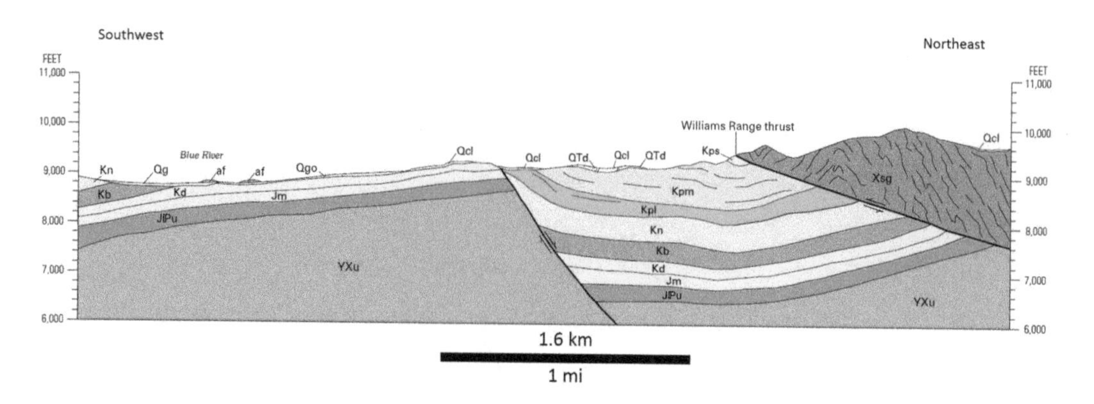

Cross section along I-70 from Williams Range Thrust to Silverthorne. Kp, Kpm, and Kpl are all Pierre Shale; Kn = Niobrara Formation; Kb = Benton Shale; Kd = Dakota Sandstone; Jm = Morrison Formation; JPu = Entrada Sandstone (Middle Jurassic), Chinle Formation (Upper Triassic), and Maroon Formation (Lower Permian to Middle Pennsylvanian), undivided; YXu = Middle? and Early Proterozoic rocks, undivided. (From Kellogg, 2002.)

Precambrian to Pierre Shale: *Continue west on I-70 for 2.7 km (1.7 mi; 2 min) and pull over on the right shoulder. This is* ***Stop 9, Pierre Shale*** *(39.639506, −106.046246).*

STOP 9 PIERRE SHALE (MILE 206.4)

Between the last outcrop and this one, you crossed over the Williams Range Thrust. This west-directed thrust moved Precambrian Rocks up and over the Cretaceous Pierre Shale. In the process, the Pierre Shale near the thrust was dragged along and overturned so that it is dipping east. The greenish-gray Pierre Shale is laterally continuous with the Mancos Shale farther west. Both are deep marine mudstones.

East-dipping, overturned Pierre Shale, equivalent to the Mancos Shale of western Colorado and Utah. The Pierre Shale is adjacent to and below the Williams Range Thrust on the north side of I-70 at Mile 206.4. Google Street View looking west.

Continuing west from Stop 9, in half a kilometer (0.3 mi), you pass an outcrop of thrusted Dakota Formation sandstone on the right (north) side of the freeway at Mile 206.1 (39.635480, −106.051704). A small, west-directed thrust fault with perhaps 1 m of displacement can be seen in the Dakota Sandstone near the base of the outcrop. The narrow shoulder and heavy traffic make it difficult and possibly unsafe to stop here.

Small west-directed thrust in the Dakota Sandstone at Mile 206.1. Google Street View north.

Pierre Shale to Dillon Scenic View: *Continue west on I-70 to* ***Stop 10, Scenic View*** *pullout and rest area (39.601577, −106.089031) on the right, for a total of 6 km (3.8 mi; 4 min).*

STOP 10 DILLON RESERVOIR SCENIC OVERLOOK (MILE 203.5)

You are in the Blue River Valley. The valley occupies a half graben (downfaulted block) between the roughly north-south, down-to-the-east Blue River normal fault near Frisco and the Williams Range Thrust east of Dillon. The valley is the northernmost expression of the Rio Grande Rift, a 20–35 million-year-old rift extending south through Colorado and New Mexico to Chihuahua, Mexico. The earliest basin fill in the graben is middle Oligocene (27 Ma). In a few tens of millions of years, this may be ocean-front property.

View south over Dillon Reservoir, a major source of Denver's water. Mt. Guyot is on the left; Bald Mountain on the right.

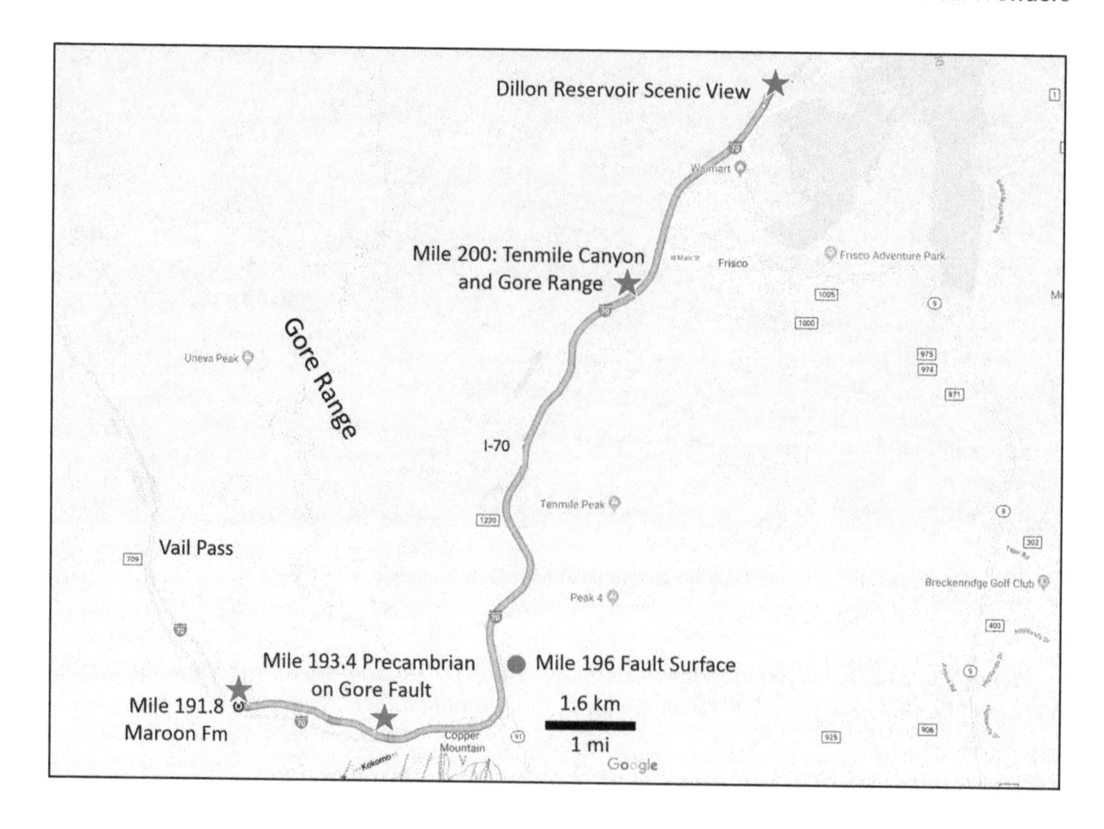

Dillon Lake Scenic Overlook to Tenmile Canyon: *Continue west on I-70 to* ***Stop 11, Tenmile Canyon*** *at Mile 200 (39.572348, −106.114401), for a total of 5 km (3 mi; 4 min). Pullout on the right is just west of the outcrop.*

STOP 11 TENMILE CANYON AND THE GORE RANGE (MILE 200)

At Exit 201 I-70 crosses the Blue River Fault and enters Tenmile Canyon in the Gore Range. The Gore Range is a Precambrian-cored mountain range that was carried west by the Gore Thrust. Tenmile Canyon is at least partly controlled by erosion along a fault. It cuts through near-vertical Precambrian metasediments that include hornblende gneiss, quartz-biotite gneiss, schist, and migmatitic (partially melted) biotite gneiss.

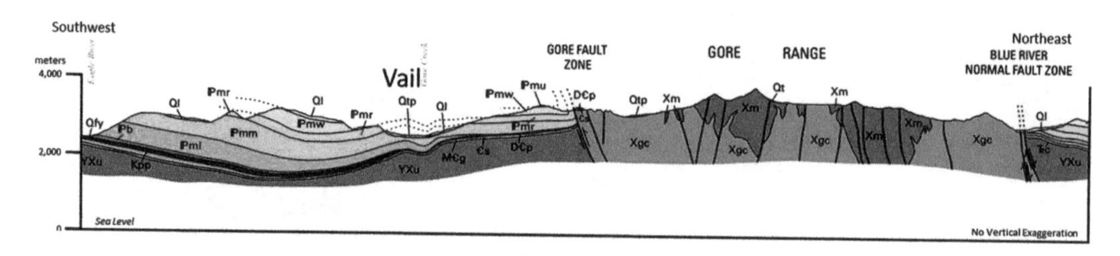

Southwest-northeast cross section through Vail and the Gore Range about 20 km north of Tenmile Canyon. Xgc = Cross Creek Granite; Xm = migmatitic biotite gneiss. (From Kellogg, et al. 2011.)

Precambrian metasediments in a roadcut at the east entrance to Tenmile Canyon, Mile 200.

Excelsior claim and mine, Mile 200.

The Excelsior Mine claim is located just west of the Blue River Fault. You can see it from this stop on the north side of the freeway. This inactive silver and gold mine was one of the most important producers in Ten Mile Canyon from the 1890s through the 1910s.

As you continue west on I-70, a fault is exposed as a smooth, "slickensided" (polished and striated) rock surface on the east side of Tenmile Canyon above Curtain Ponds at Mile 196. As the offset is down-to-the-west, this is probably a northern extension of the Mosquito Fault (Kellogg et al., 2011). The Mosquito Fault is a major regional structure that defines the east side of the Rio Grande Rift south of Leadville.

Fault surface (arrow) above Curtain Ponds, east side of valley at Mile 196, Tenmile Canyon.

Tenmile Canyon to Precambrian on Gore Fault: *Continue west on I-70 for 10.2 km (6.4 mi; 6 min) to Mile 193.4 and pull over on the right shoulder. This is* ***Stop 12, Precambrian on Gore Fault*** *(39.502767, −106.170412).*

Side Trip 3, Tenmile Canyon to Climax Mine: *Continue west on I-70 to Exit 195, Colorado 91 South to Copper Mountain/Leadville; head south on CO-91 to pullout on the right opposite the entrance to the Climax Mine. This is the* ***Side Trip 3 Stop, Climax Mine*** *(39.370632, −106.190621), for a total of 25.1 km (15.6 mi; 20 min).*

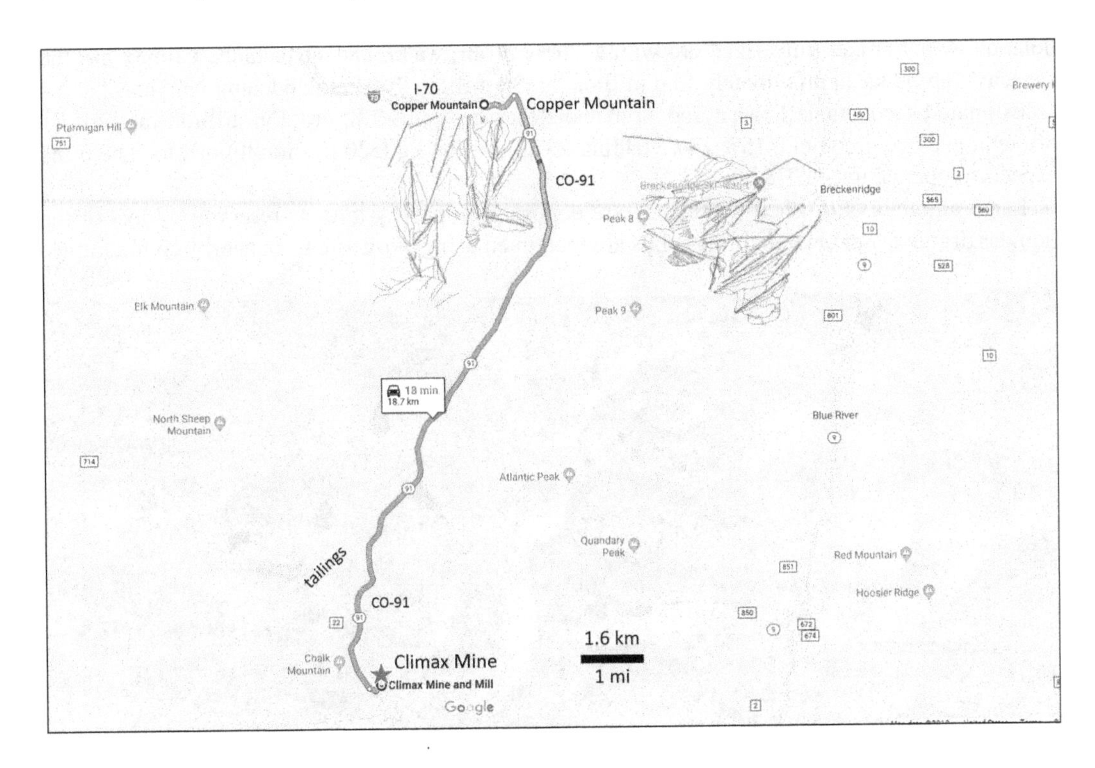

SIDE TRIP 3 CLIMAX MINE OVERLOOK

At one time, Climax was the world's largest molybdenum mine. The underground workings are not visible from the road, but you can see extensive tailings ponds consisting of crushed waste rock, and you can see some of the open-pit mine workings and the altered country rock. The mine is owned by Climax Molybdenum Company, a subsidiary of Freeport-McMoRan.

Travelers along CO-91, between Copper Mountain and Leadville, can tune their car radios to a special AM station to learn about the mine, and interpretive signs near the parking lot describe the geology of the ore body.

HISTORY

Charles Senter discovered molybdenite (molybdenum sulfide) veins here in 1879, during the Leadville Silver Rush, but had no idea what the mineral was. He maintained his claims, convinced that they had value. In 1895, a chemist finally identified the gray mineral as molybdenum ore. At the time, there was no market for the metal.

During World War I, the U.S. army learned that the Germans were using molybdenum to strengthen steel and increase the durability of weapons. A mine was opened and produced its first ore in 1915. In 1918, Senter was paid $40,000 for his claims and retired to Denver.

Demand for molybdenum fell dramatically after World War I. And yet today, molybdenum is used in jet engine turbines, electrical filaments, missile parts, nuclear power plants, and as a catalyst in petroleum refining.

The mine closed in 1995 because of low molybdenum prices. It reopened in May 2012. Historically, mining was by block caving, a method that involves undermining an ore body and allowing the broken rock to collapse under its own weight into a void where it can be gathered and hauled off to a crusher.

Current mining operations at Climax are by open-pit mining. The mine includes a 25,000 metric ton/day mill. The ore is crushed, and the molybdenite is separated from waste material by froth flotation, which mixes pulverized ore with a slurry of air, water, and surfactants. Climax has the capacity to produce approximately 13.6 million kg (30 million lb) of molybdenum per year.

Estimated recoverable Proven and Probable Reserves, end 2018, are 168 million metric tons of ore at average grade of 0.15% Mo, yielding 235.9 million kg (520 million lb) of Mo. The mine expects to operate until 2038.

Large amounts of crushed waste rock are deposited in nearby tailings reservoirs. Most of the liquid is drained, and the remaining solids are reclaimed using bio-waste from nearby communities.

Climax open-pit mine as seen looking east from CO-91.

Google Earth view of the Climax open-pit molybdenum mine at Fremont Pass. Imagery © 2019 Landsat/Copernicus; Maxar Technologies; USDA Farm Service Agency.

Geology

The deposit at Climax consists of many intersecting small veins of molybdenite. This "stockwork" occurs in an altered quartz monzonite porphyry intrusive (quartz-rich igneous rock that solidified deep underground). The ore is low-grade, but the ore body is large. Beside molybdenum, the mine has also produced minor amounts of tin (from cassiterite) and tungsten (from hübnerite).

Climax-type moly deposits are thought to originate in extensional tectonic settings inland from shallow-angle subduction zones (Colorado Geological Survey, online; Wikipedia, Climax Mine; Freeport-McMoRan, 2018).

Side Trip 3—Climax Mine to Precambrian on Gore Fault: *Return north to I-70 and enter the freeway westbound. Drive to* ***Stop 12, Precambrian on Gore Fault*** *(39.502767, −106.170412), for a total of 20.4 km (12.7 mi; 14 min).*

STOP 12 PRECAMBRIAN CARRIED ON THE GORE FAULT (MILE 193.5)

The Gore Fault is a high-angle east-dipping reverse fault that bounds the west side of the Gore Range and puts Precambrian gneiss on the east over Pennsylvanian Minturn Formation on the west. The fault is covered where it crosses I-70.

The outcrops immediately west of this stop are Minturn Formation, over 1,900 m (6,200 ft) of red sandstone, conglomerate, and shale deposited in the Eagle Basin that existed between Frontrangia and the Uncompahgre Uplift. The Minturn grades laterally into the Eagle Valley Formation farther west in the basin.

Precambrian carried on the Gore Fault outcrops between Copper Mountain and Vail Pass.

Precambrian on Gore Fault to Vail Pass Maroon Formation: *Continue on I-70 west 2.4 km (1.5 mi; 1 min) to Mile 191.7 (39.506498, −106.197601) and pull off on the right shoulder. This is* ***Stop 13, Maroon Formation, Vail Pass****.*

STOP 13 MAROON FORMATION, VAIL PASS (MILE 191.7)

The striking red color of the Maroon Formation here is due to oxidized iron in the sedimentary rock. These layers were originally laid down as braided channel sands and alluvial conglomerates. They are the western equivalent to Fountain Formation seen at Red Rocks Park and record uplift and erosion of the Ancestral Rockies.

Pennsylvanian-Permian Maroon Formation exposed in a roadcut east of Vail Pass.

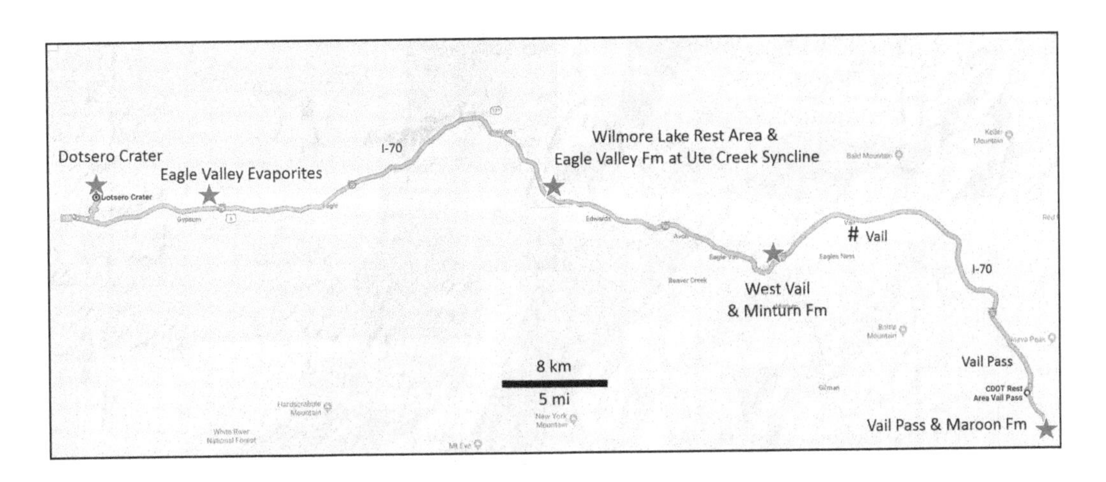

Vail Pass Maroon Formation to West Vail Minturn Formation: *Continue west on I-70 for 30.6 km (19 mi; 19 min) to* ***Stop 14, West Vail*** *(39.614630, −106.442536) and pull over on the right shoulder beyond the guardrail.*

STOP 14 MINTURN FORMATION, WEST VAIL (MILE 171.9)

This stop highlights the transition from red, alluvial Minturn shed west off the flank of *Frontrangia* (seen east of Vail) to light gray marine Minturn shale and thin sandstone and limestone layers here that were deposited on the east side of the Pennsylvanian Eagle Basin. As you go farther west in the Eagle Basin, the unit becomes rich in gypsum, with some salt, indicating evaporation of the shallow Pennsylvanian sea. The salt has mostly been dissolved away.

Marine Minturn Formation sandstone and shale west of Vail. These rocks grade laterally into the terrestrial Minturn (red sandstones) seen east of Vail.

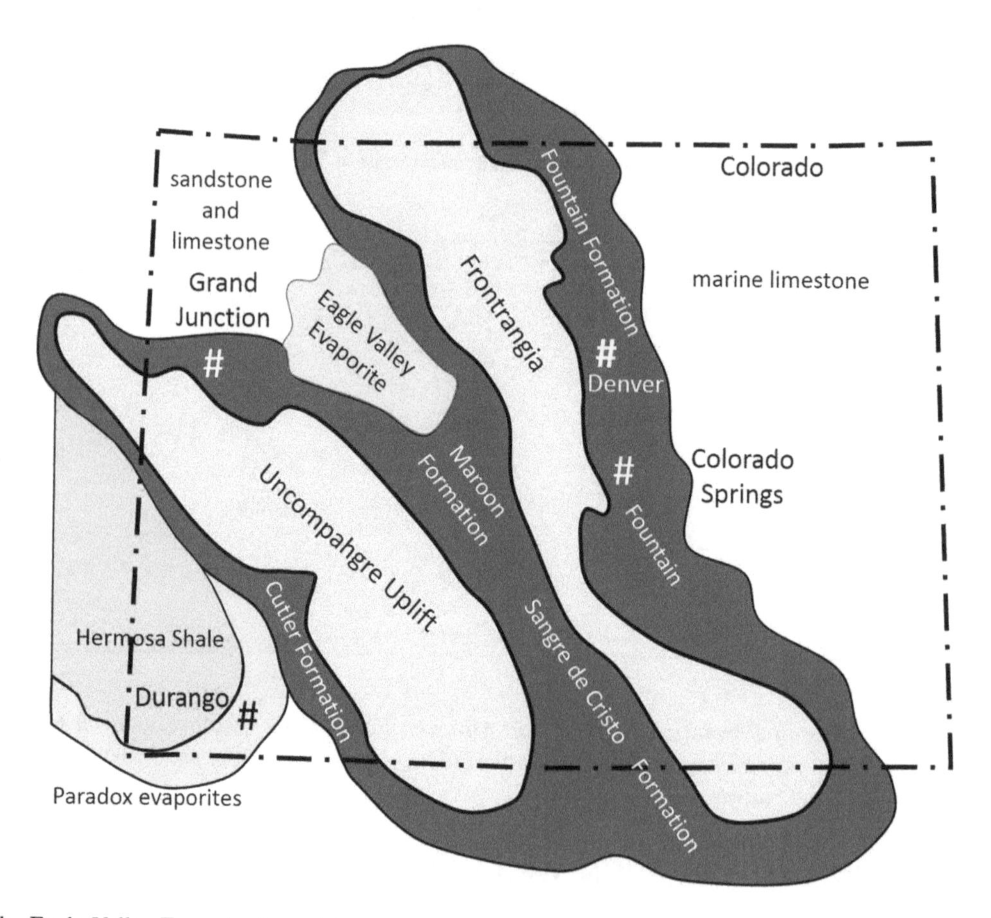

The Eagle Valley Evaporite formed in an evaporating inland sea on the flanks of the Ancestral Rockies. The Maroon, Minturn, and Fountain formations were shed off the Pennylvanian uplifts. (Data derived from Lindsey, 2010)

West Vail Minturn to Eagle Valley Formation: *Continue on I-70 west for 18.6 km (11.6 mi; 11 min) to* ***Stop 15, Edwards Rest Area at Wilmore Lake*** *(39.658401, −106.632150) on the right.*

STOP 15 EAGLE VALLEY FORMATION AT WILMORE LAKE, EDWARDS REST STOP (MILE 160.3)

The Edwards pullout at Wilmore Lake, Mile 160.3, exposes west-dipping Eagle Valley/Minturn Formation on the east limb of the Ute Creek Syncline.

Steeply west-dipping Eagle Valley Formation. This is near the transition from Minturn to Eagle Valley.

At Mile 159.3, we again drive through outcrops of the Maroon Formation. Since the units are getting younger, we are driving into a syncline (a fold with the youngest units in the center). The formations will continue to get younger until we pass through the axis of the Walcott Syncline.

Maroon Formation outcrop at Mile 159.3. Google Street View looking west.

Near Exit 157–156, we pass through the axis of the Walcott Syncline and the center of the Eagle Basin, having driven through Pennsylvanian, Permian, Jurassic, and Cretaceous units.

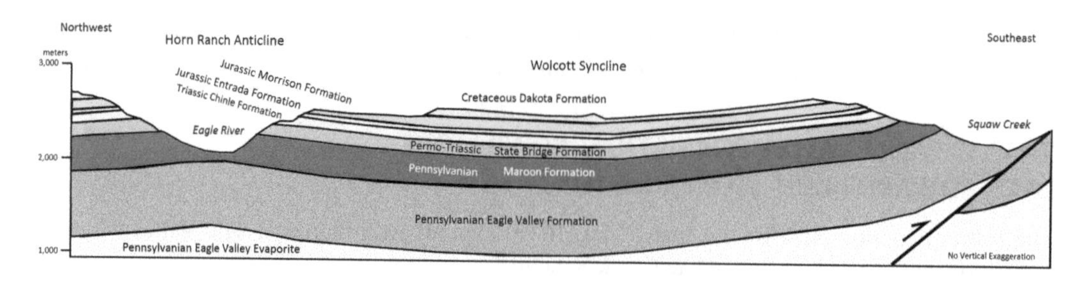

Northwest-southeast cross section through the Walcott Syncline south of I-70. (Modified after Lidke, 1998.)

Just east of the Eagle River crossing, near Mile 154, we go from Jurassic Morrison to Triassic Chinle redbeds. Since the units are now getting older as we drive west, we are driving into an anticline, in this case the Horn Ranch Anticline.

Eagle Valley Formation to Eagle Valley Evaporites: *Continue west on I-70 for 30.3 km (18.9 mi; 16 min) to* ***Stop 16, Eagle Valley Evaporites*** *(39.654487, −106.921796), and pull over on right shoulder beyond the west end of the guardrail.*

STOP 16 DEFORMED EAGLE VALLEY EVAPORITES (MILE 141.5)

The Eagle Valley Formation type section (where the unit was first described) is near the town of Eagle. The formation is 2,743 m (9,000 ft) thick here. This unit, which locally contains salt as well as gypsum, indicates evaporation of a shallow inland sea within the Ancestral Rocky Mountains. Because these ductile units flow readily under pressure, and the formation had been buried under thousands of meters of rock now eroded away, deformed Eagle Valley Evaporites can be seen along I-70 between Mile 139 and Mile 143.

The Eagle Valley Formation outcrops north of I-70 until you reach Dotsero, where you begin to enter the White River Uplift.

Gypsum outcrops on the north side of I-70 at Mile 141.5. Deformed Eagle Valley Evaporites are exposed between Mile 139 and 143, but most of it is fenced off.

The town of Gypsum and the American Gypsum drywall plant can be seen at Mile 140.

Between Gypsum and Dotsero, I-70 follows the axis of the east-west-trending Eagle River Anticline. Layers can be seen dipping away from the river both to the north and south.

Eagle Valley Evaporites to Dotsero Crater: *Continue west on I-70 to Exit 133 at Dotsero; exit freeway and turn right (east) on Frontage Road 8460; continue on Road 8460 to* ***Stop 17, Dotsero Crater and Flow Viewpoint*** *(39.647899, −107.041864), for a total of 15.7 km (9.8 mi; 17 min). If you would like to see the volcanic crater, continue north on unimproved Road 8460 another 2.5 km (1.5 mi) to get to the crater rim (39.659173, −107.033044).*

STOP 17 DOTSERO CRATER AND LAVA FLOW (EXIT 133)

Dotsero volcano is the youngest volcanic rock in Colorado at 4,200 years old. When it erupted, it created a 700 by 400 m (2,300 by 1,300 ft) crater, cinder cones, and a basaltic lava flow that crosses I-70 at Mile 135. The crater is located 1.4 km (0.9 mi) north of the highway. You can see the dark basalt on the hillside north of the highway between Miles 135 and 133.

Dotsero tuff and cinders (reddish units, valley bottom and center).

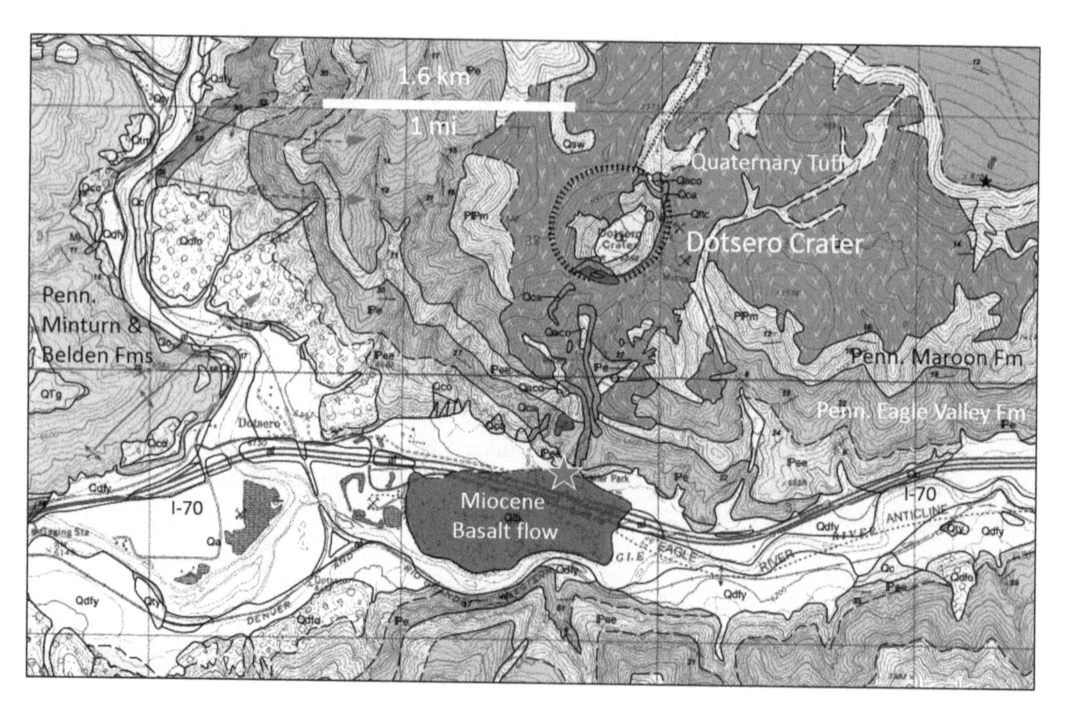

Dotsero crater and lava flow. The stop is indicated by the star. (From Streufert et al., 2008.)

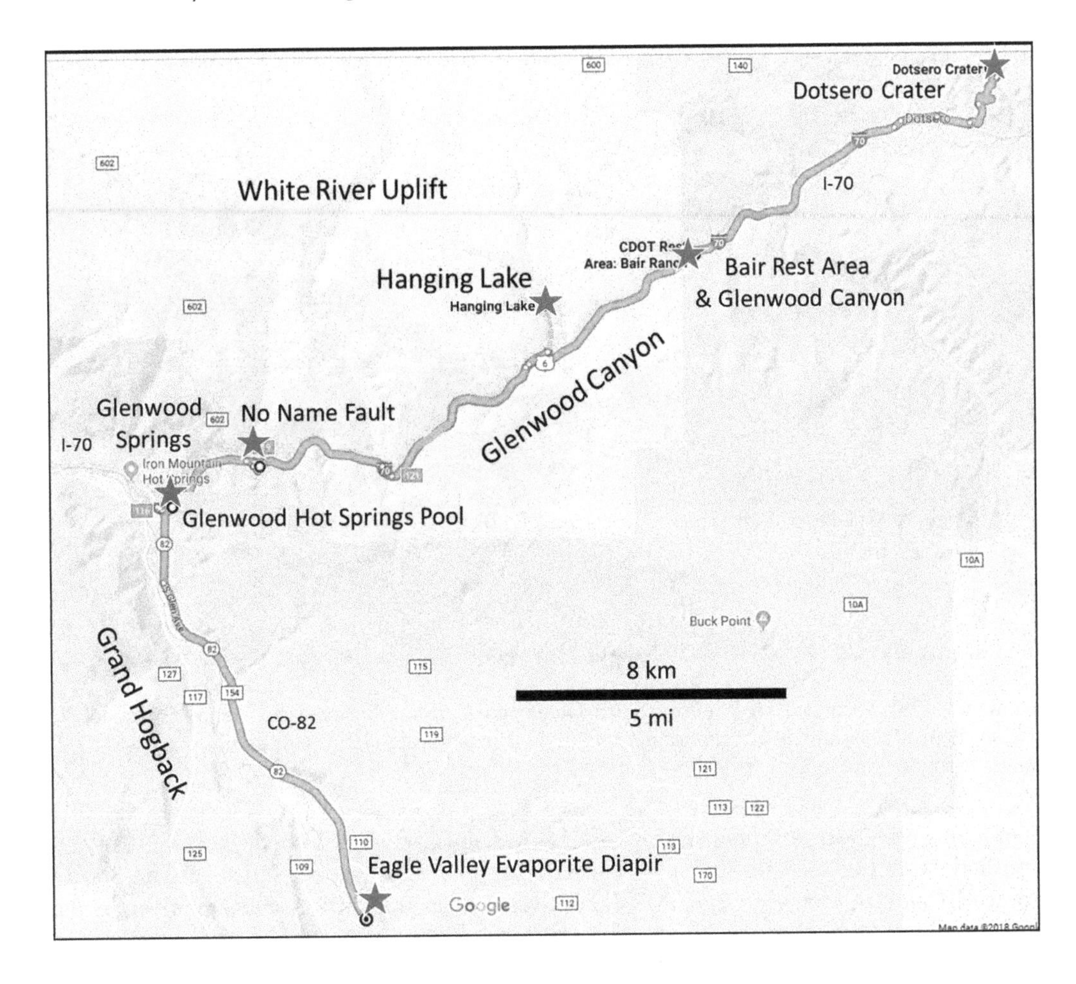

Dotsero Flow to Bair Ranch Rest Area: *Return to I-70 and continue west to Exit 129; exit the freeway and drive to* ***Stop 18, Bair Ranch Rest Area*** *(39.614187, −107.139692) and Glenwood Canyon, for a total of 13.5 km (8.4 mi; 19 min).*

STOP 18 BAIR RANCH REST AREA, GLENWOOD CANYON, AND WHITE RIVER UPLIFT (MILE 129)

Hills of Pennsylvanian-age Belden Formation limestone and shale at Mile 132 indicate you are entering the White River Uplift. Glenwood Canyon begins at Mile 131 with cliffs of the 80 m (260 ft) thick Mississippian Leadville Limestone. The Bair Ranch Rest area provides great views of the east end of Glenwood Canyon.

Bair Ranch Rest Area and view west toward Glenwood Canyon. Between here and the tunnel you drive through Mississippian to Cambrian formations, Precambrian (1.7–2 Ga) gneiss, and 1.6–1.7 Ga granitic rocks.

Prior to around 10 million years ago, there was no Glenwood Canyon and the Colorado River flowed west on a gentle surface 730 m (2,400 ft) above the present valley bottom. Starting between 10 and 5 million years ago, the White River Plateau gradually rose and the river, by then entrenched in place, carved the 20 km (12.5 mi) long, 400 m (1,300 ft) deep Glenwood Canyon through Pennsylvanian to Precambrian rocks. The steep inner canyon was cut into resistant, cliff-forming rocks of the Mississippian Leadville Limestone, slope-forming Devonian Chaffee Group (sandstone and dolomite), the cliff-forming Ordovician Manitou Dolomite, slope-forming Cambrian Dotsero Formation (limestone), the cliff-forming Cambrian Sawatch Quartzite, and cliff-forming Precambrian metamorphic rocks. The unconformity at the top of the Precambrian represents a time gap of about 1.2 billion years and is equivalent to the Great Unconformity seen in the Grand Canyon and elsewhere. You can see this unconformity both at the Hanging Lake and No Name Fault stops.

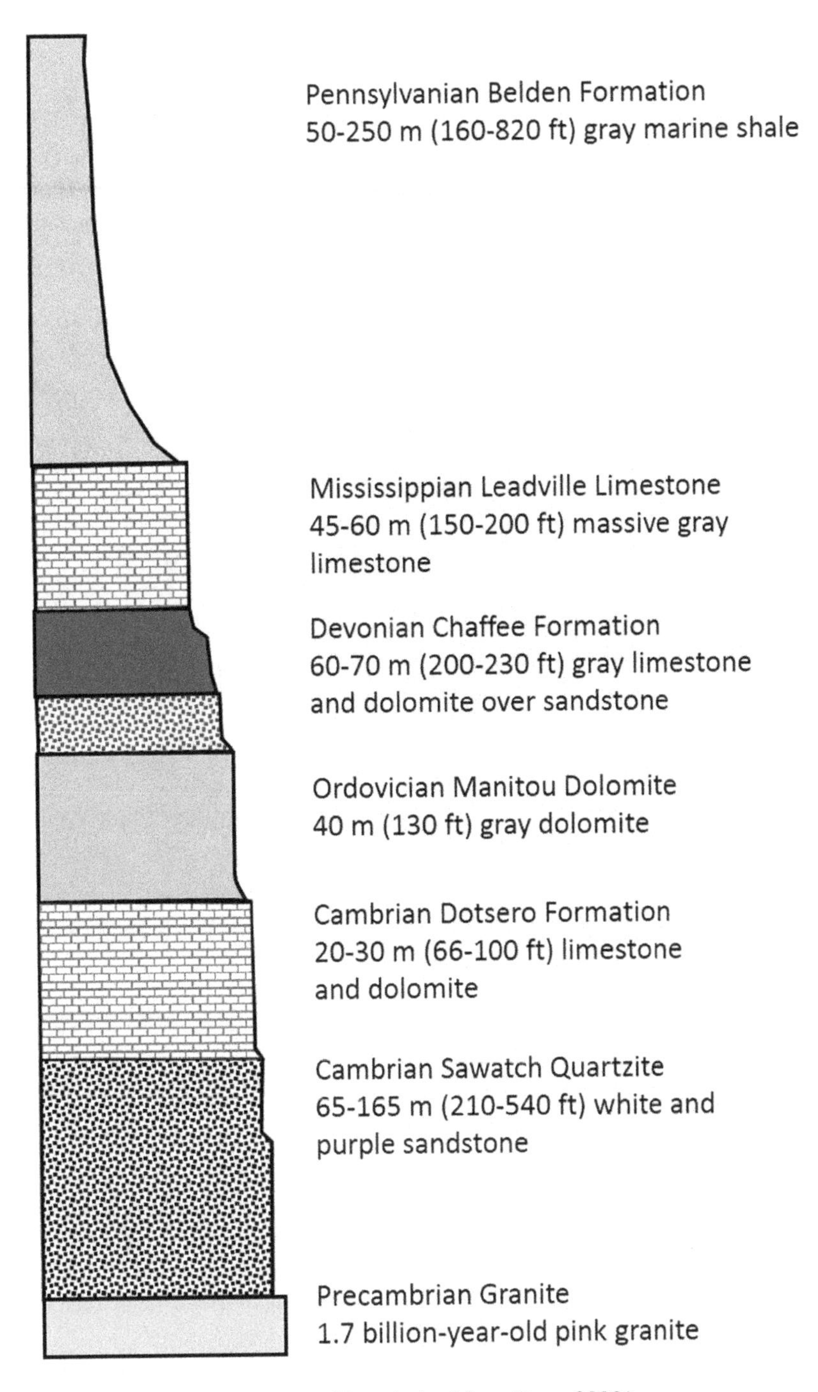

Stratigraphy of the Glenwood Canyon area. (Data derived from Baca, 2000.)

Bair Ranch Rest Area to Hanging Lake: *Continue west on I-70 to Exit 121, Grizzly Creek Rest Area; do a U-turn under the freeway and get on I-70 east; drive east to Exit 125; exit the freeway and follow the signs to* ***Stop 19, Hanging Lake*** *(39.588508, −107.192007), for a total of 32.4 km (20.2 mi; 32 min). Note: Hanging Lake parking area is under the freeway. If you miss it, you will automatically get back on the freeway westbound. It is not well marked.*

STOP 19 HANGING LAKE REST AREA (MILE 125)

A steep 3.2 km (2 mi) round-trip trail leads to the waterfall and pool at Hanging Lake. Wear appropriate shoes, and bring drinking water. Allow 3 h for the round-trip. These waterfalls emerge from the Mississippian Leadville Limestone (equivalent to the Redwall Limestone in the Grand Canyon) carrying calcium carbonate dissolved from the limestone. The travertine that rims Hanging Lake at the base of Bridal Veil Falls derives from dissolved carbonate coming out of solution and precipitating around the pool.

Hanging Lake and Bridal Veil Falls, Glenwood Canyon. (Courtesy of Joshuahicks, https://commons.wikimedia.org/wiki/File:Glenwood_Canyon.jpg.)

The trail begins in Precambrian gneiss and schist and climbs through the Cambrian Sawatch Quartzite and Dotsero Formation dolomite. Much of the quartzite was originally derived from the underlying gneiss and granite. The trail ends in the more readily eroded Ordovician Manitou Dolomite. A good deal of the water and dissolved carbonate derives from the Mississippian Leadville Limestone.

The lake was designated a National Natural Landmark in 2011. There is no swimming or fishing in the lake, nor are dogs allowed on the trail. Because of the popularity of this hike, the parking lot fills early during the summer and remains full into the evening. If the traffic alert signs on I-70 indicate that the lot is full, return another time. Parking on the entrance or exit ramps of I-70 is strictly prohibited.

Hanging Lake to No Name Fault: *Return to I-70 west and drive toward Glenwood Springs. Pull over on the right just before the tunnel (and just after Exit 119) at* ***Stop 20, No Name Fault*** *(39.562270, −107.302666), for a total of 12.2 km (7.6 mi; 9 min).*

STOP 20 NO NAME FAULT (MILE 118)

The No Name Fault is exposed on the right (north) just above the eastern entrance to the tunnel. This is a normal fault, down on the northeast (right) side. It has between 300 and 450 m (1,000 and 1,500 ft) of vertical displacement and drops the Mississippian Leadville Limestone against Precambrian rocks.

The town of No Name (exit 119) is built on a Pleistocene debris flow (landslide) that overlies alluvium of the Pinedale terrace. The Pinedale terrace was deposited by Pinedale Glaciation between 30,000 and 10,000 years ago.

No Name Fault at Glenwood Springs tunnel.

No Name Fault to Hot Springs Pool: *Continue west on I-70 to exit 116; take the exit and turn right on CO-82, then an immediate right again on North River Street; follow North River Street to* ***Stop 21, Glenwood Hot Springs Pool*** *(39.549266, −107.323170), for a total of 3.4 km (2.1 mi; 4 min).*

STOP 21 GLENWOOD SPRINGS AND HOT SPRINGS POOL

About a dozen springs flow from fault-shattered granite and through river gravels in and around Glenwood Springs. The biggest spring is used to fill a large swimming pool, and caves in the Leadville Limestone containing natural steam are used for steam baths. The water is as warm as 50°C (122°F) and contains large amounts of dissolved salt, probably derived from halite in the Eagle Valley Evaporite. Since there are Pleistocene volcanic rocks in the area, and eruptions as recently as 4,150 years ago at Dotsero volcano, it is thought that shallow, warm magma is heating groundwater that then moves upward along faults to the springs.

Glenwood Hot Springs pool. (Courtesy of J. Cipriani, https://commons.wikimedia.org/wiki/File:Glenwood Springs_HotSpringsPool_Mar2014.jpg.)

Visit

Address: 415 East 6th Street Glenwood Springs, CO 81601
Phone: 970-947-2955
Rates vary based on age, time of year, and time of day. Check the website for current entry fees.
Website: hotspringspool.com

***Hot Springs Pool to Grand Hogback:** Drive west on North River Street to the I-70 west onramp; continue west on I-70 to Exit 105; turn right (north) on Castle Valley Blvd, then left (west) onto US-6/Main Street; drive west on Main Street to the Elk Creek Elementary School turnoff on the right. This is **Stop 22, Grand Hogback** (39.570742, −107.542186), for a total of 21.1 km (13.1 mi; 17 min).*

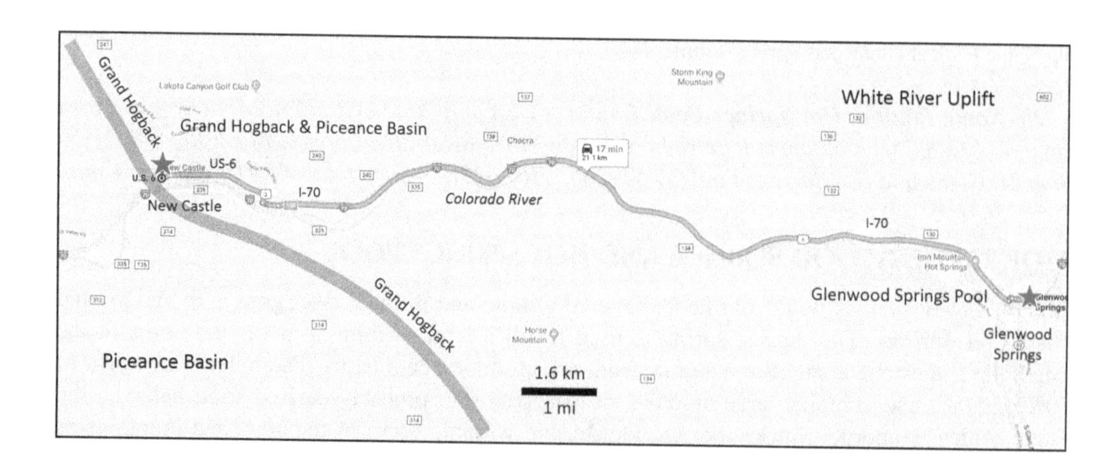

Side Trip 4—Hot Springs pool to Eagle Valley Diapir: *Drive west on North River Street to Grand Avenue/CO-82; turn left (south) on CO-82 and drive past Cattle Creek Road to a pullout on the right at Mile 9. This is* ***Side Trip 4, Eagle Valley Evaporite Diapir*** *(39.442044, −107.255547), for a total of 14.9 km (9.3 mi; 15 min).*

SIDE TRIP 4. EAGLE VALLEY EVAPORITE DIAPIR

A diaper is a dome of rock that has moved upward, breaking through the overlying layers. In order to move upward, the dome must consist of ductile rock, rock that flows under the weight and pressure of overlying strata. In this area, ductile rock is salt and gypsum of the Pennsylvanian Eagle Valley Evaporite.

Exposed on the northeast side of the valley is a diaper of gray to white Eagle Valley Evaporite that intruded and bowed up the overlying red sandstone of the Pennsylvanian Eagle Valley and Maroon formations.

Eagle Valley Evaporite gypsum diaper.

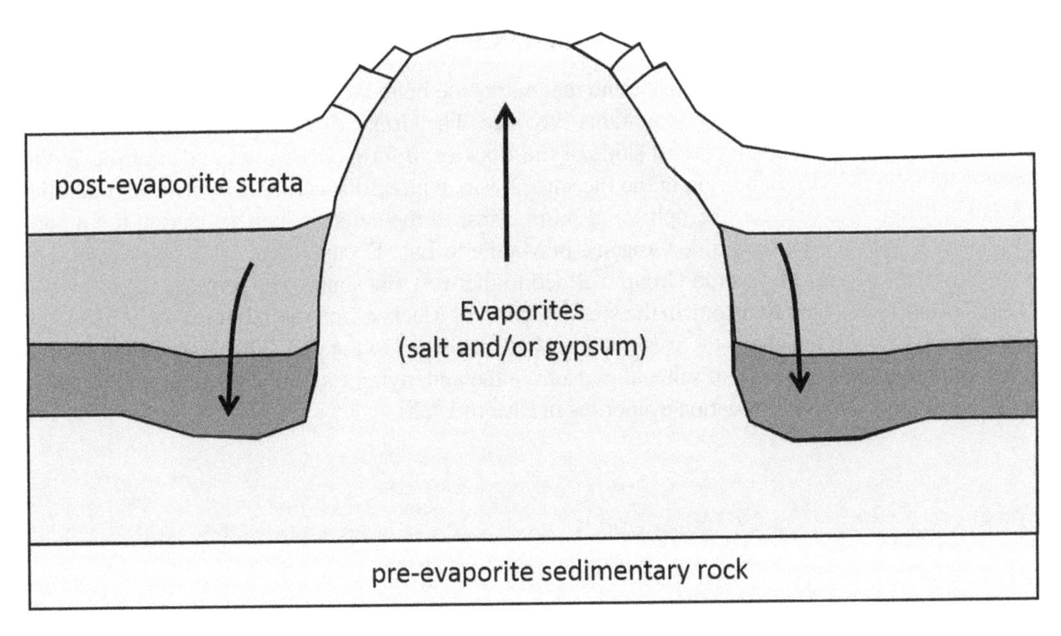

The diaper near Glenwood Springs appears to be an active piercement structure, that is, the salt and gypsum are actively flowing up into the overlying rocks due to lateral squeezing and compression. (Modified after A. Hoffmann, https://commons.wikimedia.org/wiki/File:Peircemnt_2.jpg.)

Side Trip 4—Eagle Valley diaper to Grand Hogback: *Continue south on CO-82 to the first crossover; do a U-turn and return north to I-70; drive west on I-70 to Exit 105; turn right (north) on Castle Valley Blvd, then left (west) onto US-6/Main Street; drive west on Main Street to the Elk Creek Elementary School turnoff on the right. This is* ***Stop 22, Grand Hogback*** *(39.570742, −107.542186), for a total of 37.6 km (23.3 mi; 29 min).*

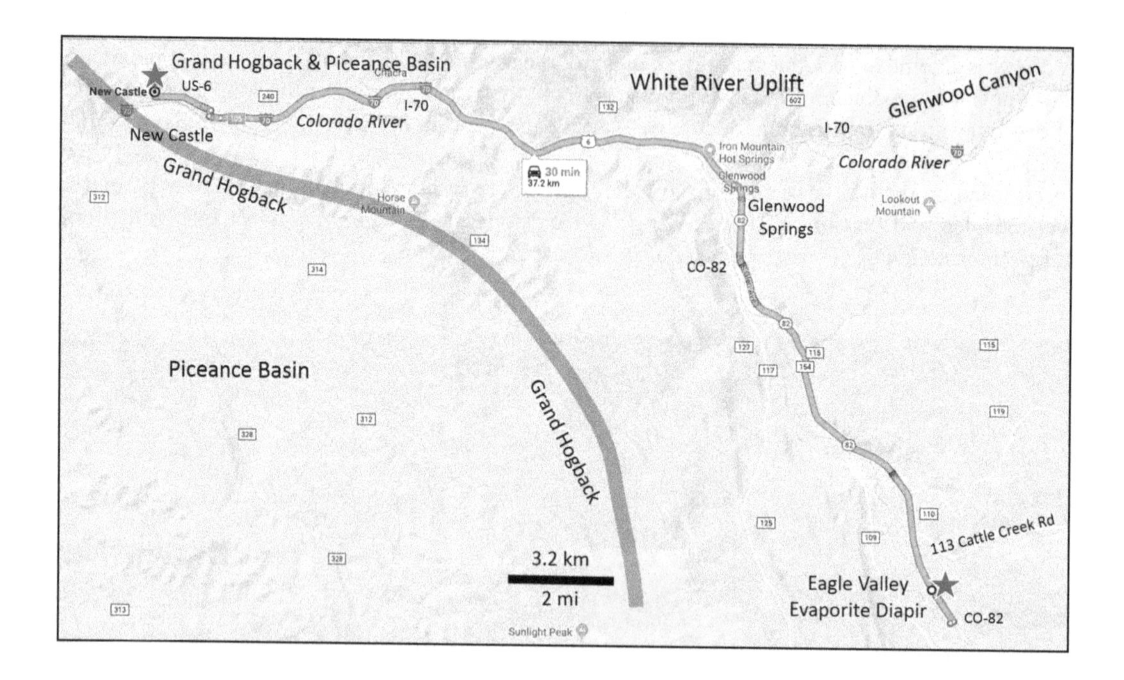

STOP 22 GRAND HOGBACK AND PICEANCE BASIN

The Grand Hogback is a regional monocline that marks the boundary between the Colorado Plateau Province and the Southern Rocky Mountains Province. The Grand Hogback is the mirror image of the Dakota Hogback along the eastern slope of the Rockies. It forms the western edge of the White River Uplift and the eastern margin of the Piceance Basin, with 6,100 m (20,000 ft) of structural relief (elevation difference) between the uplift and basin. Most of the deformation that caused the monocline occurred late in the Laramide Orogeny, in Middle to Late Eocene time. The hogback consists of layers of Cretaceous Mesaverde Group inclined to the west and south. The serrated ridges extend 112 km (70 mi) from near Redstone in the south to north of Meeker, Colorado. Interestingly, blocks of a basaltic lava flow lying above the west-inclined beds are tilted to the east. This is thought to be the result of dissolution or flowage of salt and gypsum in the underlying Eagle Valley Evaporite (roughly equivalent to the Paradox Formation evaporites in eastern Utah).

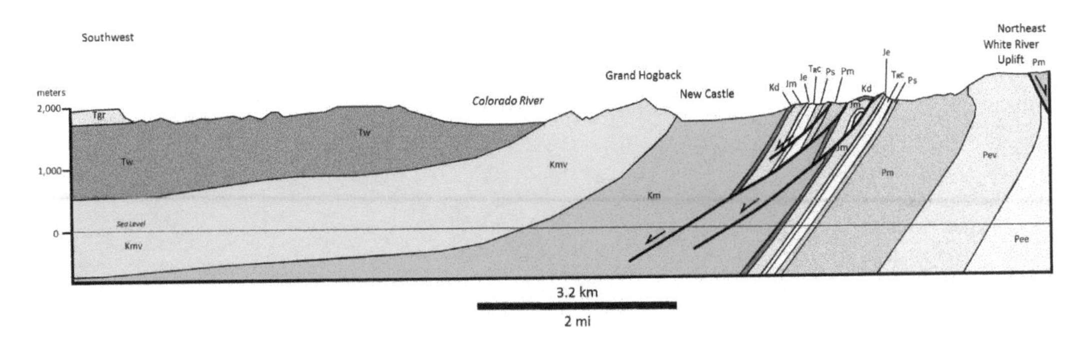

Cross section through New Castle showing the Grand Hogback and White River Uplift. Pee = Pennsylvanian Eagle Valley Evaporite; Pev = Pennsylvanian Eagle Valley Formation; Pm = Pennsylvanian-Permian Maroon Formation; Ps = Permian State Bridge Formation; T_Rc = Triassic Chinle Formation; Je = Jurassic Entrada Formation; Jm = Jurassic Morrison Formation; Kd = Cretaceous Dakota Formation; Km = Cretaceous Mancos Formation; Kmv = Cretaceous Mesaverde Formation; Tw = Eocene Wasatch Formation; Tgr = Eocene Green River Formation. (Modified after Scott and Shroba, 1997.)

The town of New Castle, as one might imagine, was named after the English town of coal mining fame. Abandoned coal mines dating to the 1880s dot the slopes of the Mesaverde Formation along the Grand Hogback here.

Continuing west on I-70 between the Grand Hogback and the Book Cliffs, we skirt the southern flank of the Piceance Basin (pronounced PEE-Ants, it is a Native American Shoshone word meaning "tall grass"). This bowl-shaped depression, half in Utah and half in Colorado, is filled with thousands of meters of sedimentary rocks ranging in age from Cambrian to Eocene. The gently north-inclined rocks on both sides of I-70 are mostly sandstone and shale of the Eocene Wasatch and Green River formations. Oil and gas are produced primarily from the Cretaceous Mesaverde Group and Mancos Shale. In addition to many oil and gas fields, the basin has coal and oil shale resources. In fact, it is the largest single source of shale oil (Taylor, 1987) in the world. The Mahogany Shale zone of the Eocene Green River Formation has an estimated 199 billion m^3 (1.525 trillion bbl) of oil-in-place locked in the shale (Wikipedia, Piceance Basin). A story, perhaps apocryphal, goes that an early settler discovered oil shale when he built his fireplace and chimney out of this attractive brown stone. Upon lighting it for the first time, the chimney caught fire and burned his cabin to the ground.

Difficult-to-extract hydrocarbons are the source of another local story. Operation Plowshare was a program of the U.S. government to find peaceful uses for nuclear explosions. In an attempt to stimulate natural gas production, in 1969 the government drilled a hole 2,560 m (8,400 ft) deep just north of the town of Parachute, Colorado, and detonated a 40 kiloton nuclear bomb. This was called Project Rulison. The test was successful in that large amounts of natural gas were released. However, the gas was radioactive and thus not fit for heating homes and cooking. Unwilling to give up (or enjoying large explosions), in May 1973, they tried again. Project Rio Blanco involved three 33 kiloton explosions in Mesaverde Formation and Fort Union Formation sandstones at depths between 1,779 and 2,039 m (5,800 and 6,700 ft) northwest of Rifle, Colorado. Ultimately, public opposition forced the projects to be abandoned.

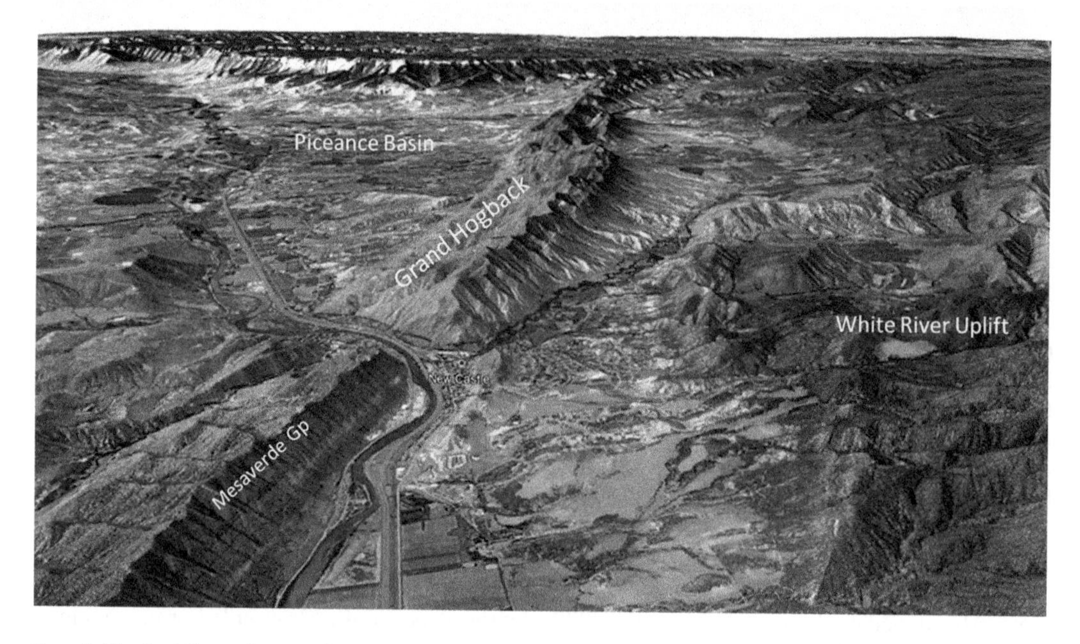

Google Earth oblique view northwest of the Grand Hogback and Colorado River near New Castle, CO. Image © 2017 Landsat/Copernicus.

Grand Hogback looking south from New Castle.

We have been driving along the Colorado River since Dotsero, but it was not always so. Prior to 1921, this was the Grand River. The Grand River originated at Grand Lake high in the Rockies, flowed past the Grand Hogback and into the Grand Valley at Grand Junction. In fact, it flowed all the way to the confluence with the Green River in central Utah. From that point onward, it was known as the Colorado River. But in 1921, the Congressional Committee on Interstate and Foreign Commerce was persuaded that the Colorado River should extend to its point of origin in Colorado, and it has been called the Colorado River ever since.

Grand Hogback to Book Cliffs: *Continue west on US-6 to 9th Street in Silt; turn left (south) on 9th and get on I-70 west; continue west on I-70 to pullout on the right at the sign that says "Grand Junction Next 4 Exits." This is* ***Stop 23, Book Cliffs*** *(39.110503, −108.419304), for a total of 107 km (66.5 mi; 1 h 2 min).*

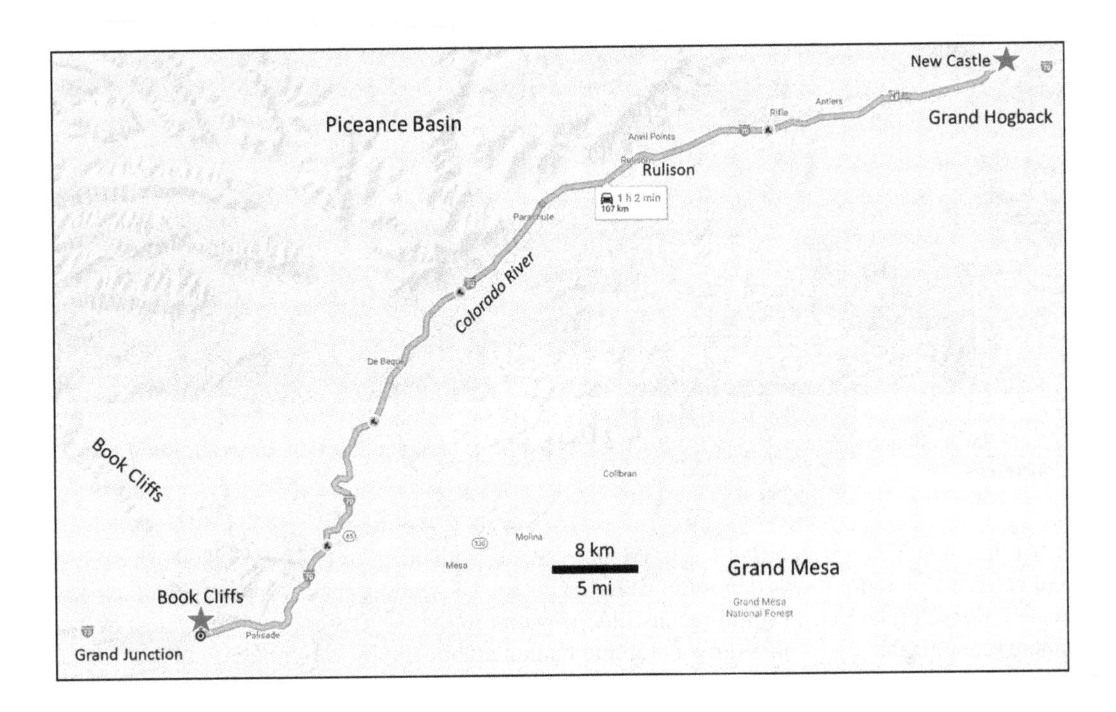

STOP 23 BOOK CLIFFS AT GRAND JUNCTION

The Book Cliffs mark the western margin of the Piceance Basin here. Major John Wesley Powell, on his 1869 trip down the Colorado River (one of the four great surveys of the west, the others being the Hayden Survey of Yellowstone, the King Survey along the 40th Parallel, and the Wheeler Survey west of the 100th Meridian), named these bluffs the "Book Cliffs" because the sandstone cliffs over shale slopes resembled books lying on their side. As with the Grand Staircase in southern Utah, these cliffs exist because the layers are gently inclined to the northeast.

Google Earth oblique view north over the Book Cliffs northeast of Grand Junction. Image © 2016 Landsat/Copernicus.

At this stop, you are standing on the Mancos Shale, equivalent to the Pierre Shale in central and eastern Colorado. It was deposited as Late Cretaceous seafloor mud. From personal experience, it doesn't take much rain to reconstitute this unit to the original muck. It is almost everywhere recognizable by its olive-gray color and badlands topography. Maximum thickness of the Mancos Shale in this area is 1,200 m (4,000 ft). The Mancos is famous for its fossil ammonites and baculites. Look for them in rusty-colored concretions sitting on the green-gray surface, especially after a rain.

During Late Cretaceous time, the thrusting and folding of the Sevier Orogeny formed highlands in Nevada and Utah. Across eastern Utah and all of Colorado was an inland sea that extended from the Gulf of Mexico to the Arctic. The interplay between uplift and sedimentation in the west, and rising and lowering of sea level to the east, led to complex interfingering of marine and non-marine sediments. The slope-forming marine Mancos Shale is interbedded with shoreline sandstone (barrier bars, beach sand) and coastal lagoon coal beds to create the ledge- and slope-forming Mesaverde Group. Thus, the Book Cliffs tell the story of multiple advances and retreats of the Cretaceous Western Interior Seaway.

The Books Cliffs and the canyons that cut into them provide exceptional three-dimensional views of rock layers. This is the area where, during the 1980s, Exxon geologists came up with the concept of "sequence stratigraphy." Sequence stratigraphy is a method of subdividing rock sequences by their bounding surfaces. These erosional, depositional, and non-depositional surfaces define the packages of sediments they enclose.

Sunset on the Book Cliffs and Mount Garfield northeast of Grand Junction.

GRAND JUNCTION

We end this transect on the eastern outskirts of Grand Junction. To the west is the Uncompahgre Uplift, the western range of the Ancestral Rockies. If you continue west on I-70, you will skirt around the northern end of this uplift staying just south of the Book Cliffs, driving mostly on the Mancos Shale until you get to the town of Green River. After that you cross the impressive San Rafael Swell of central Utah, an uplift bounded on both sides by imposing hogbacks. You then cross the Fish Lake Mountains, cross the Wasatch Front, and enter the Basin-and-Range geologic province near Salina, Utah.

If, on the other hand, you are heading to the Moab area and would like to take the scenic route, see Chapter 3 in Volume 1 of this series, Owens Valley to Colorado National Monument.

ACKNOWLEDGMENTS

Thanks to Elizabeth Prost for assisting with the field work and photography required to compile this transect.

REFERENCES AND FURTHER READING

Baca, L. (ed.). 2000. *Guide to the geology of the Glenwood Springs area, Garfield County, Colorado.* Colorado Geological Survey, 45 p.

Chronic, H., and F. Williams. 2002. *Roadside Geology of Colorado.* Mountain Press Publishing, Missoula, 398 p.

Coates, M.M., E. Evanoff, and M.L. Morgan, (eds.). 2004. *Symposium on the Geology of the Front Range, in Honor of William A. Braddock: Colorado Scientific Society*, Boulder, CO. 41 p.

Colorado Geological Survey. Climax mine. http://coloradogeologicalsurvey.org/colorado-geology/colorado-points-of-geological-interest/climax-mine/. Accessed 4 July 2019.

Delvaux, M.J. 2016. Geologic history of the Denver area (Eocene). www.slideshare.net/michaeljdelvaux/denver-area-geology. Accessed 24 June 2019.

Epis, R.C., and Chapin, C.E. 1975. Geomorphic and tectonic implications of the post-Laramide, Late Eocene erosion surface in the southern Rocky Mountains. *GSA Memoir*, v. 144, pp. 45–74.

Fisher, D.J., C.E. Erdmann, and J.B. Reeside Jr. 1960. Cretaceous and Tertiary formations of the Book Cliffs, Carbon, Emery, and Grand Counties, Utah, and Garfield and Mesa Counties, Colorado. USGS Professional Paper 332, 92 p.

Freeport-McMoRan. 2018. Annual Report. https://s22.q4cdn.com/529358580/files/doc_financials/annual/FCX_AR_2018.pdf. Accessed 4 July 2019.

Gable, D.J. 2000. Geologic map of the Proterozoic rocks of the central Front Range, Colorado. USGS Geologic Investigations Series Map I-2605, 1: 100,000.

Higley, D.K., and D.O. Cox. 2007. Oil and gas exploration and development along the Front Range in the Denver Basin of Colorado, Nebraska, and Wyoming. USGS Digital Data Series DDS-69-P, 45 p.

Hughes, T.H., G.C. Mitchell, C.S. Goodknight, and W.L. Chenoweth. 2009. Road Log for the trip from Denver to the Grand Junction along Interstate 70 in Colorado. *AIPG 46th Annual Meeting*, Grand Junction, CO. 20 p.

Kellogg, K.S. 2002. Geologic map of the Dillon quadrangle, Summit and Grand Counties, Colorado. USGS Miscellaneous Field Studies Map MF-2390, 1:24,000.

Kellogg, K.S. 2004. Western Front Range margin near Dillon, Colorado—Interrelationships among thrusts, extensional faults, and large landslides. *In* Coates, M.M., E. Evanoff, and M.L. Morgan (eds.), *Symposium on the Geology of the Front Range, in Honor of William A. Braddock: Colorado Scientific Society*, Boulder, CO. pp. 13–14.

Kellogg, K.S., P.J. Bartos, and C.L. Williams. 2002. Geologic map of the Frisco quadrangle, Summit County, Colorado. USGS Miscellaneous Field Studies Map MF-2340, 1:24,000.

Kellogg, K.S., R.R. Shroba, B. Bryant, and W.R. Premo. 2008. Geologic map of the Denver West 30' × 60' quadrangle, North-Central Colorado. U.S. Geological Survey Scientific Investigations Map 3000, 1:100,000.

Kellogg, K.S., R.R. Shroba, W.R. Premo, and B. Bryant. 2011. Geologic map of the eastern half of the Vail 30' × 60' Quadrangle, Eagle, Summit, and Grand Counties, Colorado. U.S. Geological Survey Scientific Investigations Map 3170, 1:100,000.

Kirkham, R.M., R.K. Streufert, and J.A. Cappa. 1995. Geologic map of the Shoshone quadrangle, Garfield County, Colorado. Colorado Geological Survey and U.S. Geological Survey Open File 95-4, 1:24,000.

Kirkham, R.M., R.K. Streufert, J.A. Cappa, C.A. Shaw, J.L. Allen, and J.V. Jones. 2008. Geologic map of the Glenwood Springs quadrangle, Garfield County, Colorado. Colorado Geological Survey Map Series 38, 1:24,000.

Kluth, C.F., and P.G. Coney. 1981. Plate tectonics of the Ancestral Rocky Mountains. *Geology*, v. 9, no. 1, pp. 10–15.

Lidke, D.J. 1998. Geologic map of the Wolcott quadrangle, Eagle County, Colorado. U.S. Geological Survey Geologic Investigations Series I-2656, 1:24,000.

Lidke, D.J. 2002. Geologic map of the Eagle quadrangle, Eagle County, Colorado. U.S. Geological Survey Miscellaneous Field Studies Map MF-2361, 1:24,000.

Lindsey, D.A. 2010. The Geologic Story of Colorado's Sangre de Cristo Range. U.S. Geological Survey Circular 1349, 14 p.

Lytle, L.R. 2004. Proterozoic of the central Front Range – Central Colorado's beginnings as an island arc sequence. In Coates, M.M., E. Evanofft, and M.L. Morgan (eds.), *Symposium on the Geology of the Front Range, in Honor of William A. Braddock: Colorado Scientific Society*, Boulder, CO. pp. 13–14.

Nesse, W.D. 2006. Geometry and tectonics of the Laramide Front Range, Colorado. *Mountain Geologist*, v. 43, no. 1, Rocky Mountain Association of Geologists, Denver, CO, pp. 25–44.

Pazar, C.C. 2015. Timing of southern Rocky Mountain surface uplift. www.researchgate.net/publication/274313790_Timing_of_Southern_Rocky_Mountain_Surface_Uplift.

Raynolds, R.G., K.R. Johnson, B. Ellis, M. Dechesne, and I.M. Miller. 2007. Earth history along Colorado's Front Range: Salvaging geologic data in the suburbs and sharing it with the citizens. *GSA Today*, v. 17, no. 12, pp. 4–10.

Rice, C.M., R.S. Harmon, T.J. Shepherd. 1985. Central City, Colorado; The upper part of an alkaline porphyry molybdenum system. *Economic Geology*, v. 80, no. 7, pp. 1769–1796.

Scott, G.R. 1972. Geologic map of the Morrison quadrangle, Jefferson County, Colorado. U.S. Geological Survey Map I-790-A, 1: 24,000.

Scott, R.B., and R.R. Shroba. 1997. Revised Preliminary Geologic Map of the New Castle Quadrangle, Garfield County, Colorado. U.S. Geological Survey Open File Report 97-737, 1:24,000.

Scott, R.B., D.J. Lidke, and D.J. Grunwald. 2002. Geologic Map of the Vail West quadrangle, Eagle County, Colorado. U.S. Geological Survey Miscellaneous Field Studies Map MF-2369, 1:24,000.

Simms, P.K., A.A. Drake, and E.W. Tooker. 1963. Economic geology of the Central City district, Gilpin County, Colorado. U.S. Geological Survey Professional Paper 359, 241 p.

St. John, J. n.d. Dinosaur ridge. www.jsjgeology.net/Dinosaur-Ridge.htm. Accessed 25 July 2019.

Stracher, G.B., N. Lindsley-Griffin, J.R. Griffin, S. Renner, P. Schroeder, J.H. Viellenave, M.N.-N. Masalehdani, and C. Kuenzer. 2007. Revisiting the South Cañon Number 1 Coal Mine fire during a geologic excursion from Denver to Glenwood Springs, Colorado. In Raynolds, R.G. (ed.), *Roaming the Rocky Mountains and Environs: Geological Field Trips: Geological Society of America Field Guide 10, Appendix 1.* Geological Society of America, Boulder, CO, 38 p.

Streufert, R.K., R.M. Kirkham, T.J. Schroeder II, and B.L. Widmann. 2008. Geologic map of the Dotsero quadrangle, Eagle and Garfield Counties, Colorado. Colorado Geological Survey Open File Report 08-14, 1:24,000.

Sweet, D.E., and G.S. Soreghan. 2010. Late Paleozoic tectonics and paleogeography of the Ancestral Front Range: Structural, stratigraphic, and sedimentologic evidence from the Fountain Formation (Manitou Springs, Colorado). Geological Society of America Bulletin; March/April 2010; v. 122; no. 3/4; p. 575–594; doi: 10.1130/B26554.1.

Taylor, O.J. 1987. Oil shale, water resources, and valuable minerals of the Piceance Basin, Colorado: The challenge and choices of development. U.S. Geological Survey Professional Paper 1310, 143 p.

Tweto, O., R.H. Moench, and J.C. Reed Jr. 1978. Geologic map of the Leadville 1° × 2° quadrangle, Northwestern Colorado. U.S. Geological Survey Map I-999, 1:250,000.

Waagé, K.M. 1955. Dakota Group in Northern Front Range Foothills, Colorado. U.S. Geological Survey Professional Paper 274-B, pp. 15–51.

Wikipedia. Central City. https://en.wikipedia.org/wiki/Central_City,_Colorado. Accessed 2 July 2019.

Wikipedia. Climax Mine. https://en.wikipedia.org/wiki/Climax_mine. Accessed 4 July 2019.

Wikipedia. Piceance Basin. https://en.wikipedia.org/wiki/Piceance_Basin#cite_note-4. Accessed 25 July 2019.

4 Colorado Rocky Mountains
Colorado Springs, Colorado, to Aspen, Colorado

Pikes Peak from near Garden of the Gods, Manitou Springs, Colorado. (Courtesy of Ahodges7, https://commons.wikimedia.org/wiki/File:Pikes_Peak_from_Garden_of_the_Gods.JPG.)

OVERVIEW

Traverse 4 starts just outside of Colorado Springs and examines the geology of the Colorado Rockies from Pikes Peak to Aspen. Starting at Garden of the Gods, spectacular upturned red sandstone that represents initial uplift of the Ancestral Rockies, we visit a Precambrian granitic intrusion at Pikes Peak, pass though the unique fossil beds at Florissant National Monument, cross South Park, and enter the Rio Grande Rift where the continent is pulling itself apart. We traverse the Colorado Mineral Belt between Leadville and Aspen and explore the Leadville and Aspen silver mining districts. The trip ends at Maroon Bells in the Maroon Bells-Snowmass Wilderness, iconic symbol of the Colorado Rockies.

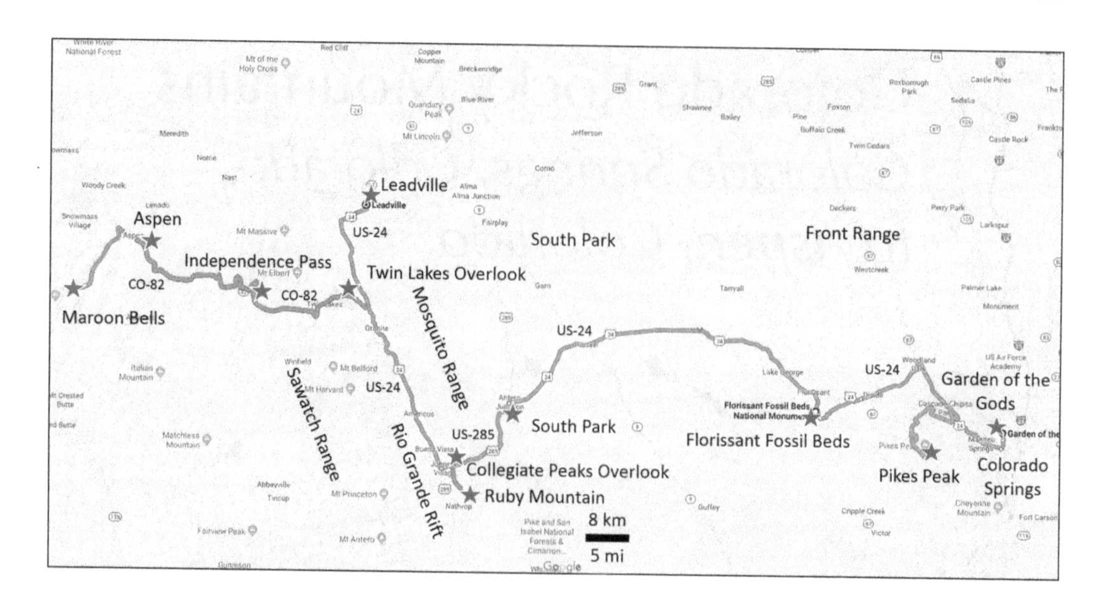

ITINERARY

Stop 1 Garden of the Gods, Colorado Springs
Side Trip 1 Pikes Peak
Stop 2 Florissant Fossil Beds National Monument
Stop 3 South Park Overview
Stop 4 Collegiate Peaks Overlook and the Rio Grande Rift
Side Trip 2 Ruby Mountain
5 Leadville Mining District and the Colorado Mineral Belt
  Stop 5.1 National Mining Hall of Fame and Museum
  Stop 5.2 Hopemore Mine
Stop 6 Twin Lakes Overlook
7 Aspen Mining District
  Stop 7.1 Smuggler Mine
  Stop 7.2 Holden Marolt Mining and Ranching Museum
Stop 8 Maroon Bells

Begin—Garden of the Gods Visitor Center, *Colorado Springs (38.878370, −104.870030)*

STOP 1 GARDEN OF THE GODS, COLORADO SPRINGS

Garden of the Gods Park, owned and operated by the City of Colorado Springs, is a popular location for hiking, rock climbing, road and mountain biking, and horseback riding. There are 24 km (15 mi) of trails to explore in the Garden of the Gods: the main trail in the park, Perkins Central Garden Trail, is a paved, wheelchair-accessible 1.8 km (1.1-mi) trail through the heart of the park's largest and most scenic red rocks. The trail begins at the North Parking lot, the main parking lot off of Juniper Way Loop. The park was designated a National Natural Landmark in 1971.

Fountain Formation spires in Garden of the Gods. View south toward Cheyenne Mountain. (Courtesy of and copyright by Robert Corby, https://commons.wikimedia.org/wiki/File:Garden_of_the_Gods.JPG.)

Geology

The spectacular red sandstone that forms the main ridges and fins in Garden of the Gods is the Pennsylvanian Fountain Formation, formed during wearing away of the Ancestral Rocky Mountains. Time-equivalent to the Maroon Formation seen in central Colorado, this coarse-grained sandstone and conglomerate was eroded off the mountains known as *Frontrangia.* These mountains were formed during the collision of South America/Gondwana against proto-North America during the assembly of Pangea, the Paleozoic supercontinent. The same formation seen at Garden of the Gods can be seen along the mountain front at Roxborough State Park southwest of Denver, in Red Rocks Park west of Denver, and in the Flatirons west of Boulder. Formed around 296–290 million years ago, these sediments were deposited as alluvial fans and braided river channel sands near the ancient mountain front. The red color comes from pink feldspar grains and oxidized iron minerals.

Above and to the east of the Fountain Formation in the park is the cream-colored Permian Lyons Sandstone, deposited as sand dunes roughly 250 million years ago.

During the Laramide uplift of the current Rocky Mountains, between 75 and 40 million years ago, these layers were folded and tilted to almost vertical. Erosion removed the softer rocks, leaving the hard, well-cemented ridges, fins, and spires seen today. Balanced Rock, made of Fountain Formation, is a combination of coarse sand and gravel. The largest outcroppings in the park, "North Gateway," "South Gateway," "Gray Rock," and "Sleeping Giant" consist primarily of Lyons Formation.

History

Prehistoric people are thought to have visited the Garden of the Gods area as early as 1330 BC. The Utes' oral history tells of their creation at the Garden of the Gods, and early Ute petroglyphs have been found in the park.

Spanish explorers visited the area in the 16th century, and later European and American explorers and trappers traveled through the area.

In 1859, two surveyors from Denver came upon the area. One of them suggested that it would be a great place for a beer garden. His companion exclaimed, "Beer Garden! Why it is a fit place for the Gods to assemble. We will call it the Garden of the Gods."

Charles Perkins, the head of the Burlington Railroad, purchased 97 ha (240 ac) in the Garden of the Gods for a summer home in 1879. He never did build there, but after he died in 1907, his children gave his then 194 ha (480 ac) to the City of Colorado Springs with the provision that it would be a free public park. Having purchased additional surrounding land, the City of Colorado Springs' park has now grown to 552 ha (1,364 ac).

Visitor Center

The Visitor Center has over 30 educational exhibits including the movie, *How Did Those Red Rocks Get There?* shown every 20 min. Natural history exhibits include geology, plants, wildlife, and Native American history in the area. Programs include nature hikes, talks, a Junior Ranger program, narrated bus tours, and educational programs.

Visit

Both the Park and Visitor and Nature Center are FREE and open to the public.
Park Hours: 5:00 am–10:00 pm
Garden of the Gods Visitor and Nature Center Hours:
8:00 am–7:00 pm, summer (Memorial Day weekend to Labor Day weekend)
8:00 am–5:00 pm, winter Months
Closed Thanksgiving, Christmas, and New Year's Day
Address: 1805 N. 30th St., Colorado Springs, CO 80904
Phone: 719-634-6666
Website: https://gardenofgods.com/

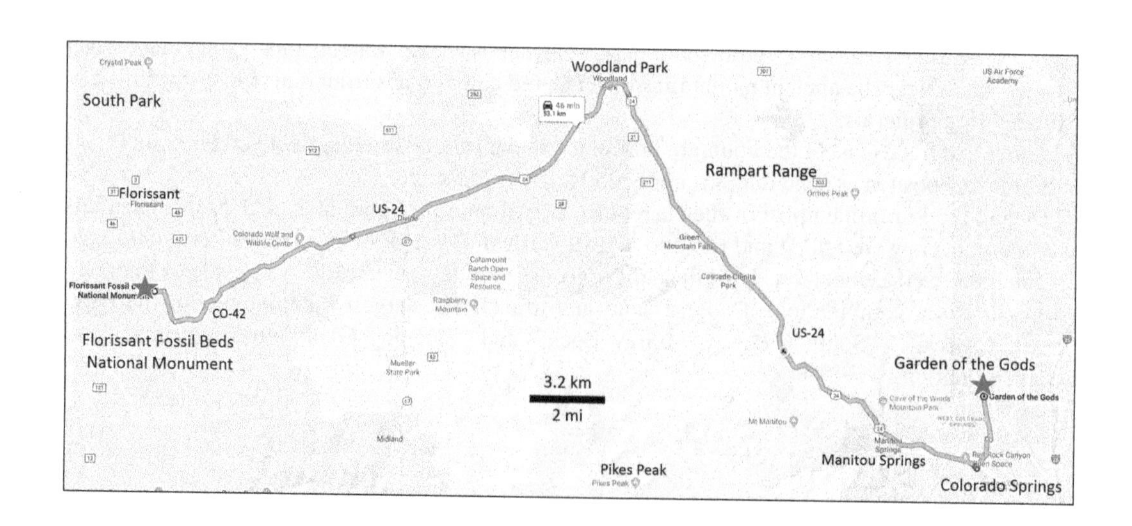

Garden of the Gods to Florissant Fossil Beds: *Drive south on N 30th St to Water St.; turn right (west) on Water to N 31st St; turn left (south) on N 31st St and drive to US-24/Midland Expressway; turn right (west) on US-24 and drive to the town of Divide; about 2.5 km (1.6 mi) west of Divide bear left onto CO-42/Twin Rocks Road at the fork; continue on CO-42 to Teller County-1; turn right (north) on Teller County-1 and drive to* ***Stop 2, Florissant Fossil Beds*** *(38.913560, −105.284769), on the left for a total of 53.1 km (33 mi; 46 min).*

Side Trip 1—Garden of the Gods to Pikes Peak: *Drive south on N 30th St to Water St.; turn right (west) on Water to N 31st St; turn left (south) on N 31st St and drive to US-24/Midland Expressway; turn right (west) on US-24 and drive to Pikes Peak Highway in the town of Cascade; turn left (southwest) onto Pikes Peak Highway and drive to* ***Side Trip 1, Pikes Peak*** *summit (38.840303, −105.042859), for a total of 45.6 km (28.3 mi; 1 h 8 min).*

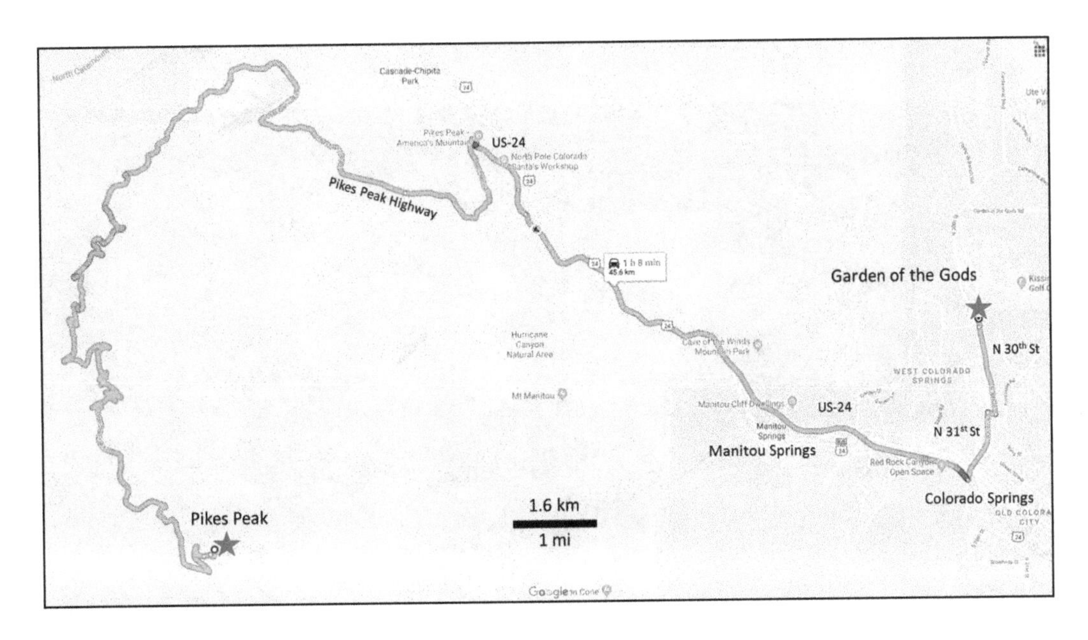

SIDE TRIP 1 PIKES PEAK

The original Ute inhabitants of the area referred to Pikes Peak as "Tava," or "Sun," and called themselves "Tabeguache," the people of Sun Mountain. Spanish explorers called the mountain "*el capitán*" or "captain." The American mountain man Zebulon Pike saw the 4,302 m (14,115 ft) peak in 1806 and called it "The Highest Peak." Thereafter, it was called Pike's Highest Peak and then just "Pikes Peak" after 1890.

Pikes Peak is famous to many because of the expression "Pikes Peak or bust." But who said that, and why? The phrase was coined during the Pikes Peak Gold Rush of the 1860s. The gold rush wasn't actually at Pikes Peak, but rather in the foothills and mountains to the north. Rumors of gold had been circulating for years when the first confirmed discovery, about 622 grams (20 troy ounces) was announced in 1859 at Little Dry Creek in what is now the Denver suburb of Englewood. From there prospectors followed the trail of placer gold to the mountains and founded mining camps along the South Platte River and Clear Creek, camps that became Idaho Springs, Central City, Georgetown, Breckenridge, Fairplay, Alma, and Como, among others. The slogan "Pikes Peak or bust" referred to the prominent peak that guided prospectors across the prairies to the Rockies. On a clear day you can see Pikes Peak from Limon, Colorado, about 125 km (78 mi) to the east.

Geology

The 1.08 billion-year-old Pikes Peak magma intruded into 1.7 billion-year-old metamorphic gneiss and schist. It is actually one of several igneous bodies that include the Tarryall Creek Pluton, the Redskin Stock, and the Lake George Pluton, that together form the Pikes Peak Batholith. The batholith is about 130 km (80 mi) long north-to-south and about 40 km (25 mi) wide east-west. The dominant rock is a coarse-grained pink to reddish-orange granite with large white plagioclase feldspar crystals. The granite weathers to rounded hills and forms a coarse, gravelly soil called *grus*. The pink color comes from microcline feldspar and weathered iron minerals.

The batholith formed deep underground and cooled slowly, leading to the large mineral crystals. In some places, pegmatites, dikes in the granite containing especially large minerals, contain spectacular crystals of smoky quartz, topaz, and amazonite, a rare blue-green microcline feldspar. Many amazonite claims are located north of Florissant (see, e.g., The Collector's Edge, https://collectorsedge.com/pages/two-point-amazonite-mine-teller-county-colorado).

Microcline feldspar (the variety called Amazonite) with smoky quartz from the Pikes Peak area. From the Halpern Mineral Collection, Colorado. (Courtesy of E. Hunt, https://commons.wikimedia.org/wiki/File:Microcline_feldspar_variety_amazonite.jpg.)

The batholith was first exposed at the surface during Pennsylvanian time when the Ancestral Front Range, known as *Frontrangia*, was uplifted on the order of 1,300 m (4,265 ft). Much of the material eroded from the uplift forms the coarse red Maroon Formation sandstone in central Colorado and coarse red Fountain Formation sandstone in eastern Colorado. The Fountain

Formation is on display in Garden of the Gods Park near Colorado Springs and at Red Rocks Park west of Denver. After being eroded to near sea level by Early Cretaceous time, the entire area was again uplifted during the Late Cretaceous to Eocene Laramide Orogeny (75–40 million years ago). Laramide uplift and faulting formed the modern Rocky Mountains and brought Pikes Peak to nearly its current elevation. Faults such as the Rampart Fault and Ute Pass Fault along the eastern margin of the mountain range are gently to steeply inclined to the west, much like the Golden Fault near Denver, indicating that the uplift is bounded by a reverse or high-angle thrust fault (Jacob, 1983; Rowley et al., 2002). These mountain-bounding faults show evidence of a prior life, originally active during the Pennsylvanian Ancestral Rocky Mountains. Drill cores show Precambrian granite over Cretaceous Pierre Shale (Temple et al., 2007). In fact, when the Army Corps of Engineers was carving out Cheyenne Mountain for NORAD headquarters they found Precambrian granite faulted above overturned Late Cretaceous Niobrara Limestone and Pierre Shale (Theodosis and Skehan, 1965). The granite has been uplifted in excess of 6,000 m (20,000 ft) during Laramide mountain building, allowing more sedimentary rock to be eroded off Pikes Peak and the surrounding mountains. These faults may have Neogene and even Recent movement.

During the Eocene (57–37 million years ago), the Front Ranges had been eroded to a nearly flat surface gently tilted 1° or 2° to the east. The Late Eocene Erosion Surface (also called the Rocky Mountain Erosion Surface) today is an undulating, nearly flat surface with a few high peaks rising above it. The surface was formed by mountain front erosion (coalescing pediment surfaces) and was protected from further erosion by onlapping alluvial fans and volcanic cover (such as the Wall Mountain Tuff) along the eastern margin of the range.

Pikes Peak from Colorado Springs showing the Eocene Erosion Surface in the foreground (arrows). (Courtesy of Aravis, https://commons.wikimedia.org/wiki/File:Pike%27s_Peak4.jpg.)

Following the Eocene, and ongoing today, there has been broad regional uplift of the Rocky Mountains and mid-continent. A minimum of 700m (2,300ft) of surface uplift has been estimated to account for the present surface elevation in Colorado.

About 3.7km (12,000ft) of sedimentary rock and granite has been eroded off Pikes Peak since Laramide uplift began. At least 300m (1,000ft) of that has been eroded just since the Eocene (in the last 37 million years).

Over that past 2 million years, glaciers have carved cirques and gauged out valleys on the mountain and deposited moraines and outwash debris along its flanks.

Side Trip 1—Pikes Peak to Florissant Fossil Beds: *Backtrack to US-24 and turn left (west); drive on US-24 to the town of Divide; about 2.5 km (1.6 mi) west of Divide bear left onto CO-42/ Twin Rocks Road at the fork; continue on CO-42 to Teller County-1; turn right (north) on Teller County-1 and drive to* ***Stop 2, Florissant Fossil Beds*** *(38.913560, −105.284769), on the left for a total of 73.0km (45.4 mi; 1h 27min).*

STOP 2 FLORISSANT FOSSIL BEDS NATIONAL MONUMENT

As you approach Florissant, you begin to see light gray shale alongside the road. These are late Eocene-Oligocene (34 million-year-old) Florissant Formation lakebed sediments. The name Florissant is from the French for "flowering." The area was first visited and described by Ferdinand Hayden as part of his "Geological and Geographical Survey of the Territories" in the 1873. Edward Drinker Cope, the famous dinosaur collector, described fossil fish from Florissant as part of the Hayden Survey. The classic description of the site was published by MacGinitie in 1953. The National Monument was established in 1969 to "preserve and interpret for the benefit and enjoyment of present and future generations the excellently preserved insect and leaf fossils and related geologic sites and objects." It is one of the most famous and important fossil plant and insect locations in North America.

The oldest rocks in the monument are pink, 1.08 billion-year-old Pikes Peak Granite. The granite forms most of the rounded hills in the area. All of the Paleozoic and Mesozoic rocks that had been deposited here were eroded off due to the Laramide uplift. The paleo-valleys eroded into the granite would be filled by the Late Eocene (36.7 million-year-old) Wall Mountain Tuff, a rhyolitic welded tuff or ignimbrite. This unit was the result of a hot ash cloud erupting from near Mt. Princeton in the Collegiate Range 80km (50 mi) to the west.

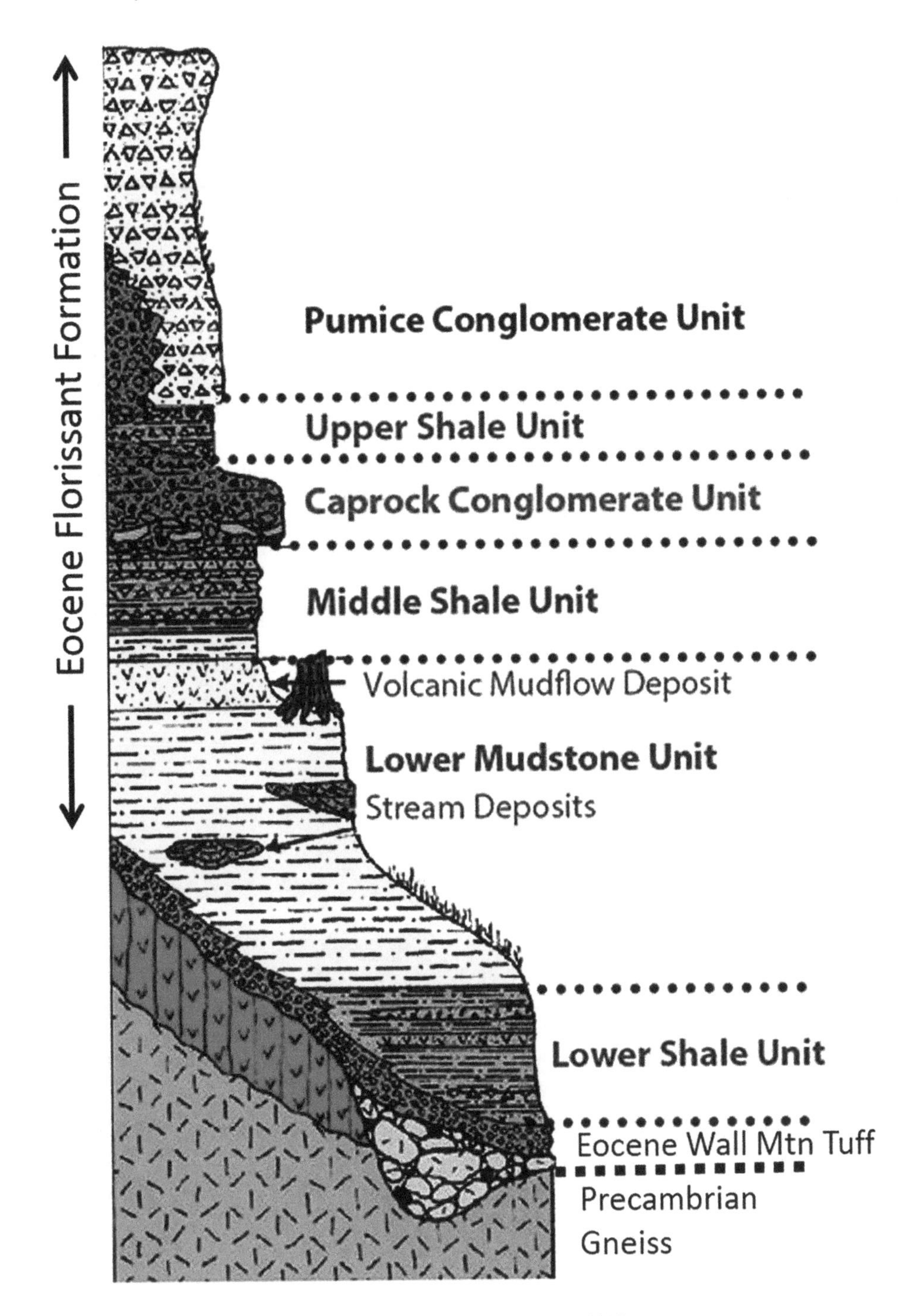

Strata in the Florissant area. (Modified after National Park Service, 2015.

During the Oligocene, roughly 34.9 million years ago, the Guffey Volcanic Center, part of the Thirtynine Mile Volcanic Field, began erupting 30 km (18 mi) southwest of Florissant. Volcanic mudflows, called lahars, covered the area and dammed the Florissant valley to form a lake. Thin layers of clay and volcanic ash were deposited in these lakes to form the "paper shales" that contain most of the fossils here. Over 40,000 specimens have been recovered, including more than 140 plant species. The mudflows buried leaves from birch, willow, maple, beech, and hickory trees; fir needles; fruits, seeds, cones, and flowers; and stumps from 500 to 700-year-old giant redwoods. The Florissant Formation consists of lava flows, massive tuffs, river gravels, volcanic conglomerates (agglomerates), and lakebed mudstone. At least three lakes occupied the valley over a period of 2 million years.

The lakebed shale at Florissant is also famous for arthropod fossils including spiders, millipedes, and insects. Insects include exquisitely preserved mosquitos, mayflies, dragonflies, grasshoppers, aphids, flies, beetles, wasps, ants, bees, and butterflies. Over 10,000 species of insects have been described from the Florissant Formation.

In addition to plants and bugs, fossils found here include mollusks (clams, snails), fish, birds, and mammals. Curiously, no reptiles or amphibians have been found.

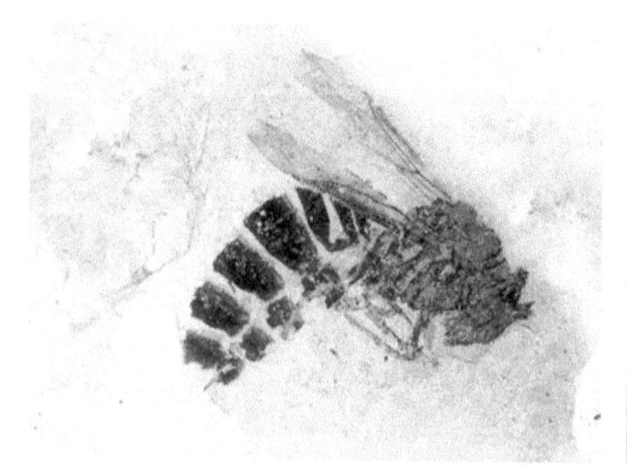

Fossil wasp and fossil flower found in "paper shale" of the Florissant Formation. (Wasp courtesy of National Park Service, https://commons.wikimedia.org/wiki/File:Palaeovespa_florissantia.jpg; Flower courtesy of Slade Winstone (Sladew), https://commons.wikimedia.org/wiki/File:Eocene_fossil_flower,_Clare_Family_Florissant_Fossil_Quarry,_Florissant,_Colorado,_USA_-_20100807.jpg.)

The fossils were preserved in fine volcanic ash that washed into the lake or fell out of the air. Silica dissolved out of nearby rock moved with groundwater into the lake and slowly replaced the organic material with quartz. Fine details such as leaf veins, insect antennae, legs, and hairs, and flower petals were preserved.

Over the past 2 million years, glaciers covered the higher peaks around South Park, including Pikes Peak. Although glaciers did not cover the Florissant area, most of the sediments above the Florissant Formation are glacial gravels and outwash deposits.

Florissant Fossil Beds to South Park Overview: *Drive north on Teller County-1 to US-24; turn left (west) on US-24 and drive to* ***Stop 3, South Park Overview*** *(38.922044, −105.967337) at Antero Junction for a total of 73.4 km (45.6 mi; 48 min).*

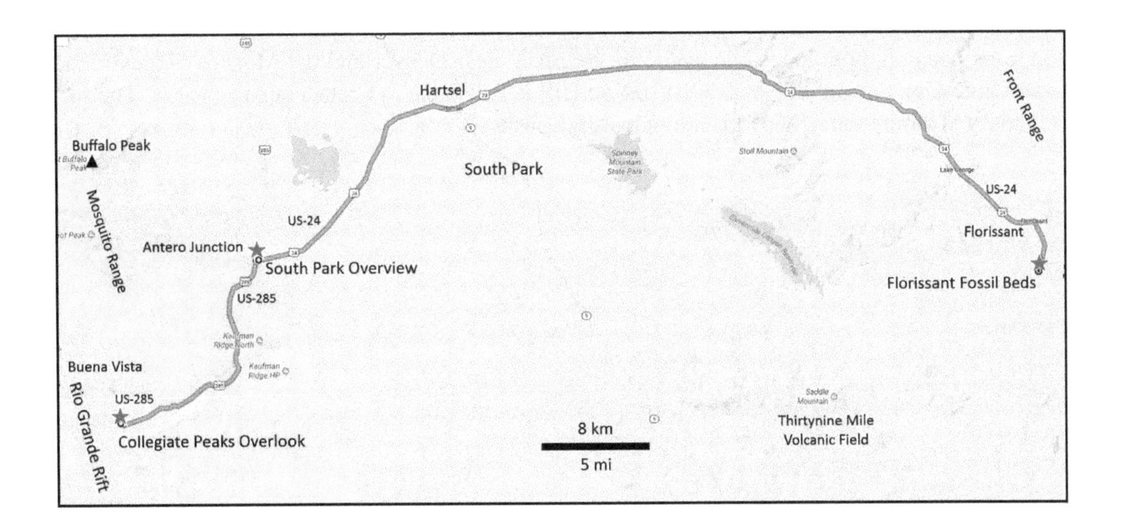

STOP 3 SOUTH PARK OVERVIEW

Between Florissant and Wilkerson Pass, you are in the Thirtynine Mile Volcanic Field. West of Wilkerson Pass, the rolling hills alongside the road consist of Precambrian metamorphic gneiss and schist. Deposited near the margin of proto-North America 1.8 billion years ago, these were metamorphosed around 1.7 billion years ago and later intruded by 1.4 billion-year-old granite.

From this stop, you can see that South Park is a large, relatively flat intermontane basin, essentially a high-altitude prairie. It is large, 2,330 km² (900 mi²), and high, with elevations between 2,400 and 3,000 m (8,000 and 10,000 ft). It is bounded by the Mosquito Range on the west, the Front Range to the east and north, and the Thirtynine Mile Volcanic Field to the south. The sedimentary rocks in the subsurface range from Pennsylvanian and Permian rocks seen in the Mosquito Range to Cretaceous rocks seen in the Denver Basin. They are inclined to the east up to 20° and terminate against the east-dipping Elkhorn Thrust Fault on the east side of the park. This Laramide thrust puts Precambrian granites over Cretaceous and early Tertiary rocks. The Mosquito Fault on the west side of the Mosquito Range is a west-dipping normal fault that separates the Mosquito Range from the Arkansas River Valley/Rio Grande Rift. Directly north from here, along the crest of the Mosquito Range, you can see the andesitic volcanic cone of Buffalo Peaks. The Thirtynine Mile Volcanic Field consists of Oligocene andesitic and basaltic flows and tuffs that blocked streams and caused a large shallow lake to form in the basin. Red Hill is a long north-south ridge (hogback) of Cretaceous Dakota Sandstone that runs through South Park. East of the ridge is mostly Cretaceous Pierre Shale (the same ocean-bottom mudstone as the Pierre Shale near Denver and the Mancos Shale around Grand Junction). Beneath the park west of Red Hill is Paleozoic to Cretaceous-age rocks. The surface today is covered mostly by glacial outwash deposits.

South Park looking west toward the Mosquito Range from Kenosha Pass area. (Photo courtesy of Adam Prost.)

Placer gold was found near Fairplay in 1859 and mining and dredging continued well into the 20th century. Lode (vein) deposits like those at Leadville were mined all along the Mosquito Range.

Google Earth oblique view north over South Park. Image © 2016, Landsat/Copernicus.

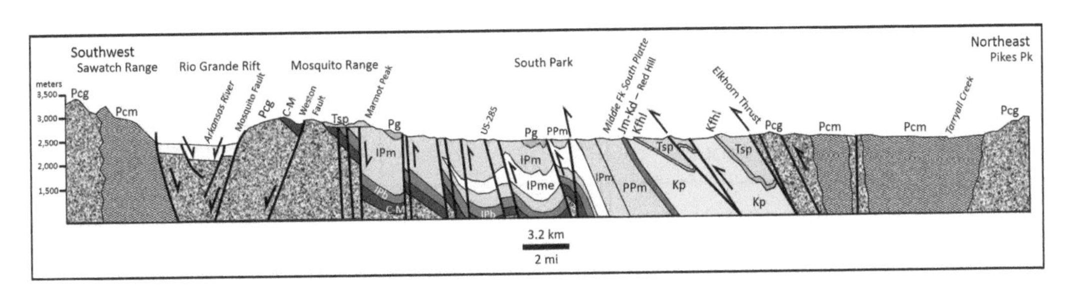

Southwest-northeast cross section through the Rio Grande Rift and South Park. Pcg = Precambrian granite; Pcm = Precambrian metamorphics; C-M = Cambrian to Mississippian sedimentary units; IPb = Pennsylvanian Belden Formation; IPm = Pennsylvanian Minturn Formation; IPme = Minturn evaporites; PPm = Pennsylvanian-Permian Maroon Formation; Pg = Permian Garo Formation; Jm-Kd = Jurassic Morrison-Cretaceous Dakota formations; Kp = Cretaceous Pierre Shale; Kfhl = Cretaceous Fox Hills-Laramie formations; Tsp = Paleocene South Park Formation. (Modified after Barkman et al., 2013; Bilodeau, 1986.)

Age		Formation	Lithology & Thickness
Cretaceous	Upper	Laramie Fm	Shale, sandstone, coal - 115 m (375 ft)
		Fox Hills Sandstone	Sandstone - 107 m (350 ft)
		Pierre Shale	Shale – 1,280-1,615 m (4,200-5,300 ft)
	Lower	Niobrara Fm	Shale & limestone 122-168 m (400-550 ft)
		Benton Fm	Shale, limestone, bentonite 76 m (250 ft)
		Dakota Sandstone	Sandstone, conglomerate, shale 76-91 m (250-300 ft)
Jurassic		Morrison Fm	Shale, sandstone, basal limestone 55-107 m (180-350 ft)
Permian		Garo Fm	Sandstone 18-70 m (60-230 ft)
		Maroon Fm	Sandstone, siltstone, shale, conglomerate up to 1,000 m (3,300 ft)
Pennsylvanian		Minturn Fm	Sandstone, siltstone, shale, conglomerate, limestone up to 1,525 m (5,000 ft) Shale with gypsum and halite up to 300 m (1,000 ft) Sandstone, siltstone, shale, conglomerate 300 m (1,000 ft)
		Belden Fm	Shale 228-260 m (750-850 ft)
Miss		Leadville Limestone	Limestone and dolomite 228-260 m (140-270 ft)
Devonian		Chaffee Fm	Quartzite, dolomite, limestone 34-79 m (110-260 ft)
Ordovician		Fremont Dolomite	Dolomite up to 33 m (to 110 ft)
		Harding Sandstone	Sandstone up to 18 m (to 60 ft)
		Manitou Dolomite	Dolomite 45-60 m (150-200 ft)
Cambrian		Dotsero Fm and Sawatch Quartzite	Quartzite and dolomitic sandstone 15-45 m (50-150 ft)
		Precambrian	Granites and Gneisses

Stratigraphy of central Colorado. Data derived from Barkman et al., 2013.

South Park Overview to Collegiate Peaks Overlook: *Take US-24 west to US-285; turn left (south) on US-285 and drive to* ***Stop 4, Collegiate Peaks Overlook*** *parking area on the right (38.816997, −106.086613), for a total of 19.8 km (12.3 mi; 15 min).*

STOP 4 COLLEGIATE PEAKS OVERLOOK AND THE RIO GRANDE RIFT

A rift is a valley or series of linked valleys where a continent (or ocean floor) is splitting apart. The Rio Grande Rift is just such a place. It is a linear, north-south depression that extends from northern Mexico to central Colorado near Leadville, almost 1,000 km (600 mi) long. In central Colorado, the valley formed by the rift is occupied by the Arkansas River. In southern Colorado and New Mexico, it occupied by the Rio Grande River.

The rift is part of the east-west crustal extension happening in the western United States. It is the same spreading that caused the Basin-and-Range province in Nevada, Utah, southern New Mexico, and Arizona over the past 20 million years and that continues today. Some geologists feel the rift began spreading even earlier, perhaps as long ago as 35 million years. The process of rifting involves uplift and spreading. This causes the center to drop along faults, leaving mountains on either side. In the Salida area, the basin-fill deposits, called the Dry Union Formation, are up to 1,500 m (4,900 ft) thick.

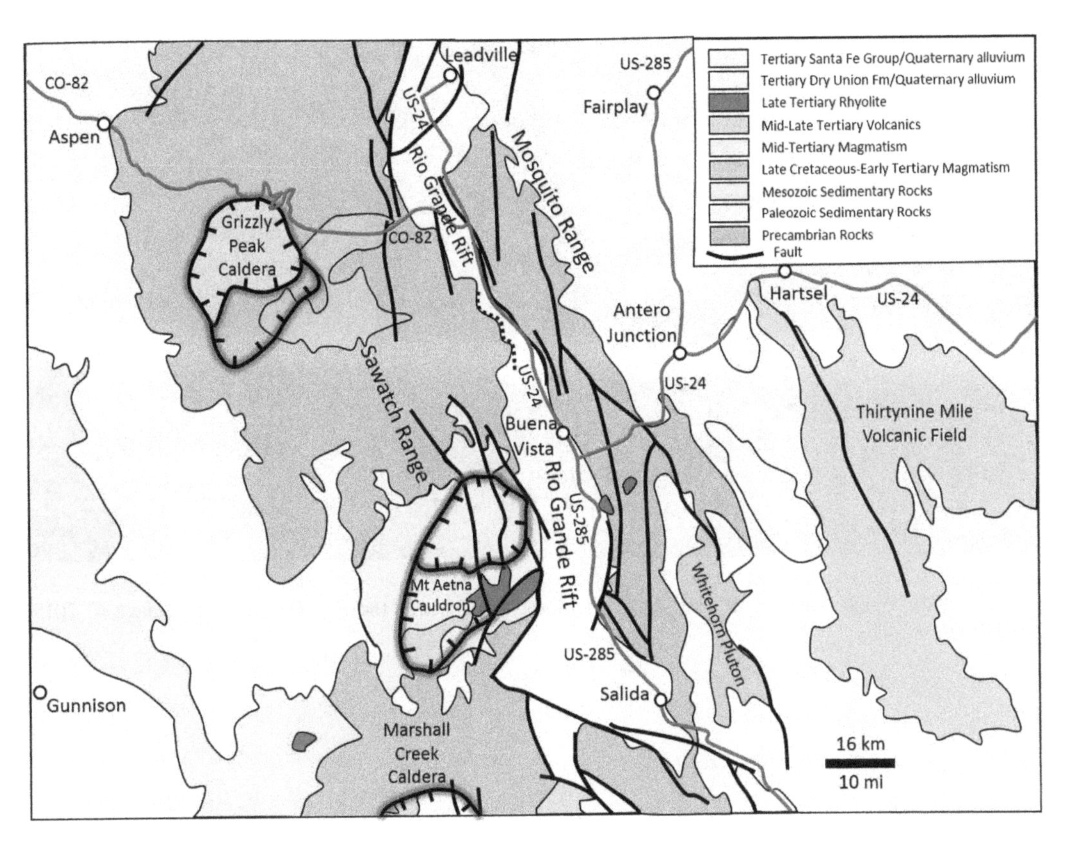

Geologic map of the Sawatch Range, Rio Grande Rift, Mosquito Range, and western South Park. (Modified after Shannon and McCalpin, 2006.)

The valleys formed by rifting are bounded on both sides by near-vertical normal faults and fault-bounded mountains that tilt away from the center of the rift. In central Colorado, these mountains are the Sawatch Range on the west and the Mosquito Range on the east. The Sawatch Range consists of 1,750 million-year-old granitic and metamorphic rocks intruded by Oligocene and Miocene igneous rocks, that is, 36–5 million-year-old granites. The Mosquito Range consists of east-tilted Cambrian to Pennsylvanian sedimentary rocks above Precambrian granitic rocks, all of which have been intruded by Tertiary granitic dikes and sills and the occasional volcano like Buffalo Peaks. Putting it all together, you have an enormous north-south anticline comprising the Sawatch and Mosquito ranges, where the center has been dropped thousands of meters in the Rio Grande Rift.

The Tertiary igneous activity was accompanied by injection of hot, metal-rich fluids along faults and fractures that formed lode ore deposits throughout the flanking mountains, as we will see at Leadville.

Today, the North American continent is actively splitting apart at this rift, about 2 mm/year (8 in/100 year). Evidence can be seen in satellite measurements that show spreading, in continued seismic activity, hot springs, and recent lava flows. Given enough time, the Gulf of Mexico will flood the depression and there will be ocean-front property in central Colorado.

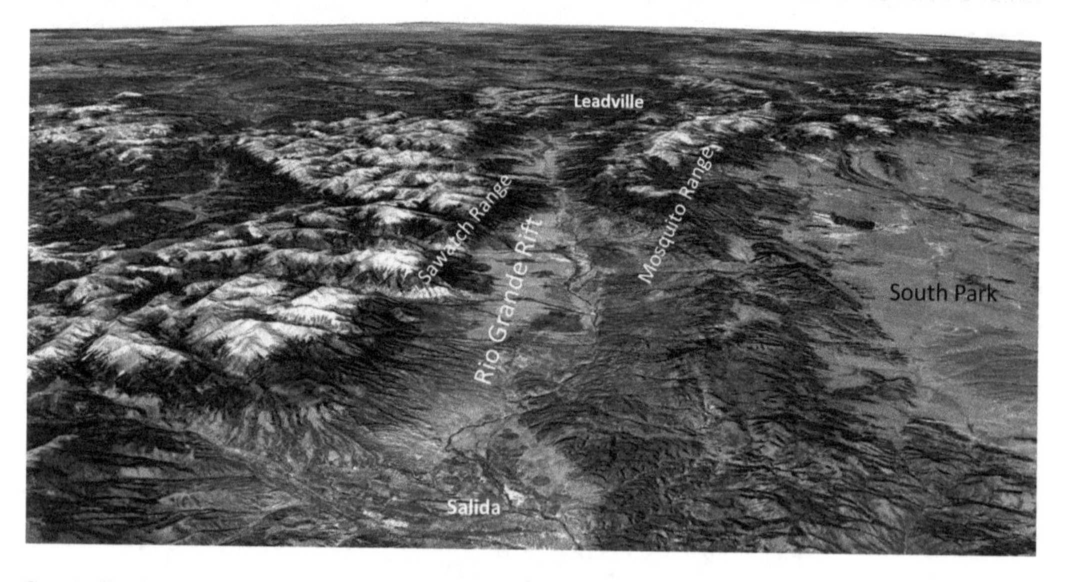

Google Earth oblique view north over central Colorado portion of the Rio Grande Rift. Image © 2015, Landsat/Copernicus.

Collegiate Peaks of the Sawatch Range looking west from overlook. (Photo courtesy of Adam Prost.)

The Collegiate Peaks are part of the Sawatch Range that extends roughly 140 km (90 mi) along the west side of the Upper Arkansas Valley. "Sawatch" comes from a native word meaning "blue earth." The Continental Divide runs along the crest of this range. Fifteen peaks in the range rise over 4,300 m (14,000 ft). During the last ice age, glaciers carved U-shaped valleys, carved cirques at the head of the valleys, and deposited moraines at the mouth of the valleys. Stream channel sediment, glacial outwash, and unsorted glacial debris including silt, sand, and conglomerate can be seen in roadcuts as you drive through the Upper Arkansas River Valley.

Side Trip 2—Collegiate Peaks Overlook to Ruby Mountain: *Continue west on US-285 to County Road 301 (poorly marked intersection); turn left and take County Road 301 south to County Road 300 and follow it to* ***Stop ST 2, Ruby Mountain*** *parking area (38.752091, −106.065553; sec. 13, T. 15 S., R. 78 W), for a total of 11 km (6.9 mi; 15 min).*

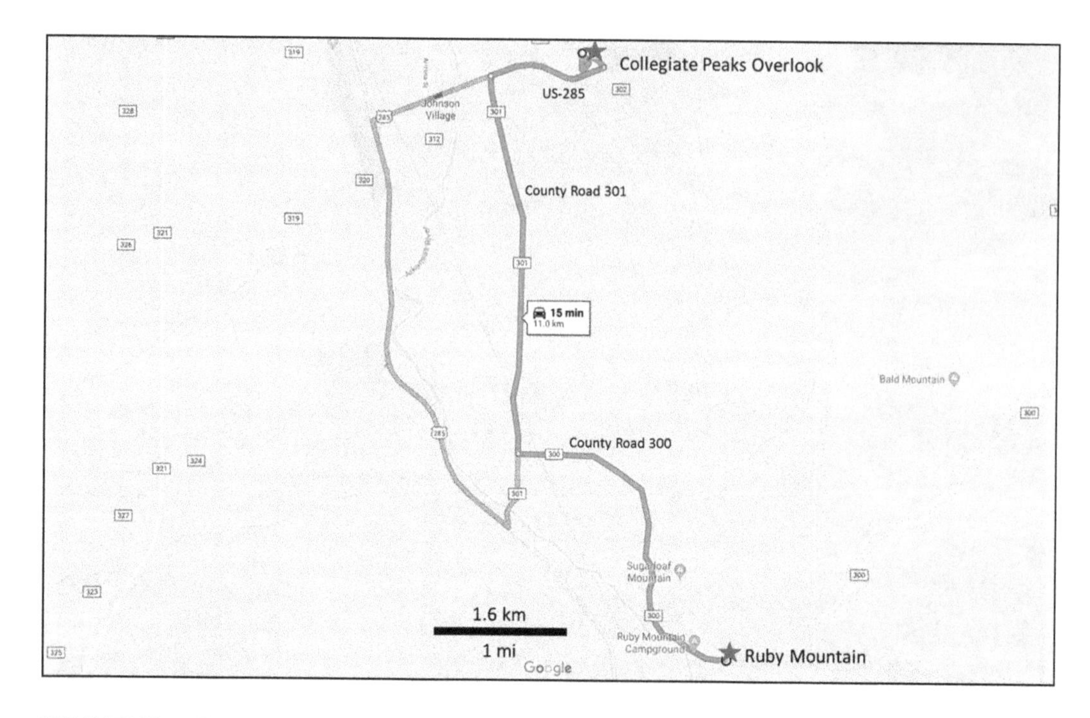

SIDE TRIP 2 RUBY MOUNTAIN

One of Colorado's prominent mineral hunting locations is just 11 km (6.9 mi) south of the Collegiate Peaks overlook. Ruby Mountain contains well-formed crystals of deep-red spessartine garnet and topaz in Oligocene rhyolite of the Nathrop volcanics. Both garnet and topaz crystals are generally less than 0.6 cm (0.25 in) in diameter, but a few specimens measure up to 1 cm (0.5 in) diameter. Topaz is less common than garnet. Some specimens are of gem quality and have been faceted into gemstones (Keller et al., 2004).

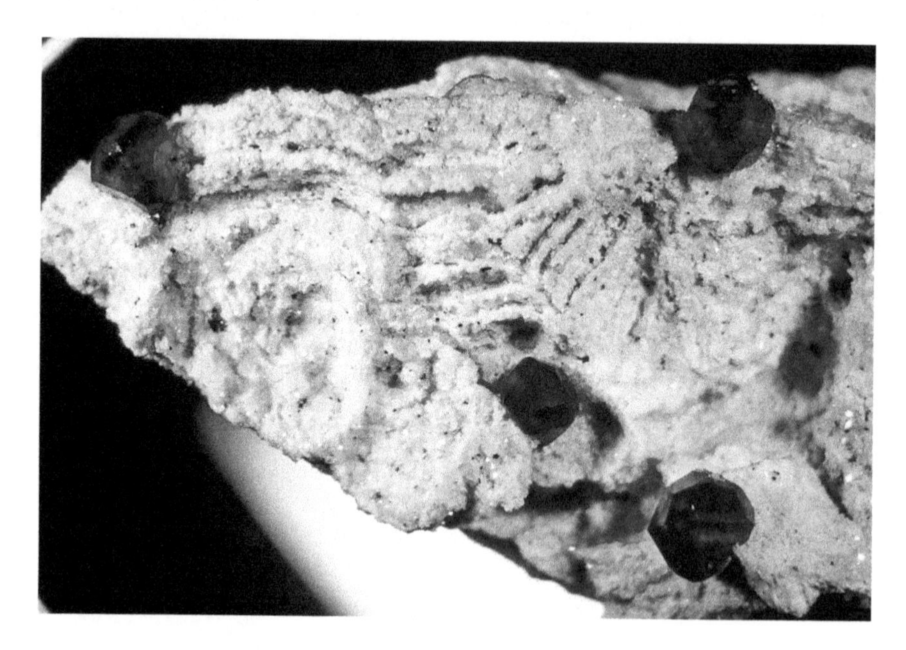

Spessartine Garnets, Ruby Mountain, Colorado. Field of view is 18.8 × 12.5 mm. (Photo courtesy of Howard Messing, www.messingminerals.com/spec_detail.php?id=1783.)

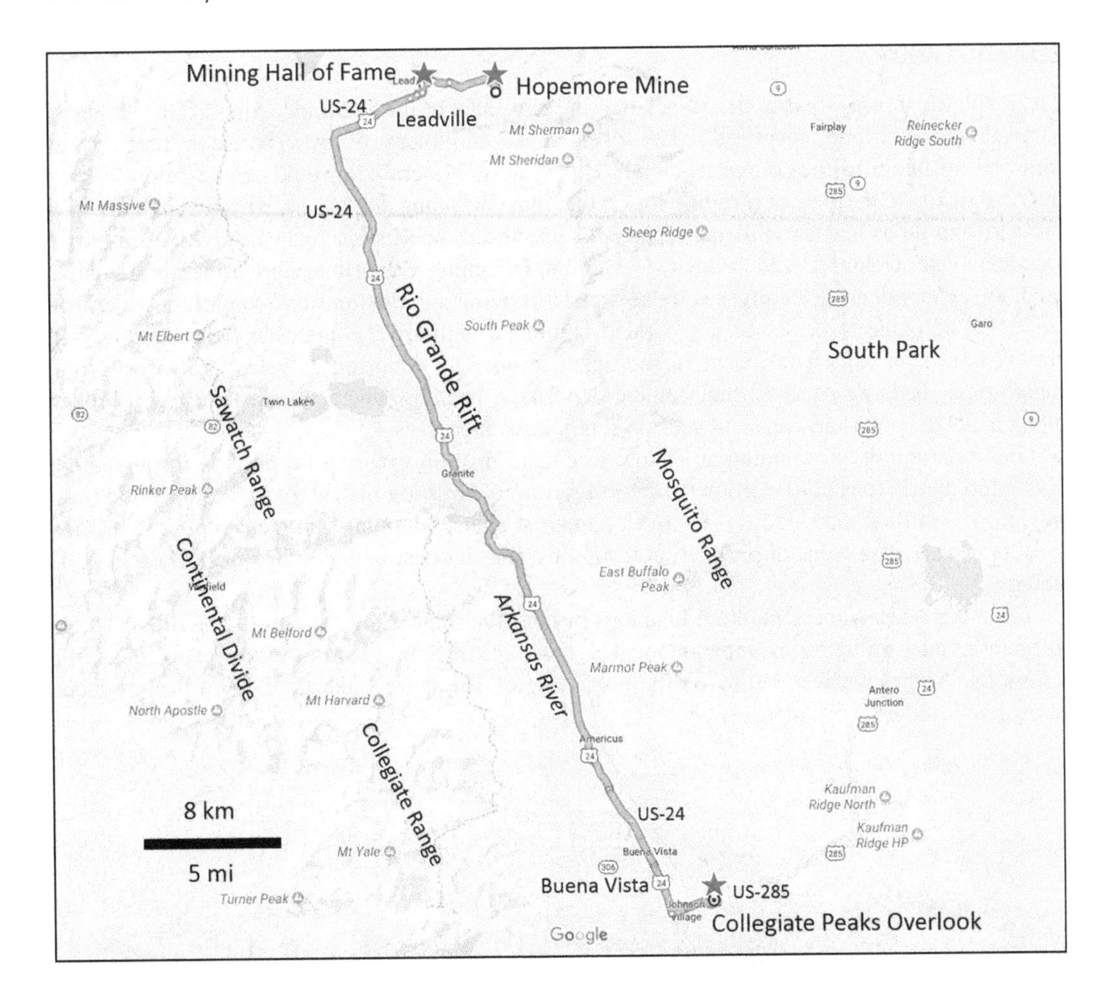

Side Trip 2—Ruby Mountain to Leadville: *Return north on County Road 300 to County Road 301; turn left (west) on County Road 301 and drive to US-285; turn right (north) on US-285; continue straight on US-24 to* ***Stop 5.1, National Mining Hall of Fame*** *at 120 W 9th St in Leadville (39.251025, −106.293942), for a total of 70.0 km (43.5 mi; 54 min).*

Collegiate Peaks Overlook to Leadville: *Continue west on US-285 to US-24 at Buena Vista; turn right (north) on US-24 and drive to* ***Stop 5.1, National Mining Hall of Fame*** *at 120 W 9th St in Leadville (39.251025, −106.293942), for a total of 67.6 km (42 mi; 57 min).*

5 LEADVILLE MINING DISTRICT AND THE COLORADO MINERAL BELT

Colorado Mineral Belt

As mentioned in the traverse from Denver to Grand Junction, the Colorado Mineral Belt (CMB) is a cluster of metallic mineral deposits that extends 400 km (250 mi) diagonally across central Colorado from the San Juan Mountains to near Boulder. It includes the mining districts of Ouray-Sneffels-Silverton-Telluride in the San Juan Mountains, the Aspen and Leadville districts near the Continental Divide, and the Idaho Springs-Central City district in the Front Range. The mineral deposits are associated with Laramide (75–42 Ma) and younger intrusions thought to be controlled by a Precambrian-age northeast shear zone. Mining districts within the mineral belt have produced prodigious amounts of gold and silver along with minor amounts of lead and zinc, between 1858 and the present.

Leadville Geology

The Leadville mining district lies smack dab in the middle of the Colorado Mineral Belt. It began as a gold placer mining operation in California Gulch a mile east of town, but soon the gold was followed upstream to the source in metal sulfide veins. Mineralizing fluids came from Tertiary quartz monzonite porphyries (granitic rocks) that intruded along and generated veins in faults. The ore-carrying fluids reacted with and replaced dolomites in the Mississippian Leadville Limestone, Devonian Dyer Dolomite, and Ordovician Manitou Dolomite. The primary ore minerals are native gold, silver in galena (lead-silver sulfide), sphalerite (zinc sulfide), and chalcopyrite (copper-iron sulfide). The oxide minerals cerussite (lead carbonate with silver), anglesite (lead sulfate), and smithsonite (zinc carbonate) occur in the upper levels. Ore occurred as veins, stockworks, and manto-type (bedding parallel) replacement deposits, especially where the host rock was broken along faults or paleo-karst features such as sinkholes and caverns.

Gold production is estimated at 93 metric tons (3 million troy oz). Production through 1963 was 7,465 metric tons (240 million troy oz) of silver, 987 million metric tons (1,088 million tons) of lead, 712 million metric tons (785 million tons) of zinc, and 48 million metric tons (53 million tons) of copper. The value of production from the mines was estimated at over $5 billion in 2007 dollars.

There are a few mineral hunting locations besides the mine dumps around Leadville. The old Turquoise Chief mine, for example, is located about 14 km (9 mi; 25 min) northwest of Leadville across the Arkansas River Valley on the north side of Turquoise Lake. The claim has produced many fine turquoise specimens.

Panoramic view up California Gulch, Leadville. (From Emmons et al., 1927.)

History

Placer gold was discovered where California Gulch empties into the Arkansas River in late 1859. The following year, a rich discovery of placer gold was found in California Gulch 1.6 km (1 mi) east of Leadville. A gold rush followed and Oro City was founded at the site. Soon there were 10,000 prospectors in town.

Black sand was found in the creeks while panning for gold, and it was hindering recovery of the gold. In 1874, someone had an assay done on the heavy black mineral and found it was cerussite, an ore rich in silver. By 1876, they had traced the cerussite to its source in a number of vein deposits. Thus began a silver boom. Because much of the lode silver was in the mineral galena, and galena also carries lead, the town that grew up around the lode deposits was named Leadville. Leadville, founded in 1877 by mine owners Horace Tabor and August Meyer, by 1880 had a population of 40,000 and, spurred on by the Sherman Silver Purchase Act, became one of the largest silver producers in the world. The Act, passed by Congress in 1890, required the government to purchase 4.5 million ounces (140,000 kg) of silver each month to help support the dollar. But the repeal of the Sherman Silver Purchase Act in 1893 caused the price of silver to plummet and many of the mines went bankrupt. The mines that remained produced mainly lead and zinc. The last active mine in the district, the Black Cloud Mine, closed in 1999.

More famous than Tabor was his wife, Elizabeth McCourt Tabor, known as Baby Doe. Born in Wisconsin, she moved to Colorado in the 1870s with her first husband, Harvey Doe. She quickly divorced Harvey for drinking, gambling, and whoring, and moved to Leadville where she met Horace Tabor. Tabor was already married and twice her age, but he left his wife of 25 years and married Baby Doe. After a brief time in Washington D.C. while Horace was a senator, they moved to Denver. Tabor lost his fortune with repeal of the Sherman Act and they became destitute. After Horace died in 1899, Baby Doe, once known as "the best dressed woman in the West," moved back to Leadville where she lived her last three decades in poverty in a shack at the Matchless Mine. She froze to death in a blizzard in 1935, age 81. The story of her life inspired the movie Silver Dollar (1932), and later an opera, The Ballad of Baby Doe (1956).

James Joseph "J. J." Brown, son of an Irish immigrant, was a self-taught prospector and geologist. He learned placer mining in the Dakotas before moving to Colorado, eventually settling at Leadville in the 1880s. In 1886, he married 19-year-old Margaret Tobin. Starting at the bottom, he quickly rose to timberman, shift boss, and foreman. He was working as superintendent of Ibex Mining Company properties when, in 1893, he came up with a method to stop cave-ins by using baled hay and timbers. Their Little Johnny Mine was one of the richest gold strikes in the district. The owners awarded J.J. 12,500 shares, or 12.5% of company stock, and a seat on the board. He had become a very wealthy man. His wife, however, is the one who became famous. Called "Maggie" by her friends, she survived the sinking of the Titanic and became known as the "Unsinkable Mrs. Brown." Her life became the basis of a 1960 Broadway musical, and the 1964 movie *The Unsinkable Molly Brown* made her a household name.

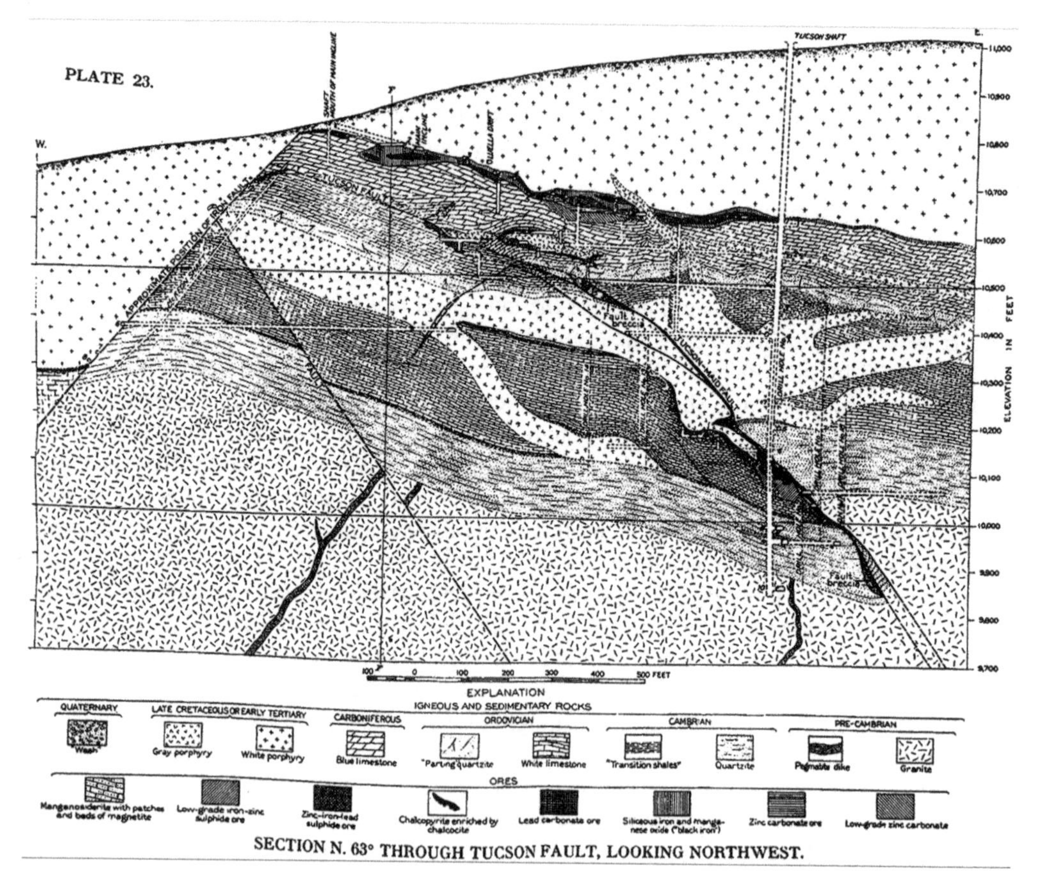

Cross section through the Leadville District showing the disposition of strata, porphyry dikes, and ore bodies. (From Emmons et al., 1927.)

STOP 5.1 NATIONAL MINING HALL OF FAME AND MUSEUM

The National Mining Hall of Fame and Museum, housed in an 1899 building in central Leadville, has 6,000 m^2 (71,000 ft^2) of interactive exhibits and collections of mining equipment, artifacts, and specimens as well as photographic archives, gem and mineral collections, and recreated mines and caves. There is a Gold Rush room, a "Railroads and Mining" exhibit, and special displays of gold, coal, and molybdenum, among others. Get a panoramic view of the Rockies from the fourth-floor terrace.

The museum is dedicated to the men and women who built our mines and to educating the public about the mining industry. It is the only national mining museum with a federal charter, which was signed by President Reagan in 1988.

National Mining Hall of Fame. Photo courtesy of C. Talbert, https://commons.wikimedia.org/wiki/File:National_Mining_Hall_of_Fame_and_Museum_(front).jpg

Visit

Hours:

The museum is open Tuesday through Sunday from 9:00 am to 5:00 pm. It is closed Thanksgiving, Christmas, and New Year's days.

See the website for current admission fees.
Website: mininghalloffame.org
Address: 120 West Ninth St., Leadville, CO
Phone: (719) 486-1229
Email: www.mininghalloffame.org/form/contact-mining-museum

Mining Hall of Fame to Hopemore Mine: *Drive east on E 9th St to Harrison Ave; turn right (south) on Harrison and drive to E 5th St; turn left (east) on E 5th St/CO-1 and drive to Stop* ***ST1.2****, the* ***Hopemore Mine*** *(39.244724, −106.234683) on the right, for a total of 6.1 km (3.8 mi; 14 min).*

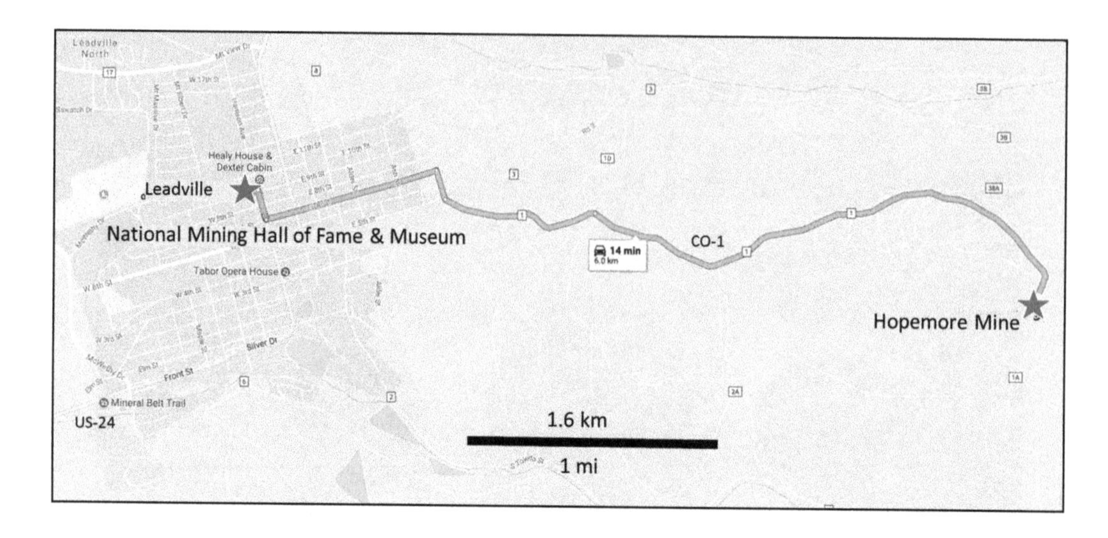

STOP 5.2 HOPEMORE MINE

Take a 1 h tour 200 m (600 ft) underground in one of Colorado's last underground mine tours. You are lowered down a vertical shaft in an old-time cage. Hear stories about the mining history of Leadville, and see demonstrations of mining tools from the mine owner, a former underground miner himself.

Reservations are required and can be made by calling ahead. Visitors should wear closed-toed shoes and bring warm clothes because the mine's temperature is 4.5°C (40°F). Children under 10 years are not allowed.

Visit

Address: 2921 County Road 1, Leadville, CO
Phone: 719-486-0301
Check the website for current entrance fees.
Website: www.leadvilletwinlakes.com/things-to-do-detail/hopemore-mine-tour/

Hopemore Mine to Twin Lakes Overlook: *Return to US-24 in Leadville and drive south on US-24 to CO-82 at Twin Lakes; turn right (west) on CO-82 and drive to County Road 10; turn right on County Road 10 and drive to Reva Ridge Road; turn right on Reva Ridge Road and drive to* ***Stop 6, Twin Lakes Overlook*** *(39.093487, −106.310870), for a total of 34.6 km (21.5 mi; 36 min).*

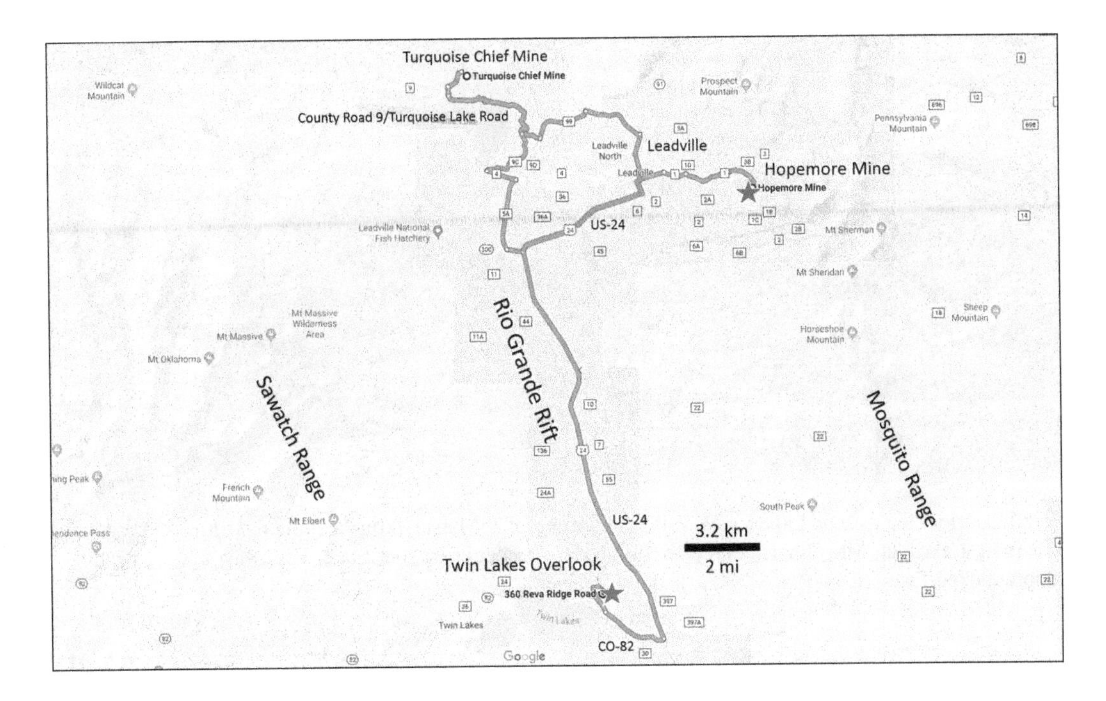

STOP 6 TWIN LAKES OVERLOOK

The Twin Lakes are hemmed in by Pinedale-age (30,000- to 10,000-year-old) lateral and terminal moraines deposited by glaciers that flowed east down Lake Creek Valley. These moraines are dated at 13,000–20,000 years old. Lee (2010) recognized at least three episodes of Pinedale glacial advance, the first extending into the Arkansas River Valley, and the second and third creating the lateral moraines seen in Lake Creek Valley (McCalpin, 2010). You will pass these moraines as you drive west on CO-82 going up the valley.

Rare evidence of earlier, Bull Lake Glaciation (125,000 to about 45,000 years ago) has been found in the lower parts of the lateral moraine on the north side of the valley. You are standing on the terminal moraine of Pinedale age. Lateral moraines flank the lakes. A recessional moraine separates the two lakes. From this vantage point, you can observe the classic glacial U-shape of Lake Creek Valley.

Gold placers were worked along Lake Creek and its tributaries from 1860 to 1951. These were hand operations until dredges were brought in 1915. Most gold was fine-grained (pinhead-size) gold flakes, although some nuggets up to 1 cm (0.5 in) were found. Altogether, the total amount of gold found in this district was estimated to be worth $1.3 million (Cappa and Bartos, 2007).

Twin Lakes looking west up Lake Creek Valley. (Courtesy of T. Matsui, https://commons.wikimedia.org/wiki/File:Twin_Lakes_at_Mt._Elbert,_entry_to_the_Independence_pass_on_State_Highway_82,_Colorado_-_panoramio.jpg.)

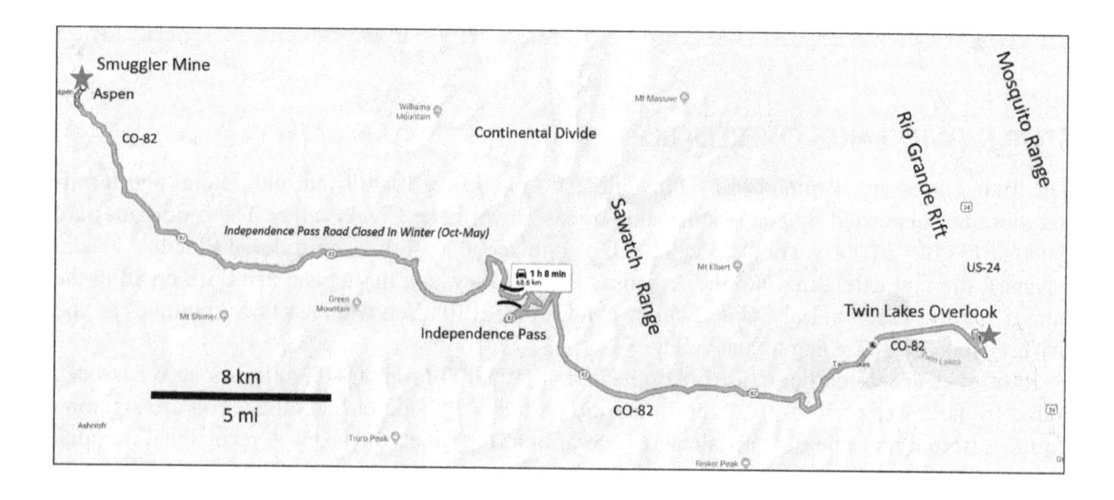

Twin Lakes Overlook to Smuggler Mine: *Return to CO-82 and drive west to Park Ave. in Aspen; turn right on Park and follow the signs to* ***Stop 7.1, the Smuggler Mine****, 110 Smuggler Mountain Rd, Aspen (39.192813, −106.808032) for a total of 68.6 km (42.6 mi; 1 h 8 min). Note: CO-82 is closed over Independence Pass from October through May.*

7 ASPEN MINING DISTRICT

As you drive west toward Aspen, you summit the Continental Divide at Independence Pass (3,686 m or 12,095 ft). You are in the Sawatch Range, a north-south mountain belt that consists primarily of 1.4 and 1.7 Ga granites and even older biotite gneiss and migmatite. These have all been intruded by Oligocene granodiorite and covered by local ash flow tuffs and andesitic lavas. Once over the pass, you are in the U-shaped glacial valley of the Roaring Fork River.

View southeast from Independence Pass to La Plata Peak. (Courtesy of N. Palmero, https://commons.wikimedia.org/wiki/File:La_Plata_Peak_from_Independence_Pass.jpg.)

Aspen, located on the Roaring Fork River, was originally the summer hunting ground of the Ute Indians. Prospecting began in the area in the 1870s when miners from Leadville headed west in search of gold and silver. Originally named Ute City, it started as a silver mining camp in 1879. Soon after, it was named Aspen for the trees on the nearby slopes. Rich silver ore was struck in 1880, and the area boomed. At one point, the district produced one-fifth of the world supply of silver. By 1893, Aspen was considered the richest silver town in the world, with production estimated at over $105M by 1947 (Vanderwilt, 1947).

Aspen grew rapidly during the 1880s, reaching almost 13,000 people, and became the seat of newly established Pitkin County. Investors built railroads, a luxury hotel, and an opera house. At one point in the 1880s, the city was the nation's leading silver producer.

After the Panic of 1893, Congress repealed the Sherman Silver Purchase Act. Deprived of its largest client, the town went into a slump that lasted until the area became a ski mecca after World War II.

STOP 7.1 SMUGGLER MINE

The Smuggler Mine is the oldest operating silver mine in the Aspen mining district and is still operating. In 1987, it was listed on the National Register of Historic Places. There are 38 underground levels, the lower half of which are flooded. The mine tour takes you 610 m (2,000 ft) underground in cars behind an electric locomotive, through some of the largest underground chambers in the world.

Smuggler Mine, Aspen. (Courtesy of bigweasel, https://commons.wikimedia.org/wiki/File:Smuggler_Mine,_Aspen,_CO.jpg.)

The prospector who discovered the Smuggler in 1879 sold it for $50 and a mule in 1881, according to *The Aspen Times* (Abraham, 2006). By 1886, it was producing $1,500 in silver *a day.* At its peak, this mine was responsible for 20% of the world's silver production. As in Leadville, prosperity ended with the Panic of 1893, when Congress repealed the Sherman Silver Purchase Act. The Smuggler was forced to close most operations. And yet in 1894, the largest silver nugget ever mined anywhere and 93% pure silver, was found at the Smuggler mine. It weighed 1,060 kg (2,340 lb) and was too large to bring up in one piece, so it was broken into three pieces.

The main ore minerals are galena, a lead-silver sulfate, and tennantite, a copper arsenic sulfosalt in which silver often substitutes for copper. The main mineralization occurs in a zone of broken rock up to 9 m (30 ft) thick along the Silver Fault, a bedding plane fault at the contact between the Pennsylvanian Weber Formation black shale and the Mississippian Leadville Formation dolomite. Minor mineralization occurred along steeply inclined faults in all the formations from Precambrian granite to Tertiary rhyolite sills (magma injected between layers of sedimentary rock). Most ore averaged 0.625 kg silver per metric ton (20 oz/ton), although some assays ran up to 3.9 kg/metric ton (125 oz/ton).

In 1917, miners reached the bottom of the vein that had been the main source of ore. The owner decided to shut down the mine. Mining resumed after World War II, but by then, skiing was becoming the main industry in Aspen and the mine languished.

Visit

Address: 110 Smuggler Mountain Rd, Aspen, CO 81611
Phone: (970) 925-2049 or 970-925-3699
See the website for current tour costs.
Website: www.allaspen.com/history_museums/smuggler_mine.php

Smuggler Mine to Holden Marolt Mining Museum: *Return to Park Circle and turn left (south); drive south on Park Circle to King Street; turn right (west) on King, then bear right onto Gibson Ave.; drive northwest on Gibson to Neale Ave; turn left (south) on Neal and drive to Main Street; turn right (west) on Main and drive to 7th Street; turn right (north) on 7th and drive to Hallam Street/CO-82; turn left (west) on CO-82 and drive over Castle Creek and immediately turn left to* ***Stop 7.2, Holden Marolt Mining and Ranching Museum*** *(39.192888, −106.834596), for a total of 3.2 km (2 mi; 8 min).*

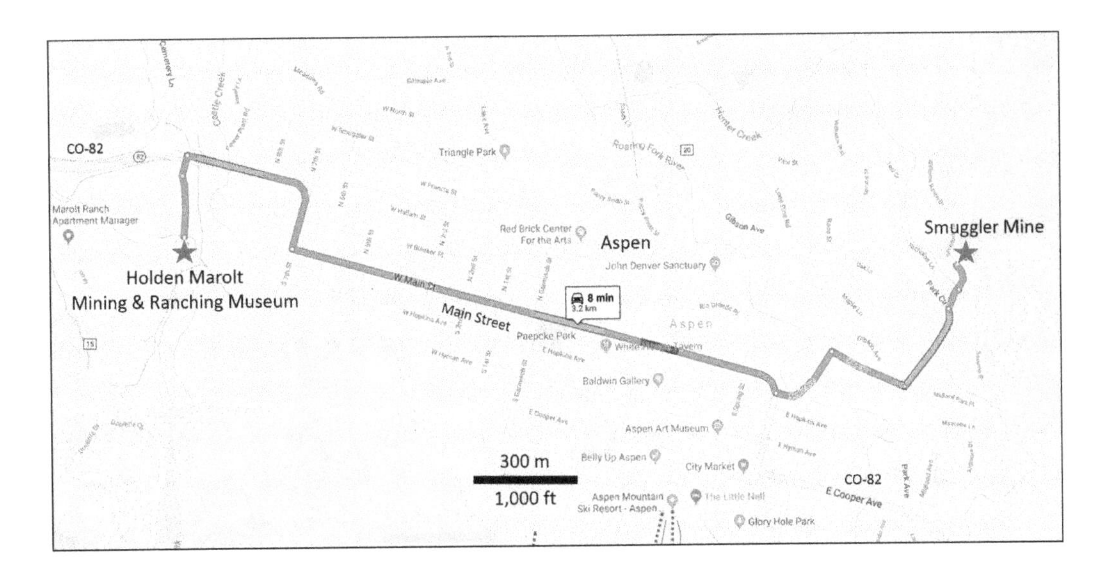

STOP 7.2 HOLDEN MAROLT MINING AND RANCHING MUSEUM

Holden Marolt Mining and Ranching Museum is on the site of the former Holden Mining and Smelting Company mill. In 1990, it was listed on the National Register of Historic Places.

In 1891, E.R. Holden formed the Holden Mining & Smelting Company with a $60,000 loan. The Holden Lixiviation Mill was built to process ore mined from the surrounding mines into silver through leaching. The Russell Lixiviation process crushed, heated, and dissolved the ore to refine silver. The mill was able to extract silver from ore with as little as 280 g (10 oz) of silver per 0.9 tonne (1 ton) at a time when most Aspen ore averaged 14-21 kg/tonne (400–600 oz/ton).

Mining went into decline after the bust of 1893. The buildings, however, remained. Mike Marolt purchased the property for a dollar in 1940 and merged it with his neighboring ranch. The museum highlights both Aspen's mining and ranching heritage.

Visit

Address: 40180 Highway 82, just over the pedestrian bridge off 7th Street.
Phone: 970.544.0820
Hours: Open Tuesday–Saturday, 11 am–5 pm (June 18–October 5, 2019). Closed July 4th.
See the website for current entrance fees.
Website: http://aspenhistory.org

Holden Marolt Mining Museum to Maroon Bells: *Return to CO-82 and turn left (west); drive to County Road 13/Maroon Creek Road and turn left (south); continue on Maroon Creek Road to* ***Stop 8, Maroon Bells*** *trailhead (39.098584, −106.940529), for a total of 15.8 km (9.8 mi; 20 min).*

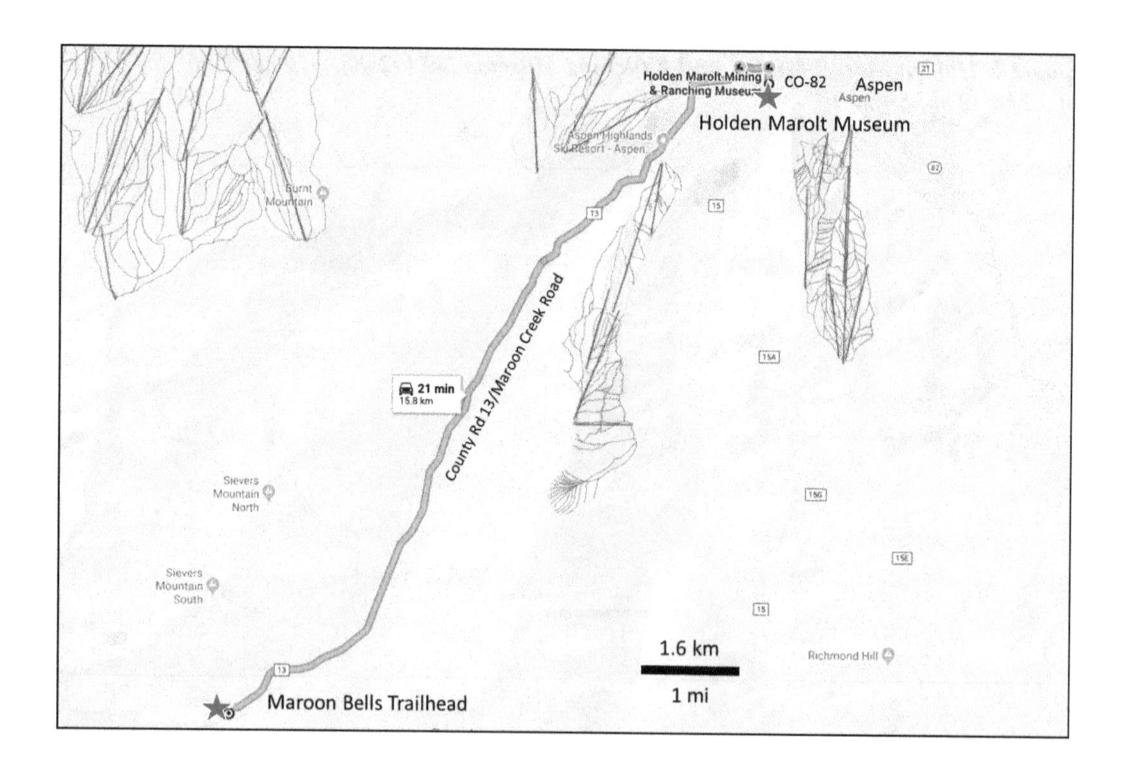

STOP 8 MAROON BELLS

We end this traverse of the Rockies at Maroon Bells. These iconic peaks are named for their distinctive color and bell shape. They consist of Middle Pennsylvanian Maroon Formation, up to 1,200 m (4,000 ft) of interbedded red sandstone, siltstone, mudstone, conglomerate, and rare thin gray limestone. The limestone was deposited in a shallow marine environment, whereas the conglomerates, sands, silts, and muds were eroded off the Ancestral Rocky Mountains and deposited in river valley, delta, and coastal plain environments. This area was the Central Colorado Trough: there was a range to the west, the Uncompahgre Uplift, and one to the east, the Ancestral Front Range. Both contributed sediments. The Maroon Formation is time-equivalent and effectively the same rock type as the Fountain Formation we saw at Garden of the Gods. Both units were eroded off the Ancestral Rockies.

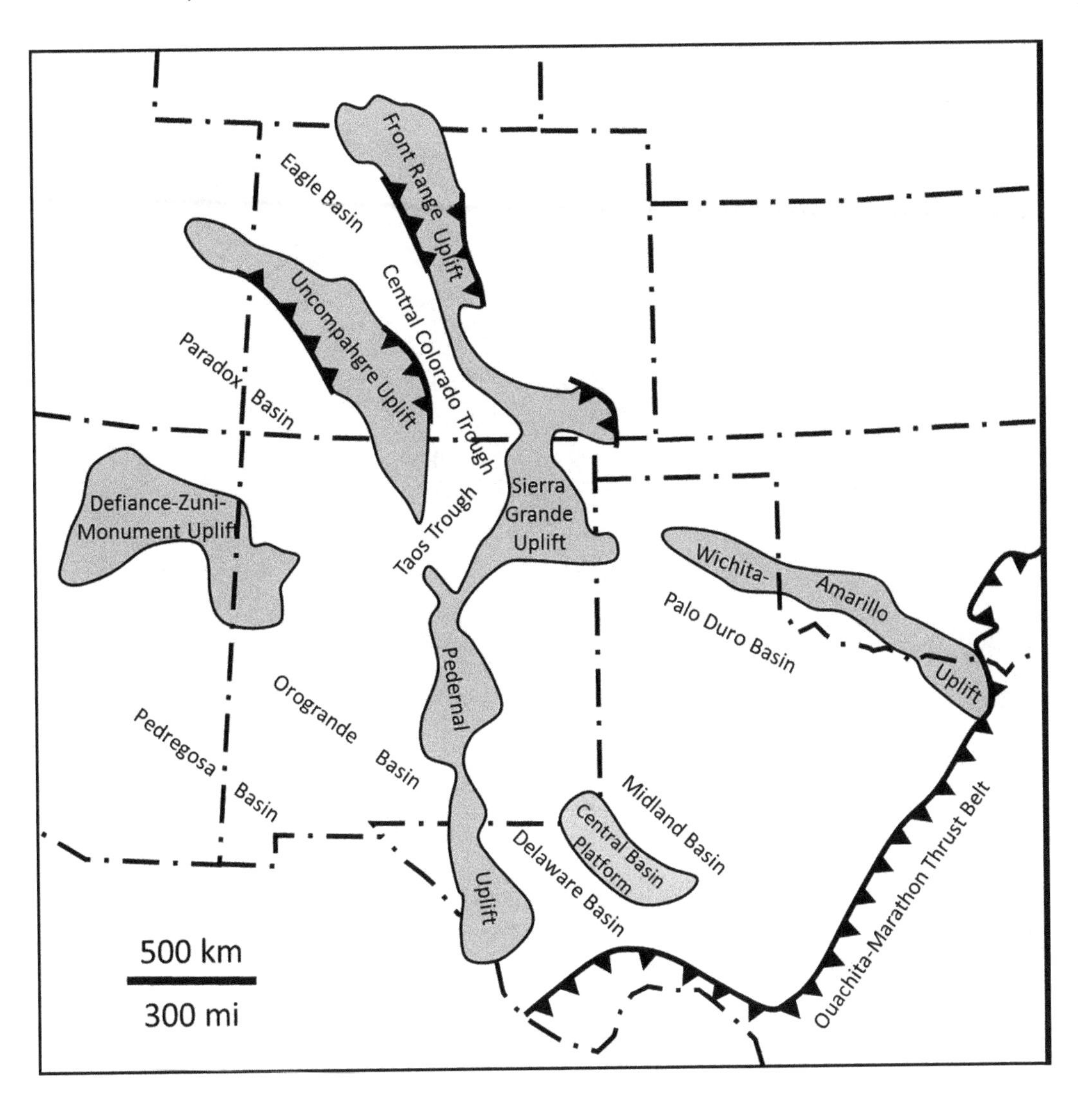

Pennsylvanian uplifts and basins in Colorado and south-central United States. The Ouachita-Marathon Thrust Belt in Texas and Oklahoma is continuous with the Alleghanian thrust belt that formed the Appalachian Mountains. Both are a result of the collision of Gondwana, the southern continent, with Laurasia (proto-North America) to form the supercontinent Pangea in Pennsylvanian-Permian time. (Modified after Lindsey et al., 1986; Lindsey, 2010.)

The area was uplifted a second time during the Laramide Orogeny in Late Cretaceous through Eocene time. There was a pause in the Eocene, when a widespread erosion surface developed in the Rockies, then uplift resumed about 34 million years ago. Erosion of the uplift was enhanced starting around 2 million years ago when glaciers began to form due to global cooling. Glaciers put the finishing touches on the peaks and carved the U-shaped valley leading up to them. The lake in the foreground formed after a landslide dammed the valley.

This stop, as well as Maroon and North Maroon peaks (both 'fourteeners'), are in the Maroon Bells—Snowmass Wilderness, designated in 1964.

Sunrise on Maroon Bells as seen from Maroon Lake. (Courtesy of R. Schuster, https://commons.wikimedia.org/wiki/File:Sonnenaufgang_an_den_Maroon_Bells.jpg.)

We end this transect near western margin of the Rocky Mountains. Just west of here, at the Grand Hogback (New Castle on I-80), we enter the Colorado Plateau Province, a region of gentle, broad uplifts and basins.

REFERENCES AND FURTHER READING

Abraham, C. 2006. Aspen's mining past. *Aspen Times*, January 6.

Baldridge, W.S., and K.H. Olsen. 1989. The Rio Grande Rift: What happens when the earth's lithosphere is pulled apart? *American Scientist*, v. 77, no. 3, pp. 240–247.

Barkmann, P.E., A. Moore, and J. Johnson. 2013. South Park groundwater quality scoping study. Colorado Geological Survey and Colorado School of Mines, 58 p.

Bilodeau, W.L. 1986. Reassessment of Post-Laramid Uplift and Tectonic History of the Front Range, Colorado. *In* Rogers, W.P., and R.M. Kirkham (eds.), Contributions to Colorado Seismicity and Tectonics – a 1986 Update. Colorado Geological Survey Special Publication, Denver, 28, pp. 3–16.

Bryant, B. 1979. Geology of the Aspen 15-minute quadrangle, Pitkin and Gunnison Counties, Colorado. U.S. Geological Survey Professional Paper 1073, 146 p.

Cappa, J.A., and P.J. Bartos. 2007. Geology and mineral resources of Lake County, Colorado. Resource Series 42, Colorado Geological Survey, 69 p.

Chronic, H., and F. Williams. 2002. *Roadside Geology of Colorado*. Mountain Press Publishing, Missoula, 398 p.

Emmons, S.F., J.D. Irving, and G.F. Loughliln. 1927. Geology and ore deposits of the Leadville mining district, Colorado. U.S. Geological Survey Professional Paper 148, 368 p.

Heylmun, E.B. 2000. The Colorado mineral belt. *ICMJ's Prospecting and Mining Journal*, CMJ Inc., v. 69, 3 p.

Jacob, A. 1983. Mountain front thrust, southeastern Front Range and northeastern Wet Mountains, Colorado. In Lowell, J.D. (ed.), *Rocky Mountain Foreland Basins and Uplifts*. Rocky Mountain Association of Geologists, Denver, CO, pp. 229–244.

Keller, J.W., J.P. McCalpin, and B.W. Lowry. 2004. Geologic map of the Buena Vista East quadrangle, Chaffee County, Colorado. Colorado Geological Survey Open-File Report 04-4, 65 p.

Lee, K. 2010. Catastrophic outburst floods on the Arkansas River, Colorado. *The Mountain Geologist*, v. 47, pp. 35–47.

Lindsey, D.A. 2010. The geologic story of Colorado's Sangre de Cristo range. U.S. Geological Survey Circular 1349, 14 p.

Lindsey, D.A., R.F. Clark, and S.J. Soulliere. 1986. Minturn and Sangre de Cristo Formations of Southern Colorado: A Prograding Fan Delta and Alluvial Fan Sequence Shed from the Ancestral Rocky Mountains: Part IV. Southern Rocky Mountains. American Association of Petroleum Geologists Memoir 41: Paleotectonics and Sedimentation in the Rocky Mountain Region, United States, pp. 541–561.

MacGinitie, H.D. 1953. Fossil plants of the Florissant beds, Colorado. Carnegie Institute of Washington Publication 599, pp. 1–198.

Maughan, E.K. 1990. Summary of the Ancestral Rocky Mountains epeirogeny in Wyoming and adjacent areas. U.S. Geological Survey Open-File Report 90-447, 8 p.

McCalpin, J.P. (ed.) 2010. *GSA Annual Meeting Quaternary Geology and Geochronology of the Uppermost Arkansas Valley; Field Trip 405*, October 29–30, 2010, Glaciers, Ice Dams, Landslides, Floods. 49 p.

Messing, H. The messing mineral collection. www.messingminerals.com/spec_detail.php?id=1783. Accessed 25 July 2019.

Moore, D.W., A.W. Straub, M.E. Berry, M.L. Baker, and T.R. Brandt. 2002. Surficial geologic map of the Pueblo 1°×2° quadrangle, Colorado. U.S. Geological Survey Miscellaneous Field Studies Map MF-2388, 1:250,000.

National Park Service. 2015. Geologic history of florissant. www.nps.gov/flfo/learn/nature/geologic-history-of-florissant.htm. Accessed 30 June 2019.

Rowley, P.D., J.W. Himmerlich Jr., and D.H. Kupfer. 2002. Geologic map of the Cheyenne Mountain quadrangle, El Paso County, Colorado. Colorado Geological Survey Open-File Report 02-5, 55 p.

Shannon, J.R., and J.P. McCalpin. 2006. Geologic Map of the Maysville Quadrangle, Chaffee County, Colorado. Colorado Geological Survey Open File Report 06-10, 225 p.

Share, J. 2011. Written In Stone... seen through my lens. http://written-in-stone-seen-through-my-lens.blogspot.com/2011/02/ancestral-rocky-mountains-and-their.html. Accessed 15 June 2019.

Temple, J., R. Madole, J. Keller, and D. Martin. 2007. Geologic map of the Mount Deception quadrangle, Teller and El Paso Counties, Colorado. Colorado Geological Survey Open-File Report 07-7, 49 p.

The Collector's Edge. Two point amazonite mine, Teller County, Colorado. https://collectorsedge.com/pages/two-point-amazonite-mine-teller-county-colorado. Accessed 25 July 2019.

Theodosis, S.D., and J.W. Skehan. 1965. Geology of the North American Air Defense Combat Operations Center, Near Cheyenne Mountain and peripheral area, Colorado Springs, El Paso County, Colorado. Project 4600, Scientific Report No. 2, Air Force Cambridge Research Labs, Bedford, MA, 40 p.

Tweto, O., R.H. Moench, and J.C. Reed. 1978. Geologic map of the Leadville 1°×2° quadrangle, Colorado. U.S. Geological Survey Miscellaneous Investigation Series Map I-999, 1:250,000.

Vanderwilt, J.W. 1947. *Mineral Resources of Colorado.* Colorado Mineral Resources Board, Denver, CO, 547 p.

Veatch, S.W. 2012. The Rio Grande Rift. http://coloradoearthscience.blogspot.com/2012/12/the-rio-grande-rift.html. Accessed 3 July 2019.

Wilson, A.B., and P.K. Sims. 2003. Colorado mineral belt revisited – An analysis of new data. U.S. Geological Survey Open-File Report 03-046, 7 p.

5 Gulf Coast to Rio Grande Rift

Austin, Texas, to Las Cruces, New Mexico

Organ Needle, at 2,740 m (8,990 ft), is the highest point in Organ Mountains-Desert Peaks National Monument. (Courtesy of B. Wick, Bureau of Land Management.)

OVERVIEW

Texas is a big place. But contrary to what you might have heard, it is not all flat. In this chapter, we start at the Balcones Fault zone, the boundary between the Gulf Coast Province and the Edwards Plateau, and drive west to the Rio Grande Rift in New Mexico. Along the way, we discuss the petroleum geology of productive basins, visit classic localities for Cretaceous and Permian units, explore Enchanted Rock State Park, examine Paleozoic rocks deformed by the Ouachita Orogeny (also called the Ouachita-Marathon Orogeny) in the Marathon Uplift, take a side trip to Big Bend National Park and Big Bend Ranch State Park, and visit Carlsbad Caverns National Park and World Heritage Site. We look at reef successions in Guadalupe Mountains National Park and the Hueco Mountains, see gypsum dunes at White Sands National Monument, and rifted and tilted fault blocks of the Rio Grande Rift in the Franklin Mountains and Organ Mountains National Monument. We end at Prehistoric Trackways National Monument, an exceptional site for Permian reptile tracks.

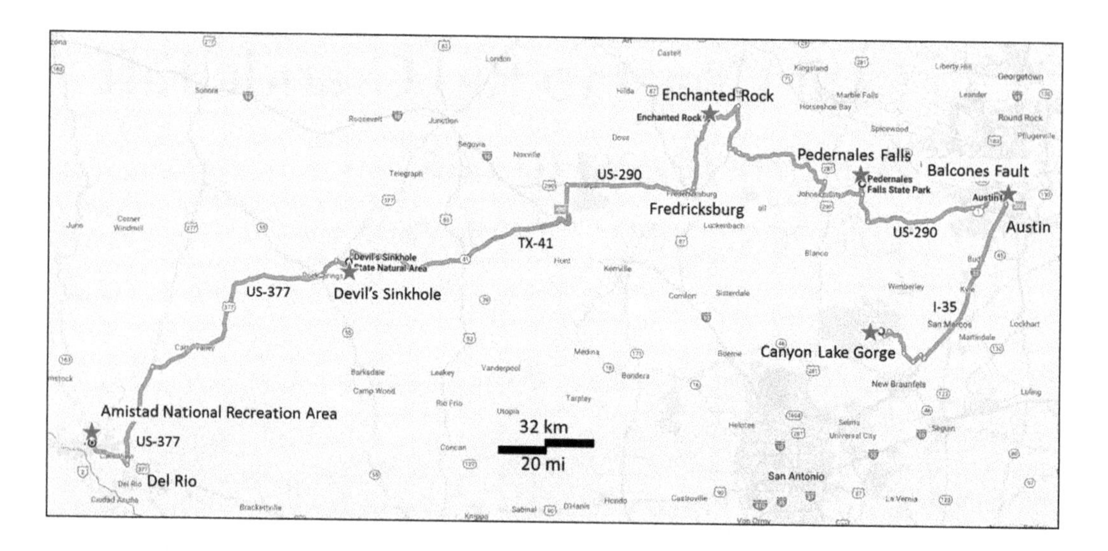
Enchanted Rock
Pedernales Falls
Balcones Fault
US-290
Fredricksburg
TX-41
US-290
Austin
US-377
Devil's Sinkhole
I-35
Canyon Lake Gorge
Amistad National Recreation Area
US-377
32 km
20 mi
Del Rio
San Antonio

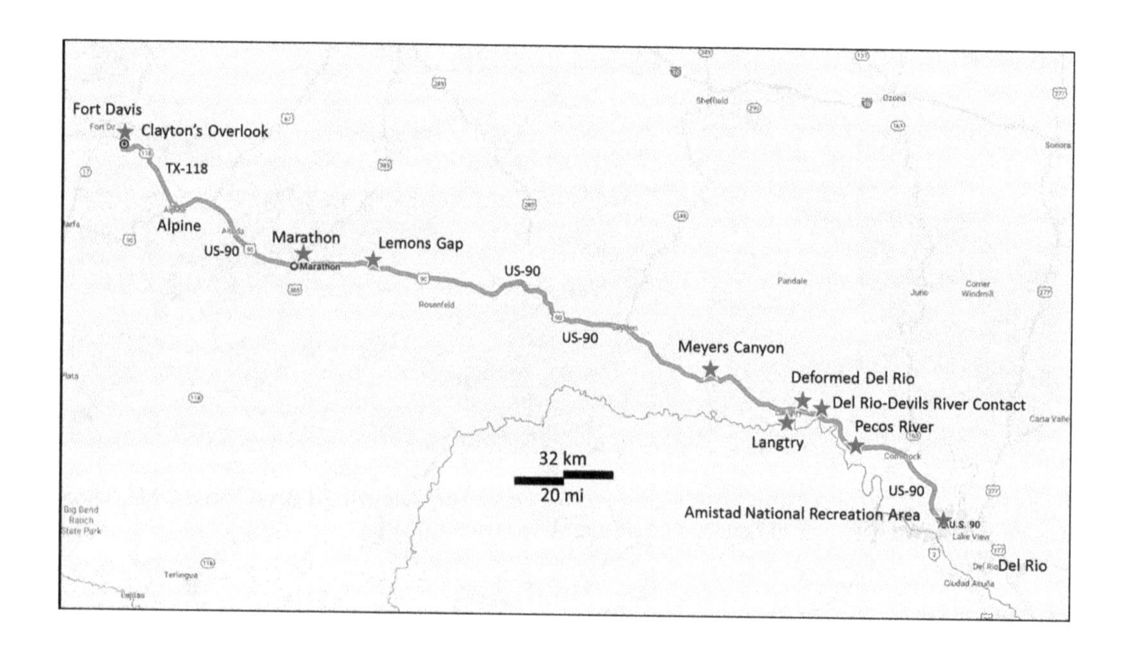
Fort Davis
Clayton's Overlook
TX-118
Alpine
US-90
Marathon
Lemons Gap
US-90
US-90
Meyers Canyon
Deformed Del Rio
Del Rio-Devils River Contact
Pecos River
Langtry
32 km
20 mi
US-90
Amistad National Recreation Area
Del Rio

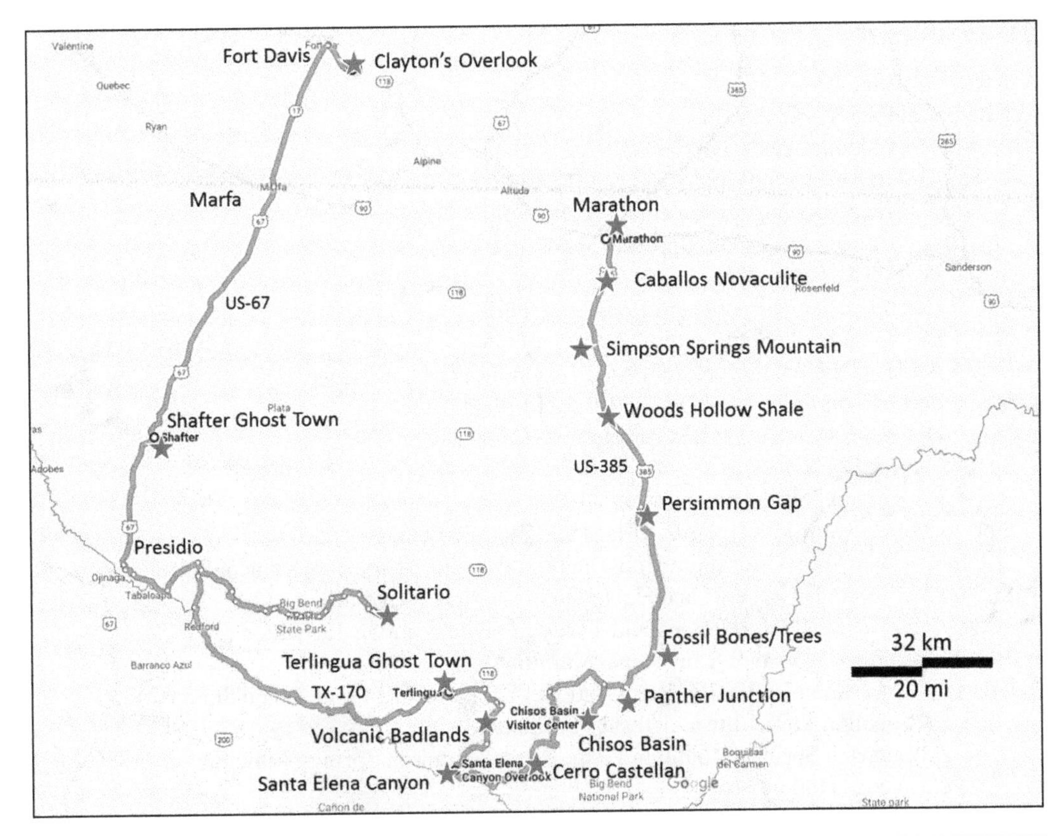
Fort Davis
Clayton's Overlook
Marfa
Marathon
Caballos Novaculite
US-67
Simpson Springs Mountain
Woods Hollow Shale
Shafter Ghost Town
US-385
Persimmon Gap
Presidio
Solitario
Fossil Bones/Trees
32 km
20 mi
Terlingua Ghost Town
TX-170
Panther Junction
Chisos Basin Visitor Center
Volcanic Badlands
Chisos Basin
Santa Elena Canyon Overlook
Cerro Castellan
Santa Elena Canyon

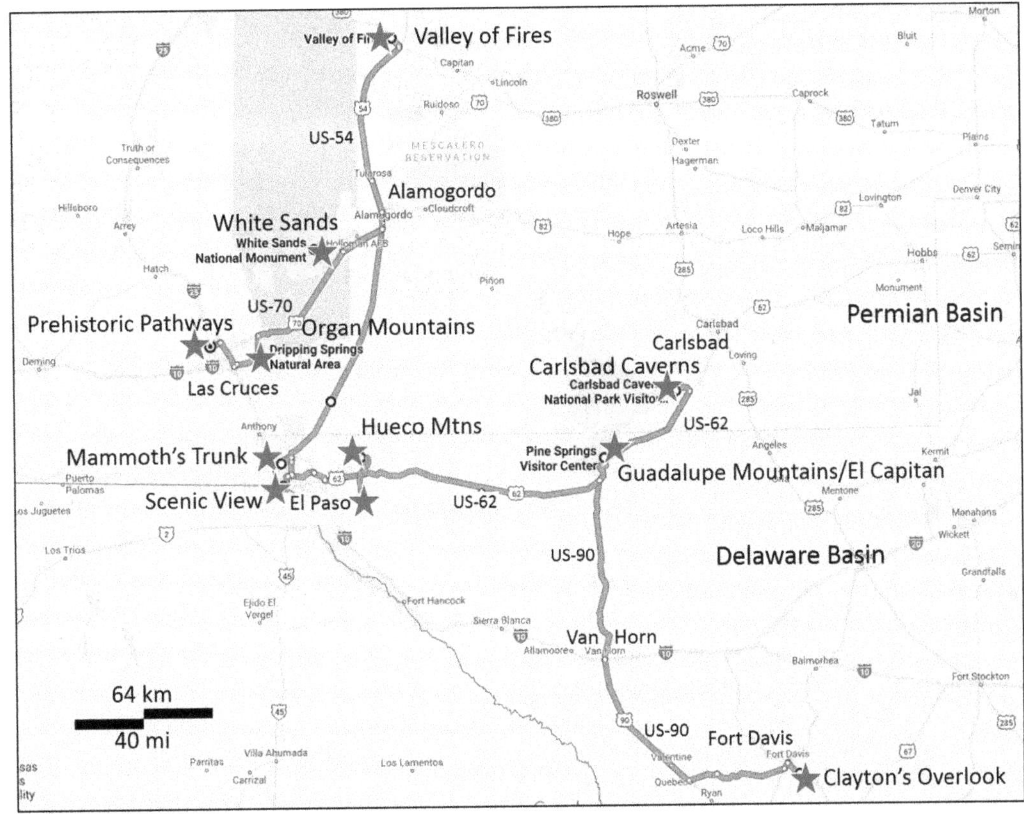
Valley of Fires
US-54
Alamogordo
White Sands
White Sands National Monument
US-70
Prehistoric Pathways
Organ Mountains
Dripping Springs Natural Area
Las Cruces
Permian Basin
Carlsbad
Carlsbad Caverns
Carlsbad Caverns National Park Visitor Center
Hueco Mtns
US-62
Mammoth's Trunk
Pine Springs Visitor Center
Guadalupe Mountains/El Capitan
Scenic View
El Paso
US-62
US-90
Delaware Basin
Van Horn
64 km
40 mi
US-90
Fort Davis
Clayton's Overlook

ITINERARY

Begin – Austin, TX
1 Balcones Fault Zone, Austin
  Stop 1.1 Balcones Fault, Perry Park
  Stop 1.2 Barton Springs and the Edwards Aquifer
  Stop 1.3 Folding along the Balcones Fault
Side Trip 1 Canyon Lake Dam Spillway and Gorge
Stop 2 Pedernales State Park
Stop 3 Llano Uplift and Enchanted Rock State Park
Stop 4 Devils Sinkhole State Natural Area
Stop 5 Amistad National Recreation Area—Salmon Peak Limestone
Stop 6 Pecos River Crossing—Devils River and Buda Limestone
Stop 7 Del Rio Clay—Devils River Limestone Contact
Stop 8 Faulted Del Rio Clay
Stop 9 Rio Grande Valley at Langtry
Stop 10 Meyers Canyon—Boquillas Flags and Buda Limestone
11 Lemons Gap Area—Marathon Uplift and the Paleozoic-Cretaceous Unconformity
  Stop 11.1 Lemons Gap East and Haymond Formation Turbidites
  Stop 11.2 Lemons Gap—Folded and Thrusted Pennsylvanian Turbidites
Stop 12 Deformed Marathon Limestone, Marathon
Side Trip 2 Marathon Uplift, Big Bend National Park, and Big Bend Ranch State Park
  ST 2.1 Caballos Novaculite and Tesnus Formation
  ST 2.2 Simpson Springs Mountain Picnic Area—Simpson Springs Anticline
  ST 2.3 Woods Hollow Shale
  ST 2.4 Persimmon Gap Overlook
  ST 2.5 Fossil Trees/Logjam
  ST 2.6 Fossil Bone Exhibit
  ST 2.7 Panther Junction Visitor Center
  ST 2.8 Chisos Basin
  ST 2.9 Volcanic Dike
  ST 2.10 Cerro Castellan
  ST 2.11 Santa Elena Canyon
  ST 2.12 Volcanic Badlands
  ST 2.13 Terlingua Ghost Town and Mercury Mining District
  ST 2.14 Big Bend Ranch State Park and Solitario Dome
  ST 2.15 Shafter Ghost Town and Silver Mining District
Stop 13 Clayton's Overlook, Chihuahuan Desert Research Institute Nature Center and Botanical Gardens
Stop 14 Carlsbad Caverns National Park and the Permian Basin
15 Guadalupe Mountains National Park
  Stop 15.1 McKittrick Canyon Trail
  Stop 15.2 Pine Springs Visitor Center
  Stop 15.3 El Capitan Lookout
16 Hueco Group, Orogrande Basin
  Stop 16.1 Southern Outlier
  Stop 16.2 Northern Outlier
17 Franklin Mountains and Rio Grande Rift
  Stop 17.1 Murchison Rogers Park Scenic Overlook, El Paso
  Stop 17.2 Mammoths Trunk Trail Roadcut, central Franklin Mountains
  Stop 17.3 Rest Area

18 White Sands National Monument
  Stop 18.1 Playa Trail
  Stop 18.2 Interdune Boardwalk
Side Trip 3 Malpais and Valley of Fires
19 Organ Mountains-Desert Peaks National Monument
  Stop 19.1 San Augustin Pass and the Organ Batholith
  Stop 19.2 La Cueva Rock
  Stop 19.3 Dripping Springs
20 Las Cruces Area
  Stop 20.1 Las Cruces Museum of Nature and Science
  Stop 20.2 Prehistoric Trackways National Monument
End –Las Cruces, NM

THE GULF COAST PROVINCE

This transect begins in Austin and traverses the Edwards Plateau and Permian Basin to end in Las Cruces. Austin sits at the confluence of the Gulf Coast Province and the Edwards Plateau. These are separated by the Balcones Fault Zone, thought to be the surface expression of the Ouachita-Marathon Fold Belt in the subsurface.

The Gulf Coast Province spans the low-lying coastal plains of Mexico and the United States from Veracruz State through south Texas to Alabama. The province contains the Gulf Coast Basin, perhaps the greatest petroleum region in North America. It is the most prolific petroleum basin in the United States, producing 31% of its oil and 48% of its gas over the past 100 years (Fails, 1990). More than 36.5 billion m^3 (230 billion barrels) of oil equivalent (BBOE) had been discovered by the early 1990s. Another 16 billion m^3 (100 BBOE) may remain to be discovered (Galloway, 2009).

Initial rifting of the Gulf and Atlantic basins began in the Triassic. Soon a restricted marine basin had amassed several kilometers of Jurassic salt as seawater alternately flooded the Gulf Basin and evaporated over a period of several million years. Since the Jurassic, the basin has accumulated on the order of 15 km (9 mi) of marine sediments. Limestones were deposited on shallow marine platforms and ramps during periods of low sediment input. Sandstones and shales were deposited by rivers that drained most of North America, particularly during periods of uplift in the Rocky Mountains and mid-continent. Mesozoic hydrocarbon reservoir units comprise limestones and sandstones, whereas Cenozoic reservoirs are mainly sandstones. Multiple, overlapping reservoirs are characteristic of this basin. Sandy deep-water fans are overlain by turbidites (underwater landslide deposits) and mini-basin sands; these were topped by delta front, barrier bar, and beach sands, and in turn overlain by river channel sand deposits. The large amounts of mudstone in the system effectively sealed the carbonate (limestone, dolomite) and sandstone reservoir units, preventing most hydrocarbons from leaking off to the surface.

Rapid, massive sedimentation and restricted circulation (anoxic bottom water) favored preservation of organic matter, leading to particularly rich and abundant hydrocarbon source rocks. Upper Jurassic, Cretaceous, and lower Tertiary organic-rich limestone, marl, and mudstone source rocks are found throughout the basin. Most hydrocarbon generation occurred during the Cenozoic. Faults and sand bodies acted as migration pathways leading to traps.

Rapid burial of water-saturated sediment led to a buildup of fluid pressure that both preserved reservoir porosity and decreased the strength of the rocks. This enabled structural deformation such as folding and faulting, and facilitated migration of hydrocarbons out of the deep basin to shallow sand bodies. Loading of deep Jurassic salt by kilometers of overlying sediments led to mobilization and flow of the salt layers. As the salt was displaced upward and basinward, it formed mounds, domes, and canopies. Mini-basins were caused by local salt evacuation. Each of these structures provides traps for oil and gas accumulation. "Growth faults" became active as sediments were deposited along the basin margin. These regional faults allowed the accumulating sediment to slip slowly basinward. At the same time, they provided space for more sediment to accumulate each time the normal fault moved. Dip reversal into these faults provided "rollover anticline" traps for hydrocarbons. Other traps include reservoirs cut off against faults, reservoir truncations against salt, upward sandstone pinchouts, and salt domes.

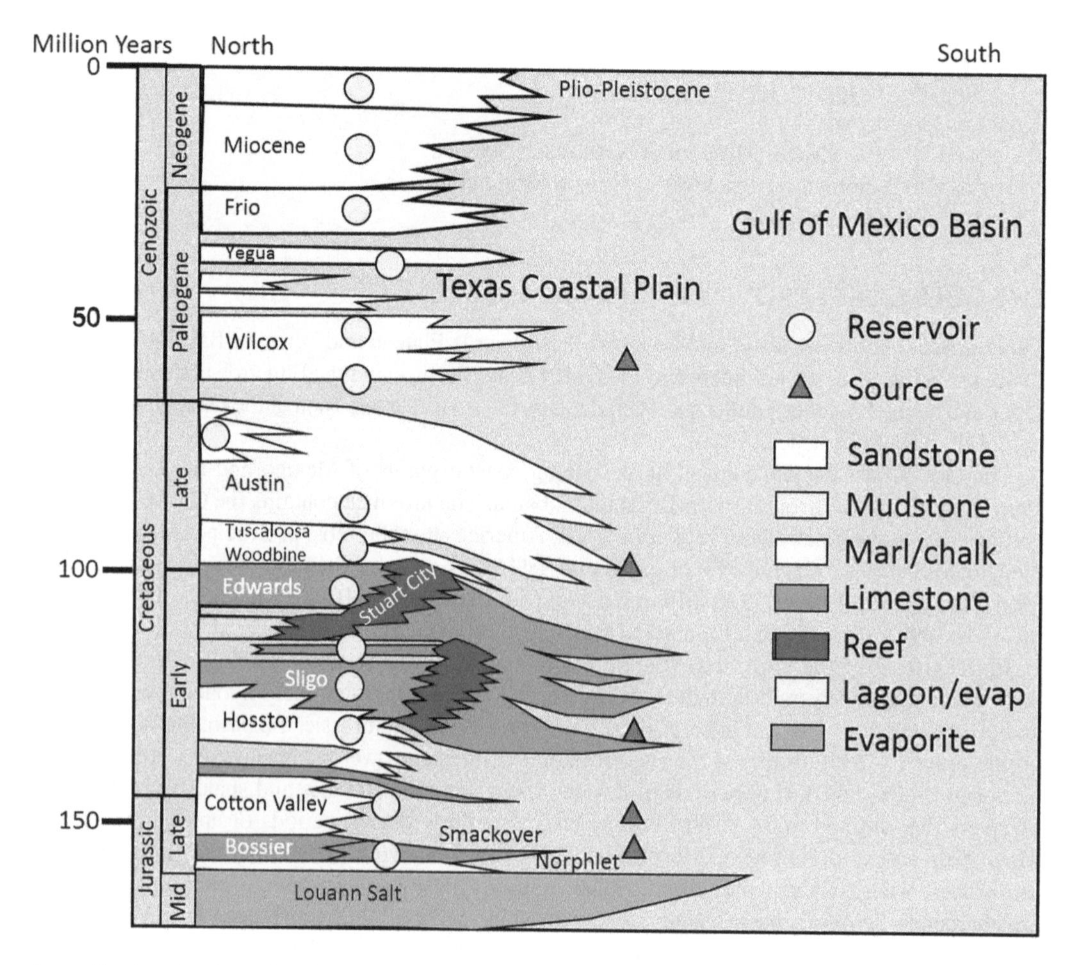

Generalized regional stratigraphy, northern Gulf of Mexico Basin. Hydrocarbon source rocks and reservoir rocks are indicated. (Modified after Galloway, 2009.)

Advances in horizontal drilling and hydraulic fracturing since the late 1990s have opened new, unconventional hydrocarbon resources in tight sandstone and shale gas reservoirs along the Gulf Coast Province.

The Edwards Plateau extends across central Texas from Austin and San Antonio in the southeast to the Llano Uplift in the north and Pecos River in the west. Surface units are nearly flat-lying Cretaceous-age limestones. The thin soil and rough terrain of the Hill Country are best suited for grazing cattle, sheep, and goats. Elevations range from 100 to 1,000 m (300 to 3,000 ft). Caves are common and contain some of the largest bat communities in the world.

Balcones Fault Zone between Del Rio and Austin (arrows). Google Earth oblique view north. Imagery © 2019 Landsat/Copernicus.

***Begin**—Balcones Fault at **Stop 1.1, Perry Park, Fairview Drive at Sunny Lane, Austin** (30.3312898, −97.7608722). Park at the Crenshaw Athletic Club and walk west past the tennis courts to the fault outcrop.*

1 BALCONES FAULT ZONE, AUSTIN

The Balcones Fault Zone extends from near Del Rio in southwest Texas through San Antonio and Austin to near Dallas in northeast Texas. It consists of multiple faults, grabens (down-dropped fault-bounded blocks), and horsts (uplifted blocks), and contains many caves and springs. The fault zone is expressed as a topographic escarpment facing the Gulf Coastal Plain and separating it from the Edwards Plateau and "Texas Hill Country." The zone is most likely the surface expression of the Ouachita-Marathon Orogenic Belt that formed by continental collision between Laurasia (proto-North America) and South America/Gondwana during Pennsylvanian-Permian time (roughly 300–270 million years ago). The folded and thrusted mountains formed by that event are exposed in the Ouachita Mountains of Oklahoma-Arkansas and in the Marathon Uplift of west Texas, but are deeply buried in the Austin area. During the Permian (285–245 Ma), these mountains defined the Texas coastline. Since that time, especially over the past 80 Ma, uplift and erosion in the mid-continent have led to thousands of meters of sediment being deposited in the Gulf of Mexico, moving the coastline south and east to its present location. About 15 million years ago (during the Miocene), the weight of the sediments, along with dewatering and compaction, caused these rocks and sediments to collapse along the old mountain front, forming the Balcones Fault Zone. The rocks were dropped down to the south and east, a total of around 360 m (1,200 ft) across a zone up to 8 km (5 mi) wide.

The fault zone is inactive, with little or no movement since the Miocene (past 15 million years or so).

STOP 1.1 BALCONES FAULT, PERRY PARK

Our first stop is the Balcones Fault Zone at Perry Park in north Austin. The fault plane is clearly exposed and contains well-expressed slickensides (grooves on the fault surface) that indicate normal offset. The fault is down-to-the-east and juxtaposes Cretaceous Glen Rose Limestone on the upthrown side with Fredericksburg Group (Edwards Limestone) on the down side.

Just south of Perry Park, where the Colorado River crosses the fault zone, an escarpment raises the topography from 140 m above sea level to 300 m (from 450 to 1,000 ft) elevation over a distance of about 6 km (3.7 mi).

Balcones Fault at Perry Park, Austin. The downthrown side is to the left (east). Cretaceous Glen Rose Limestone is exposed as the footwall, and has clear slickensides (surface scratches that indicate the direction of fault movement).

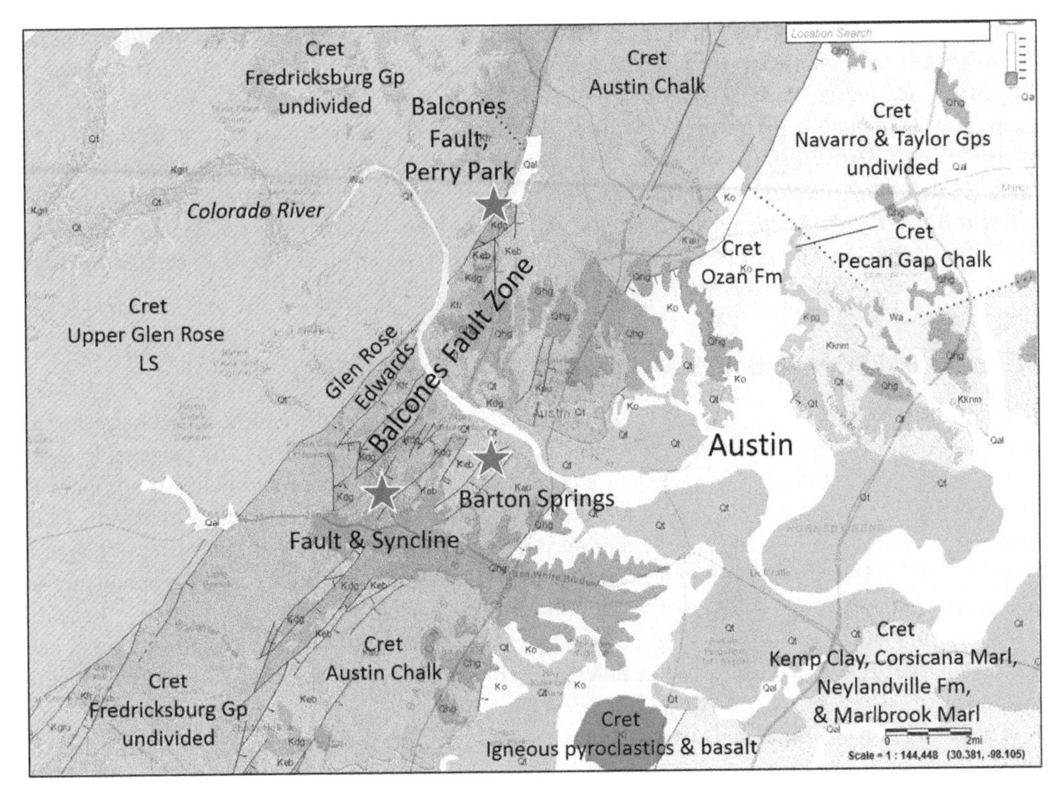

Geologic map of the Austin area showing Balcones Fault Zone. Stops are indicated by the red stars. (Courtesy of U.S. Geological Survey and Bureau of Economic Geology, https://txpub.usgs.gov/txgeology/.)

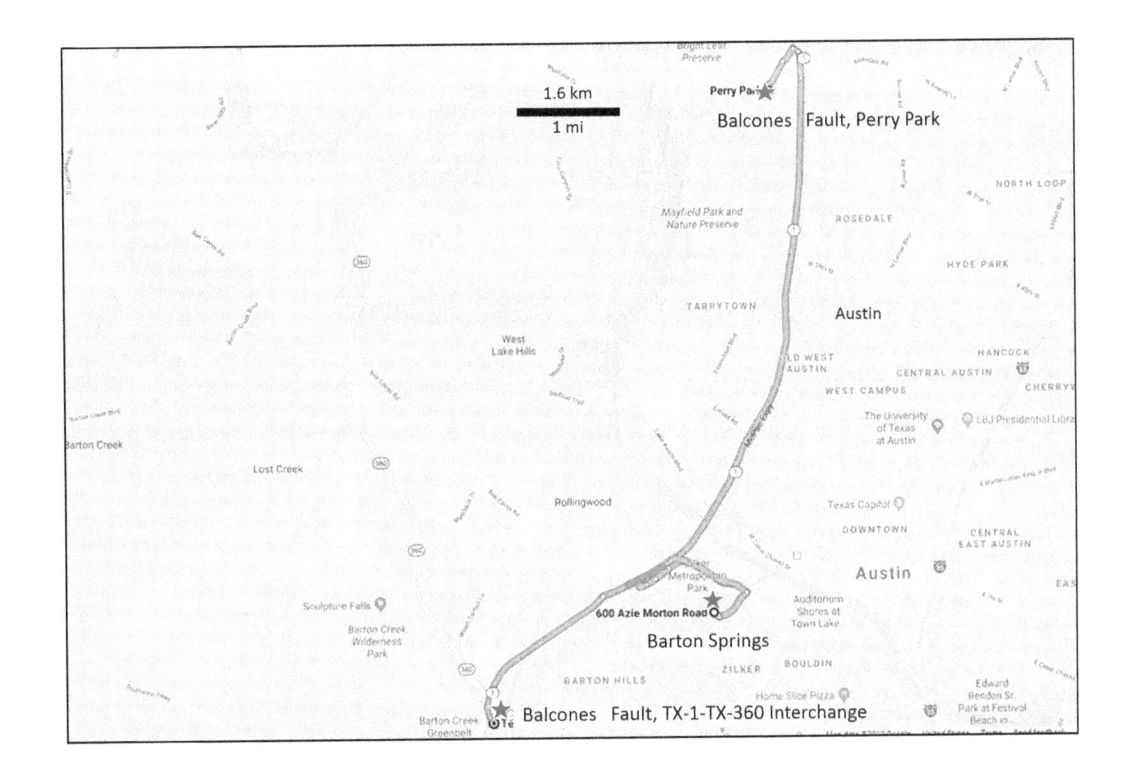

Balcones Fault to Barton Springs: *From Perry Park, take Fairview Drive south to Highland Terrace; turn left (southeast) on Highland Terrace and drive to 45th Street; turn left (south) on 45th Street and merge onto TX-1 Loop S/Loop 1 S; continue on TX-1 Loop S/Loop 1 S to W Cesar Chavez St. Take the Cesar Chavez St exit from TX-1 Loop S/Loop 1 S; drive east on Cesar Chavez St to Reynolds Drive; turn left (north) on Reynolds Drive to Lamar Blvd; turn right (south) on Lamar Blvd and drive to Barton Springs Road; turn right (west) on Barton Springs Road and drive to Azie Morton Road; turn left (south) on Azie Morton and drive to 600 Azie Morton Rd. Turn into the parking lot and walk north to Barton Springs pool. This is* ***Stop 1.2, Barton Springs*** *(30.262063, −97.770275) for a total of 10.5 km (6.5 mi; 16 min).*

STOP 1.2 BARTON SPRINGS AND THE EDWARDS AQUIFER

The Edwards Aquifer is one of the most prolific artesian aquifers in the world, supplying water to around 2 million people in central and south Texas. An aquifer is a porous and permeable layer of rock that holds groundwater. Artesian aquifers are rock layers confined beneath non-porous rock (such as shale) that contain groundwater under pressure such that springs and wells flow readily above the top of the layer. The aquifer consists of the Edwards and other limestones in a zone between 100 and 210 m (300 and 700 ft) thick. The Edwards Aquifer carries groundwater in fractures, conduits, and caves dissolved in 100 million-year-old (Cretaceous) limestone beneath the Edwards Plateau. The recharge zone is where rain and streams cross the exposed limestone. The discharge zone is where the water emerges at the surface, mainly along faults, as at Barton Springs.

The Georgetown Limestone is exposed in outcrops above the springs.

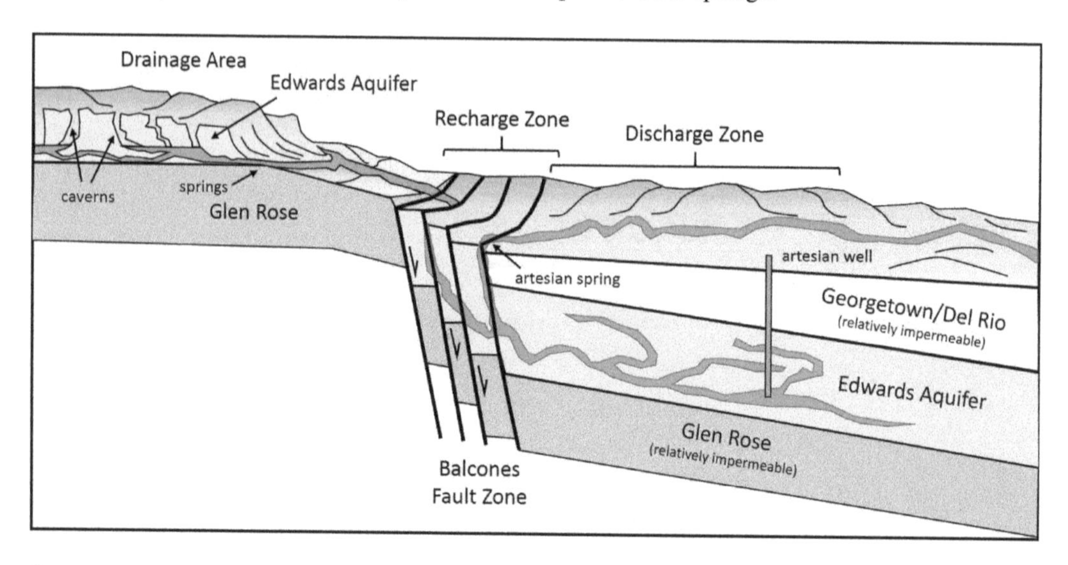

Cross section of the Edwards Aquifer near Austin. (Modified after Eckhardt, 1995–2018, www.edwardsaquifer.net/intro.html.)

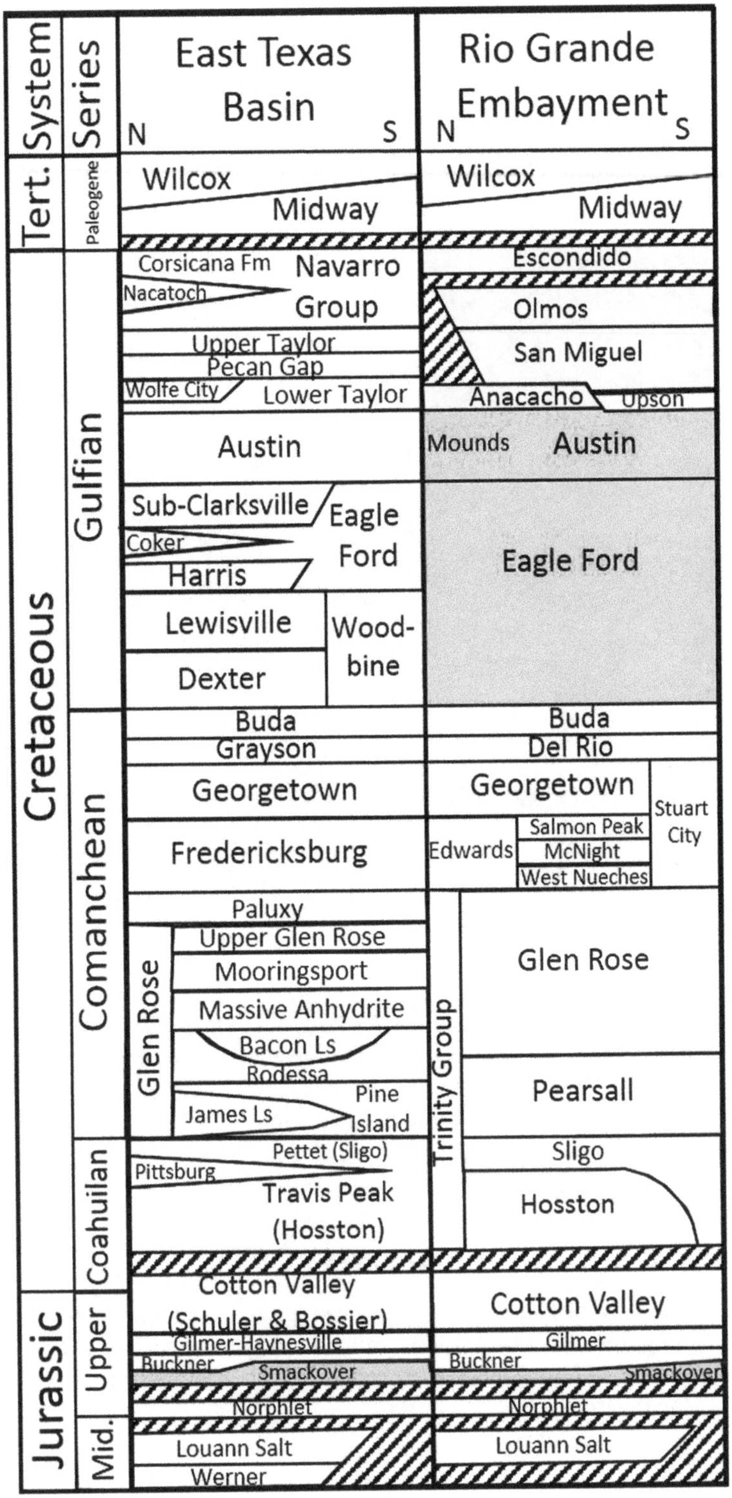

System
Series
East Texas Basin
N S
Rio Grande Embayment
N S
Tert.
Paleogene
Wilcox
Midway
Wilcox
Midway
Cretaceous
Gulfian
Corsicana Fm
Navarro Group
Nacatoch
Upper Taylor
Pecan Gap
Wolfe City
Lower Taylor
Escondido
Olmos
San Miguel
Anacacho
Upson
Austin
Mounds
Austin
Sub-Clarksville
Eagle Ford
Coker
Harris
Lewisville
Dexter
Wood-bine
Eagle Ford
Comanchean
Buda
Grayson
Georgetown
Fredericksburg
Paluxy
Glen Rose
Upper Glen Rose
Mooringsport
Massive Anhydrite
Bacon Ls
Rodessa
James Ls
Pine Island
Buda
Del Rio
Georgetown
Edwards
Salmon Peak
McNight
West Nueches
Stuart City
Trinity Group
Glen Rose
Pearsall
Coahuilan
Pettet (Sligo)
Pittsburg
Travis Peak (Hosston)
Sligo
Hosston
Jurassic
Upper
Cotton Valley (Schuler & Bossier)
Gilmer-Haynesville
Buckner
Smackover
Norphlet
Cotton Valley
Gilmer
Buckner
Smackover
Norphlet
Mid.
Louann Salt
Werner
Louann Salt

Austin area stratigraphic column. Yellow units indicate potential hydrocarbon reservoir units; green are potential source rocks. Diagonal pattern indicates unconformities (eroded rock). (Modified after Condon and Dyman, 2006.)

The four Barton Springs issue from fractures along the Balcones Fault zone in Zilker Park. They are the main outflow for the Austin section of the Edwards Aquifer.

The main spring, beneath the diving board in the pool, discharges around 120,000 m^3 (31 million gal) per day. In comparison, the average swimming pool contains around 190 m^3 (50,000 gal). The other three springs produce around 11,000 m^3 (3 million gal) per day, but sometimes dry up completely. The springs are home to the endangered Austin Blind salamander and Barton Springs salamander.

Barton Springs, Zilker Park, Austin. The springs are aligned along a splay of the Balcones Fault System.

Barton Springs to Balcones Fault and Folds: *Take Azie Morton Rd north to Barton Springs Rd; turn left (northwest) on Barton Springs and drive to Frontage Road; continue straight (west) on Frontage Road till can merge onto TX-1/MoPac Expressway southwest; take TX-1 offramp on left to TX-360;* ***Stop 1.3, Folding along the Balcones Fault*** *(30.247601, −97.805656), is on the left before you merge onto TX-360, for a total of 5.9 km (3.7 mi; 12 min). Pull well onto the shoulder and watch for traffic.*

STOP 1.3 FOLDING ALONG THE BALCONES FAULT

Deformation along the Balcones Fault System can be seen in this outcrop on the left (north) side of the interchange leading from TX-1 to TX-360 eastbound. A gentle anticline-syncline pair is developed in the Georgetown Limestone (Woodruff et al., 2017). Small bivalve fossils (clams, oysters, and such) occur in the limestone.

Low-amplitude anticline developed in the Georgetown Limestone. Roadcut at intersection of TX-1/ MoPac Expressway and TX-360. View north.

Low-amplitude syncline developed in the Georgetown Limestone. Roadcut near intersection of TX-1/ MoPac Expressway and TX-360. View northwest.

Side Trip 1—Balcones Folding to Canyon Lake Dam: *Continue southeast on TX-360; get on TX-71 E/US-290 E; continue southeast on TX-71 E to W Ben White Blvd; exit right to TX-275 Loop/S Congress Ave; turn left on Foremost Drive and get on I-35 S from I-35 Frontage Road; follow I-35 south to I-35 Frontage Road in Comal County; take Exit 195 and drive northwest on Watson Lane; turn left onto FM-1102 N; drive southwest to Hoffmann Lane and turn right onto Hoffman Lane; drive to FM-306 and turn right onto FM-306; drive northwest on FM-306 to FM-2673; turn left on FM-2673 and drive into Sattler; turn right at the traffic light onto South Access Road; drive over the bridge and immediately turn left at 16029 South Access Road; park at meeting place. This is* ***ST1, Canyon Lake Gorge*** *(29.863263, −98.187805), for a total of 87.8 km (54.6 mi; 1 h 9 min). Your (mandatory) guide will take you to the spillway to start the hike down the gorge.*

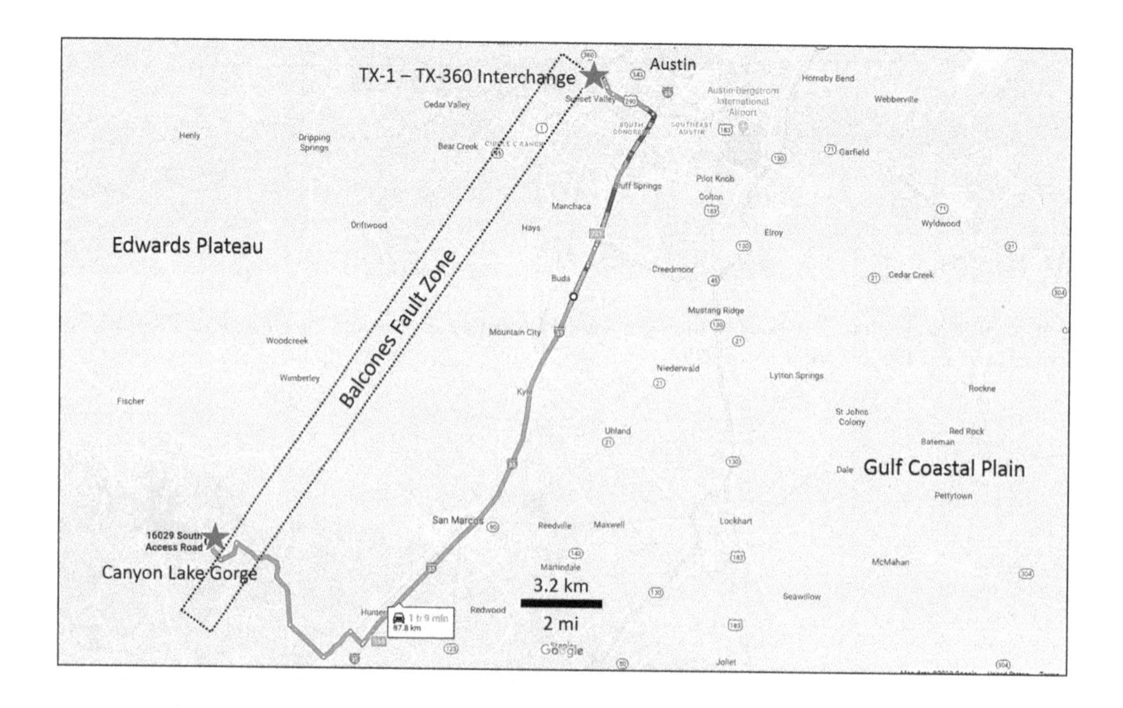

SIDE TRIP 1 CANYON LAKE DAM SPILLWAY AND GORGE

Heavy rains during July of 2002 caused the Canyon Lake Dam spillway to be overtopped, eroding the gorge below and exposing the Lower Cretaceous section. The high ground on both sides of Canyon Lake consists of the Lower Edwards Group limestone. The spillway is in Glen Rose Limestone. About 225 m (740 ft) of the Glen Rose Formation is exposed in the gorge. The limestone represents shallow marine, lagoon, reef, beach, and tidal flat environments (Woodruff et al., 2017). It is equivalent to the lower part of the Tamaulipas Superior in northeastern Mexico.

The spillway provides a view of bedding planes in the Glen Rose. You will note that there are two perpendicular fracture sets that cut the surface. Some of these are vertical joints (no offset), whereas others are small-scale faults. Some fractures show the effects of dissolution where they are widened or form small cavities. The scoured strata reveal that some layers are more resistant, while others are more easily eroded. This is the origin of the stepped topography of the Texas Hill Country.

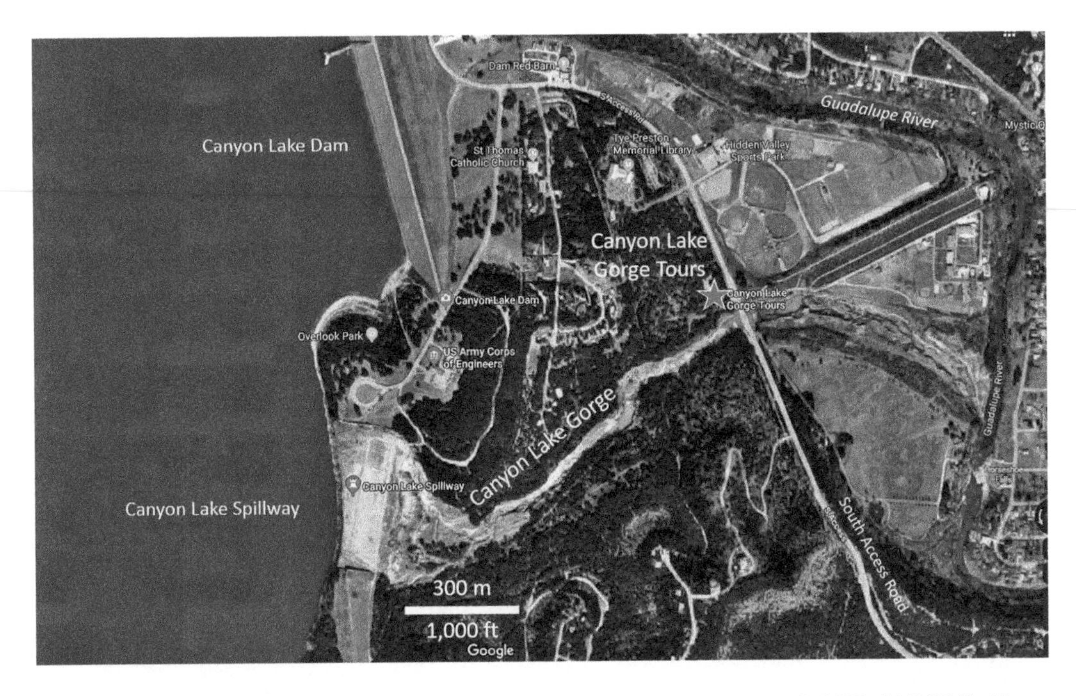

Google Earth satellite view of Canyon Lake spillway and gorge. Imagery © 2019 CAPCOG; Maxar Technologies, USDA Farm Service Agency.

The limestone contains abundant shells of single-celled marine planktonic organisms known as Orbitolina, a variety of foraminifera. The disc-shaped shells are about 1 cm (half inch) long. Weathering and solution in the limestone emphasize burrows made on bedding surfaces when this was ocean bottom lime mud.

Farther down the gorge are tracks of three-toed dinosaurs thought to be Early Cretaceous *Acrocanthosaurus*. These 10 m (30 ft) high predators roamed the tidal flats and preyed on plant-eating sauropods, whose tracks are also seen in the gorge.

Three-toed dinosaur prints thought to be from an Early Cretaceous *Acrocanthosaurus*.

Slickensides show normal offset on the Hidden Valley Fault. Offset here is about 2.4 m (8 ft).

Beyond the tracks is a near-vertical fault scarp with well-defined striations (scratches and grooves) on its surface. Known as slickensides, these grooves indicate the sense of movement along the fault. The 2.4 m (8 ft) high scarp extends about 60 m (200 ft) and trends northeast. This small-offset (about 2.4 m down to the southeast) fault is one of a series of breaks that comprise the Hidden Valley Fault. The Hidden Valley Fault is part of the Balcones Fault system in this area. Movement on the faults is thought to be around 15 million years old.

Continuing down the gorge one comes upon a resistant limestone layer with abundant rudists. These now-extinct bottom-dwelling bivalves (clams) formed reefs in warm, shallow marine waters.

Below the rudist layer, a spring discharges groundwater derived from Canyon Lake. This shows how groundwater can move through fractures as well as layers with high porosity and permeability.

At the spillway, the gorge flows southeast. The gorge abruptly turns northeast where erosion has cut along the main Hidden Valley Fault. Layers that were flat or inclined gently to the southeast are bent to near-vertical along the fault. The fault zone is about 40 m (125 ft) wide through here, with total offset around 60 m (200 ft).

Continuing down the gorge, one can find large ripple marks, evidence of tidal currents or waves flowing over a shallow, mud-bottomed sea. This resistant layer caps a 5 m (15 ft) waterfall. A "relay ramp" is exposed below the waterfall: this is a zone of tilted bedrock that separates strands of the Hidden Valley Fault and allows the displacement on one section of fault to transfer to the adjacent section.

Glen Rose Formation exposed in the gorge. The Hidden Valley Fault cuts the far outcrop at the dotted line.

Below the Bill Ward monument (to a prominent local geologist) is a half meter (1–2 ft) thick resistant layer containing the small fossil clam Corbula. About the size of grains of rice, the abundance of this clam and lack of fossil diversity probably indicates a highly saline environment. Below the Corbula Bed is a softer, easily eroded zone containing over 80 different marine fossils: these layers are characterized by a nearly spherical sea urchin, *Salenia texana*, about 2.5 cm (1 in) in diameter.

Fossils weathering out of the Glen Rose Formation near the bottom of the gorge.

The lower section of the gorge contains abundant northeast-trending joints that control erosion and form multiple, interconnected channels. This section of the gorge is known as "The Channels."

Thanks to Tim Strutz for the guided tour and providing much of the above information.

Visit

Hiking into the gorge can only be done as part of a guided tour provided by the Gorge Preservation Society. Contact canyongorge.org or call 830-964-5424.

Please check the website for current tour fees.

Website: canyongorge.org
Phone: 830-964-5424

Side Trip 1—Canyon Lake to Pedernales Falls State Park: *Continue north on South Access Road to FM-306; turn left (west) on FM-306 and drive to FM-3424; turn right (north) on FM-3424 and drive to FM-32; turn left (west) on FM-32 and drive to FM-406/Cox Road; turn right (north) on Cox Road and drive to FM-165; turn right (east) on FM-165 and drive to Highway 290; turn left (west) on Hwy 290 and drive to Ranch Road 3232 N; turn right (north) onto RR-3232 and drive to Pedernales Falls Rd; turn right (east) and immediately turn left (north) onto Park Road-6026; drive to the Falls parking and trailhead. This is* ***Stop 2, Pedernales Falls*** *(30.334394, −98.252620), for a total of 75.8 km (47.1 mi; 1 h).*

Balcones Folding to Pedernales Falls State Park: *Continue southeast on TX-360; turn right (west) onto South Lamar/US-290W Service Road; use left lane to merge onto US-290W; continue west on US-290 to Ranch Road 3232 N; turn right (north) onto 3232 and drive to Pedernales Falls Rd; turn right (east) and immediately turn left (north) onto Park Road-6026; drive to the Falls parking and trailhead. This is* ***Stop 2, Pedernales Falls*** *(30.334394, −98.252620), for a total of 64.7 km (40.2 mi; 1 h).*

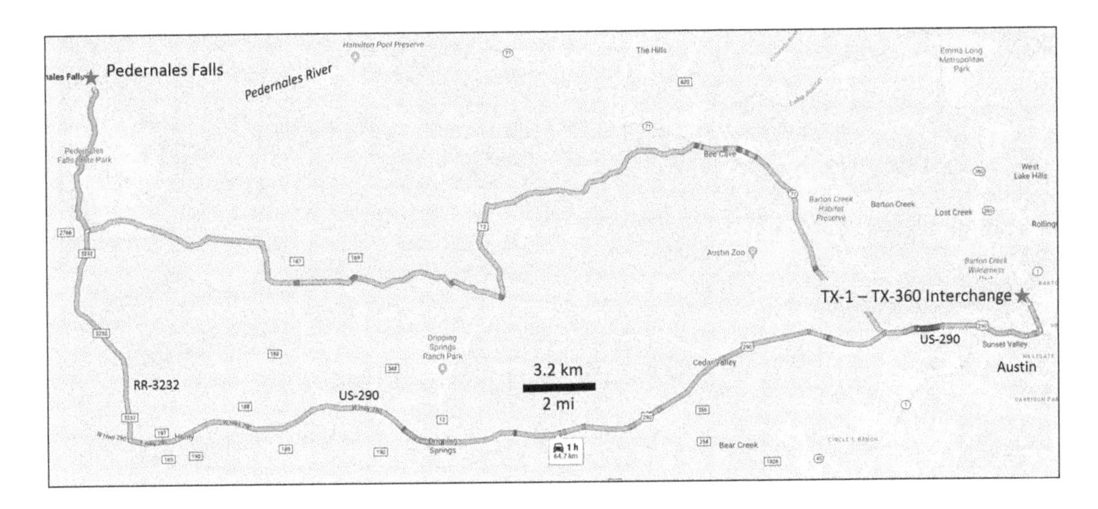

STOP 2 PEDERNALES STATE PARK

Pedernales Falls are developed in Pennsylvanian (320 Ma) Marble Falls cherty (flinty) limestone. This limestone is full of crinoid fossils, a flower-shaped animal that attached itself to the Pennsylvanian sea floor.

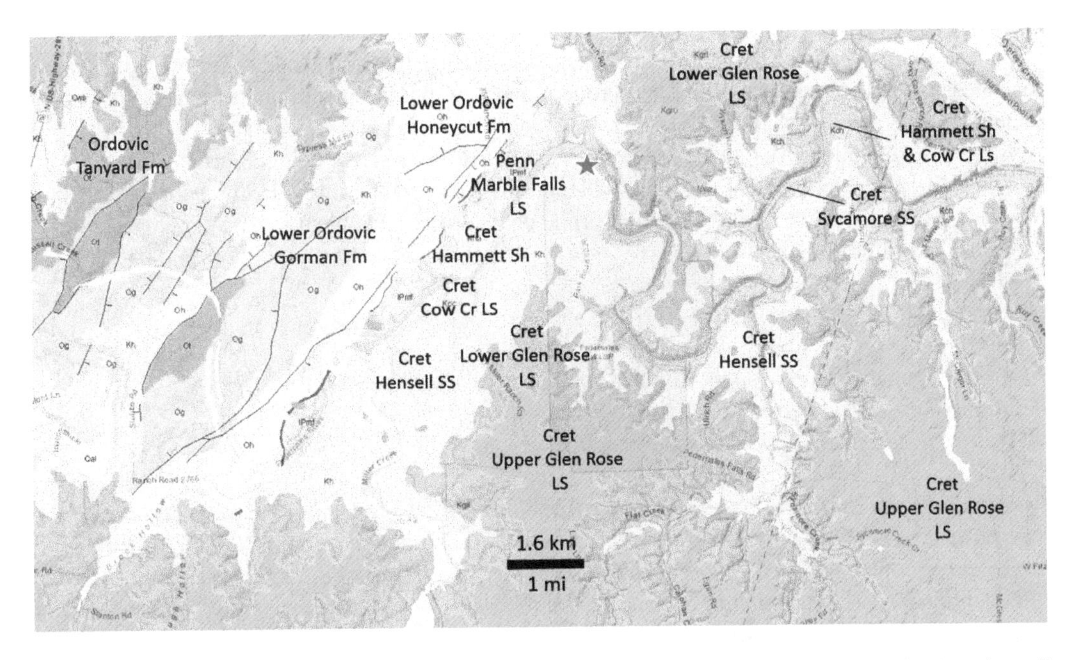

Geologic map of the Pedernales Falls area (U.S. Geological Survey and Bureau of Economic Geology, https://txpub.usgs.gov/txgeology/).

Marble Falls, Pedernales Falls State Park. The flat layers at the rim are Cretaceous Sycamore Conglomerate and Cow Creek Limestone; the inclined layers below the unconformity (dashed line) are Pennsylvanian Marble Falls Limestone.

The main geologic feature in the park is the angular unconformity between the Marble Falls Limestone, which dips (is inclined) about 10° southeast, and the nearly flat overlying Cretaceous Sycamore Conglomerate, Cow Creek Limestone, and Glen Rose Formation (105–115 Ma). The time gap at the unconformity is over 200 million years. During this time interval, more layers were deposited, the entire area was tilted and eroded, and new flat-lying sediments were deposited by ancient streams and in a shallow sea.

A lively spring gushes out of the sand on the south side of the river below Marble Falls.

Texas purchased land for the park in 1970, and the park opened in 1971. The Pedernales River was named by the Spanish for the flint cobbles and pebbles (pedernal means flint in Spanish) eroding out of the local limestone.

Visit

Hours: Open daily, year-round. Campsites and day hikes are available.
Admission: see the website for current fees.
Website: https://tpwd.texas.gov/state-parks/pedernales-falls

Pedernales Falls State Park to Enchanted Rock State Park: *Return south to Pedernales Falls Rd and turn right (W); continue onto Robinson Road/TX-2766 to US-281 N/US-290 W; turn right (north) and follow US-281 north to Ranch Road 1323; turn left (west) onto 1323 and drive to Eckert; in Eckert, turn right (north) onto TX-16 and drive to Ranch Road 965; turn left (west) onto 965 and drive to* ***Stop 3, Enchanted Rock State Park*** *parking and trailhead (30.496058, −98.820482) on the right for a total of 98.4 km (61.2 mi; 1 h 12 min).*

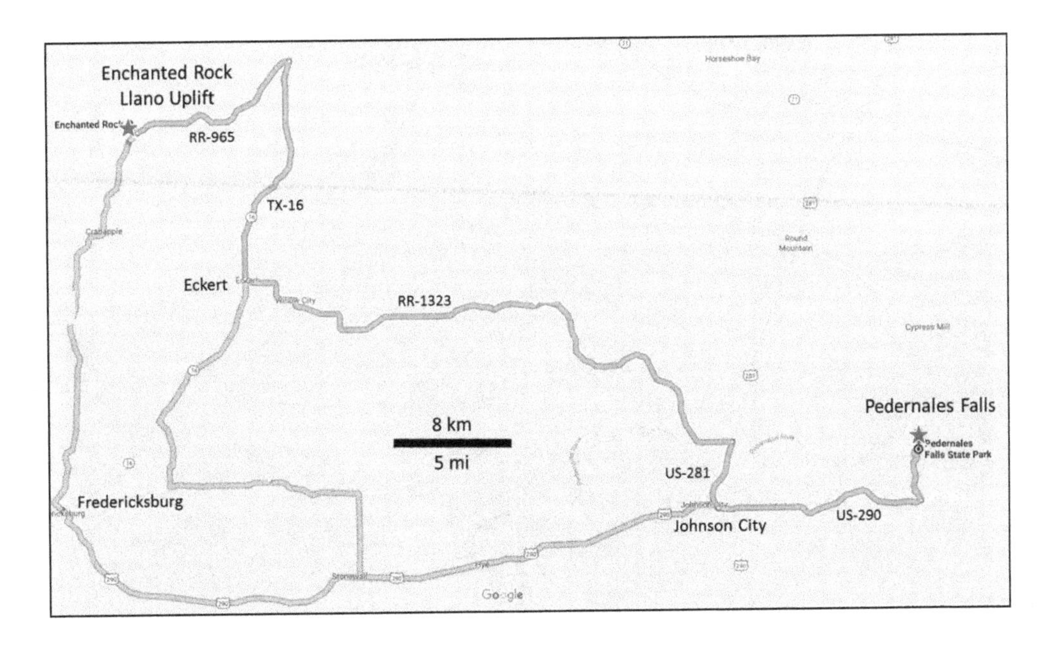

STOP 3 LLANO UPLIFT AND ENCHANTED ROCK STATE PARK

The Llano Uplift is not a structural uplift at all; it is a large Precambrian granite-cored dome, an erosional high point that already existed and influenced the location and development of the Pennsylvanian-Permian (300–270 Ma) Ouachita-Marathon Orogeny mountain belt. Enchanted Rock is now surrounded and onlapped by younger, mainly horizontal Cretaceous sedimentary layers of the Edwards Plateau. Enchanted Rock is one of several domes in the 105 km (65 mi) wide granite exposure. This distinctive pink granite is among the oldest rocks found in Texas. The 260 ha (640 ac) dome rises almost 130 m (425 ft), to 560 m (1,825 ft) above sea level. The granite is a classic location to see exfoliation, or spalling of large granite slabs. Exfoliation, a type of weathering caused by release of pressure, results from erosion and removal of overlying rocks.

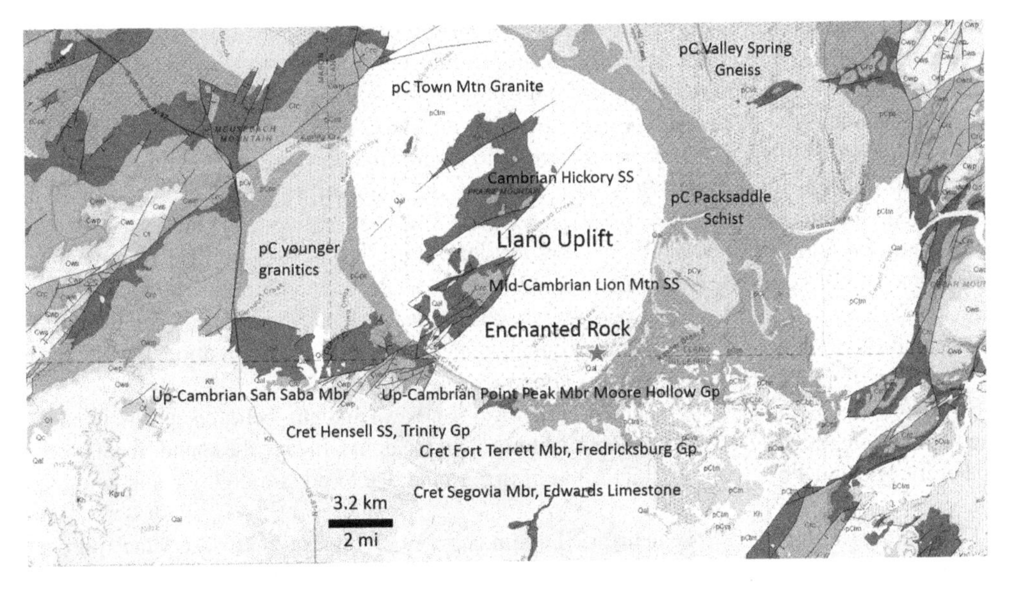

Geologic map of the Llano Uplift. Star indicates the Enchanted Rock stop (U.S. Geological Survey and Bureau of Economic Geology, https://txpub.usgs.gov/txgeology/).

Enchanted Rock, Llano Uplift. This 130 m high dome consists of 1.1 billion-year-old granite.

Precambrian rocks in the area comprise metasedimentary, metavolcanic, and meta-intrusive rocks that range in age from 1.37 to 1.23 billion years (Ga). These metamorphic rocks were later intruded and further metamorphosed by 1.13–1.07 Ga granite during the Grenville Orogeny. Erosion over the next 400 million years exposed the granite at the core of the uplift. Beginning in Cambrian time, sediments were deposited over and around the granite, continuing until the start of the Ouachita-Marathon mountain-building episode in Pennsylvanian time. The Ouachita-Marathon Orogenic Belt formed when Laurasia collided with Gondwana during Pennsylvanian-Permian time. The folded and thrusted mountains formed by that event are at the surface in Oklahoma-Arkansas and in the Marathon Uplift of west Texas. Beneath the cover rocks of central Texas, they bend around the Llano Uplift, which acted as a buttress to the north and west-directed thrusting.

Permian, Triassic, and Jurassic units were either not deposited or were eroded away in this area.

Within the uplift are remnants of Paleozoic strata that were down-dropped along northeast-trending normal faults, preserving them from erosion. These layers range in age from Cambrian sandstone to Lower Pennsylvanian limestone and shale. The flat-topped hills surrounding the uplift consist of flat-lying Cretaceous limestone, mainly the Trinity Group (Pearsall and Glen Rose formations) and Fredericksburg Group (Walnut, Comanche Peak, and Edwards formations). Since the end of Cretaceous time, the area has been subject mainly to erosion.

Enchanted Rock is said to be haunted. The local Tonkawa Indians believed ghosts lived there because of the creaking and groaning sounds coming from the dome. In fact, the sounds from the dome are a result of exfoliation (spalling, fracturing) of the granite due to pressure release and temperature changes.

The area was designated a Texas Historic Landmark in 1936. The Nature Conservancy purchased the property in 1978 and later sold it to the state of Texas. The 665 ha (1,643 ac) Enchanted Rock Park opened as a state natural area in October 1978. It is a National Natural Landmark and is on the National Register of Historic Places due to the more than 400 archeological sites found in the park.

Visit

More than 250,000 people visit the park every year. Camping is allowed, and lots of day hikes are available.

Hours: Open all day, 7 days a week, year-round.
Admission: See the website for current fees.
Website: https://tpwd.texas.gov/state-parks/enchanted-rock

Enchanted Rock State Park to Devils Sinkhole State Natural Area: *Continue southwest on Ranch Road 965 to Fredericksburg; turn right (west) on W Main St; take a slight left onto US-290 W and drive to RM-479 S; turn left (south) onto 479 and drive to I-10 westbound; drive to Exit 490 and take TX-41 southwest (left); continue on TX-41 to US-377 and turn left (southwest); take US-377 to Rock Springs; turn right on Main Street and drive 3 blocks to Devils Sinkhole Headquarters on the right at 101 N. Sweeten St., Rocksprings. This is* ***Stop 4, Devils Sinkhole*** *(30.015809, −100.208492), for a total of 183 km (114 mi; 1 h 53 min).*

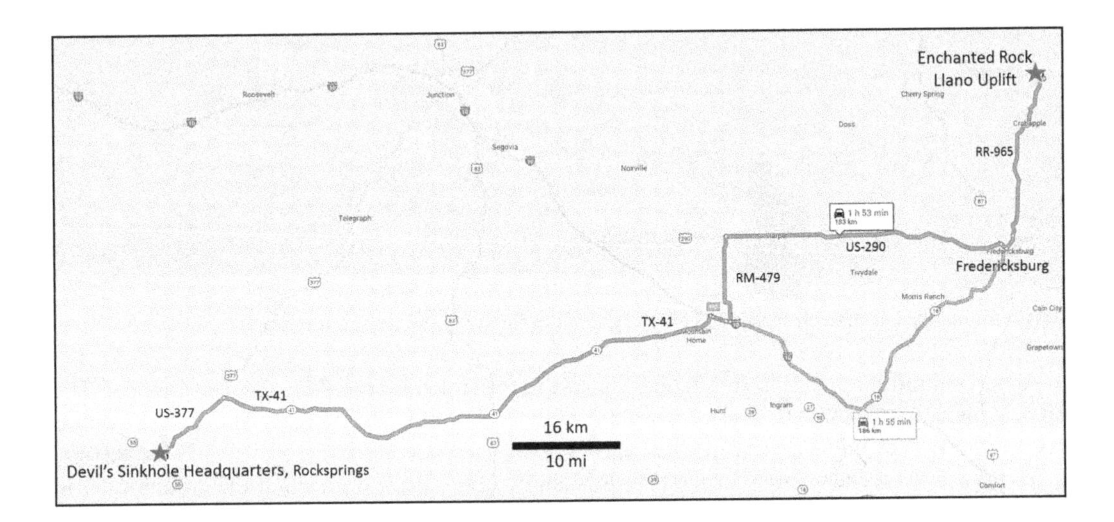

STOP 4 DEVIL'S SINKHOLE STATE NATURAL AREA

Discovered in 1876, Texas acquired the cave in 1985. This 750 ha (1,860 ac) National Natural Landmark is mainly undeveloped. The opening is a shaft approximately 12–18 m (40–60 ft) wide that drops 43 m (140 ft) into the cavern. Here, the roughly circular cavern expands to over 100 m (320 ft) and reaches a total depth of 107–122 m (350–400 ft). It contains the largest single-chamber cavern and the largest vertical drop, and it is the third deepest cave in Texas. The cavern can be seen from a viewing platform that extends over the chasm. Access to the cave system is limited to scientific researchers.

Devil's Sinkhole, a 43 m deep shaft and cavern system developed in the Buda Limestone.

About 3–4 million Mexican free-tailed bats live in the sinkhole from May through October. The bats eat about 18 tonnes (20 tons) of insects, mostly corn moths, each night. About 3,000–4,000 cave swallows occupy the cave while the bats are gone.

You are in the recharge area for the Edwards aquifer, the source of much of the groundwater in central Texas. The area gets about 56 cm (22 in) of rain per year. Over the past million years, slightly acidic rainwater slowly dissolved the Cretaceous Edwards Limestone, creating a system of caverns. As the climate became more arid the water table dropped and the water that once held up the ceiling of the cave in this area drained away, allowing the overlying rock to cave in. The roof collapsed some time during the past 10,000 years.

Stone artifacts from sites in Devil's Sinkhole State Natural Area indicate that Native Americans used the area. The chipped stone arrowheads, spear points, and scrapers from these sites were made from chert found in the Edwards Limestone.

In the early 20th century, local entrepreneurs mined the sinkhole for bat guano that was used as fertilizer. In 1968, the National Park Service designated the cave as a National Natural Landmark. The owner of the property gave it to the state of Texas in 1985, and it opened to the public as a state park in 1992.

Thanks to Mike Cox for the guided tour and providing much of the above information.

Visit

Devil's Sinkhole is open to visitors by appointment only. Access can be arranged by contacting the Devil's Sinkhole Society. Tours launch from the Rocksprings Visitor Center.

Entrance Fees: see the website for current fees.
Reservations must be made in advance.
Hours: Open Wednesday through Sunday, year-round. Closed Christmas to New Year's Day.
Contact Information: (830) 683-BATS (tel.: +1 830 683 2287)
Tours Begin at: 101 N. Sweeten St., Rocksprings, TX 78880
Website: Devil's Sinkhole Society, https://tpwd.texas.gov/state-parks/devils-sinkhole
Email: devilssinkhole@swtexas.net

Devil's Sinkhole to Amistad Reservoir: *Return to US-377 and turn left (southwest); drive to US-277/ US-377 S and turn left (south); drive to US-90 W and turn right (west); drive across the bridge over the reservoir and park on the right. This is* ***Stop 5, Amistad Natural Recreation Area—Salmon Peak Limestone*** *(29.491876, −101.037491), for a total of 169 km (105 mi; 1 h 43 min).*

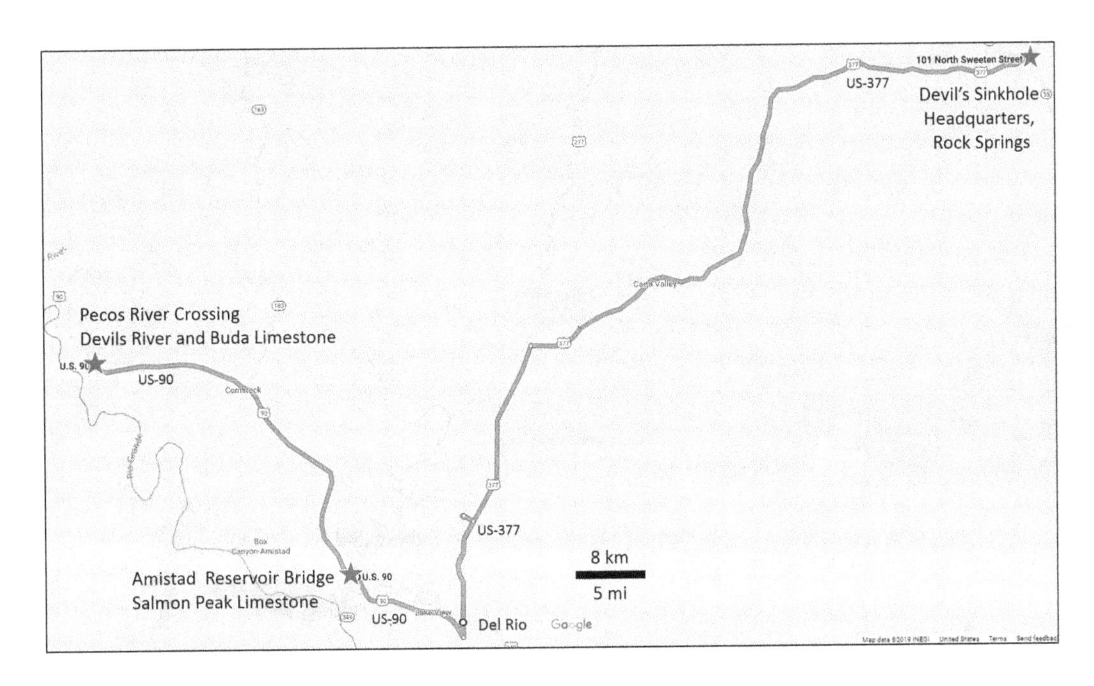

STOP 5 AMISTAD NATIONAL RECREATION AREA—SALMON PEAK LIMESTONE

As you drive west from Devil's Sinkhole, pay attention to the changing scenery. The low scrubland gradually dries out, the plants become sparse and grow thorns, and soon you are in classic west Texas desert country. There is a harsh beauty in the desert. As the filmmaker Ang Lee said, "I think the American West really attracts me because it's romantic. The desert, the empty space, the drama."

Between Devil's Sinkhole and Amistad Reservoir, you leave the Edwards Plateau and enter the Val Verde Basin. The Val Verde Basin is a foreland basin that developed north of the Ouachita-Marathon Fold-Thrust Belt and south of the Midland Basin during Mississippian to Permian time. It extends 240 km (150 mi) in a northwest direction and is 20–40 km (12–25 mi) in a northeast-southwest direction. Most of the basin fill consists of Cretaceous sediments deposited in the Western Interior Seaway. Deformation of the basin consists of minor folding. The basin is a major gas-producing area, containing an estimated 142 billion m^3 (5 trillion ft^3) of natural gas (Wikipedia).

Amistad National Recreation Area's (NRA) natural features are primarily the result of two influences: geology and climate.

The entire Amistad NRA area is underlain by 100 million-year-old Cretaceous limestone. A number of different formations make up the strata, and most include a wide variety of marine fossils. Karst landscape (dissolution features such as sinkholes and caves formed in limestone) is common in the area.

The near-surface rocks of the area consist of five Lower Cretaceous, predominantly limestone formations. From oldest to youngest they are the West Nueces Formation, the McKnight Formation, the Salmon Peak Formation, the Del Rio Clay, and the Buda Limestone.

Visitors heading west from Amistad NRA observe the transition from Gulf-influenced climate and vegetation to true desert by the time they get to the Big Bend region's Chihuahuan Desert.

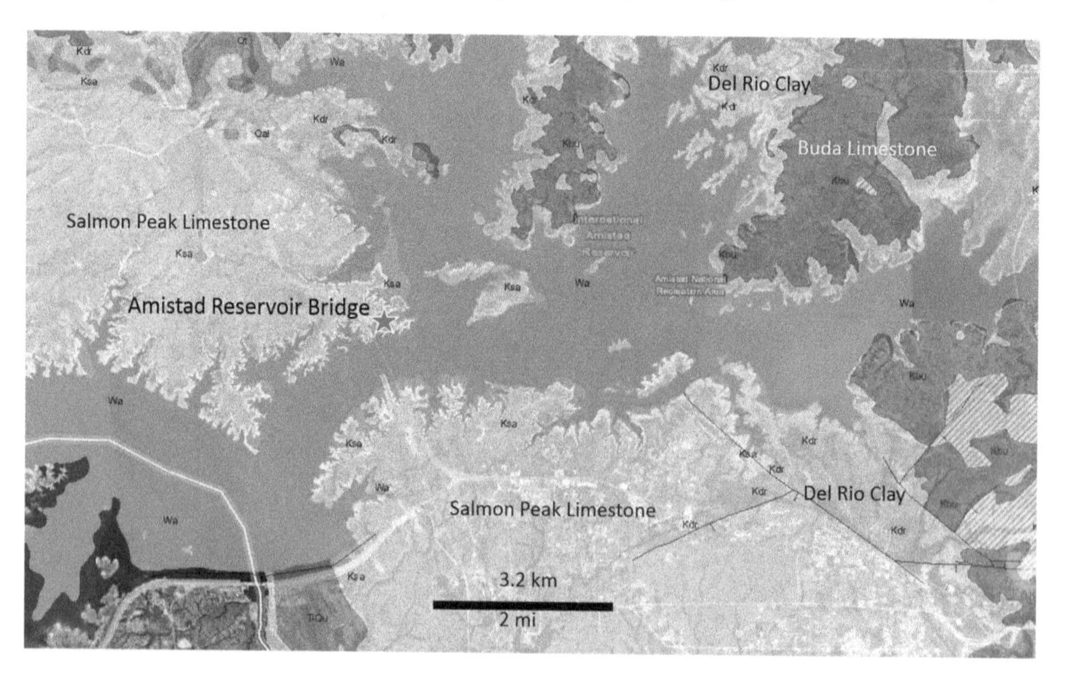

Geologic map of Amistad Reservoir bridge area (U.S. Geological Survey and Bureau of Economic Geology, https://txpub.usgs.gov/txgeology/).

Salmon Peak Limestone, north side of Amistad Reservoir Bridge looking east.

Fossil bivalves (oysters, clams, scallops) in the Salmon Peak Limestone, north-side Amistad Reservoir bridge.

Salmon Peak Formation limestone, the rock under most of Amistad Reservoir, typically has low original porosity and permeability. Secondary porosity and permeability are a result of solution enlargement of bedding planes, fractures, and faults. The amount of dissolution depends on the density of faults and joints that enhance groundwater flow.

During design of the Amistad Reservoir in the early 1960s, dissolution features such as depressions, sinkholes, and caverns in the Rio Grande valley walls, river terraces and adjacent flood plains were documented. As the reservoir began to fill in 1968, an increase in discharge of local springs was observed. Flow from Carmina Springs, near the western dam embankment, has been measured and indicates a direct correlation between the lake elevation and discharge at the springs. Thus, despite efforts to minimize seepage, data indicate a hydraulic connection between the reservoir and the shallow subsurface immediately downstream from the dam. The same connection between lake and springs has been observed at Canyon Lake Gorge.

Amistad Reservoir Bridge to Pecos River Crossing: *Continue west on US-90 for 46.6km (29.0 mi; 28min) and pull over on the right shoulder, west side of bridge. Walk to the bridge for a view of the gorge. This is* ***Stop 6, Pecos River Crossing*** *(29.709958, −101.352933).*

STOP 6 PECOS RIVER CROSSING—DEVILS RIVER AND BUDA LIMESTONE

Between the Amistad Reservoir bridge and Pecos River, you drive through several excellent roadcuts in the Salmon Peak and Buda/Del Rio formations.

Uppermost Albian (Early Cretaceous) carbonates of the Devils River-Salmon Peak formations are exposed in the Pecos River canyon. The Devils River Formation is laterally equivalent to the Georgetown Formation along the Gulf Coast and the lower part of the Cuesta del Cura Formation in northeastern Mexico. The Salmon Peak Formation is equivalent to the upper Edwards Group/lower Georgetown Formation on the Gulf Coast.

The canyon contains some of the best examples of rudist and skeletal reefs and carbonate sand complexes in the world. These complexes comprise a major petroleum system throughout the Lower Cretaceous of Texas. Along the river you can examine rudist reef systems, wave and tide-dominated carbonate grainstones (limestone lacking any mud), and complex fracture systems, solution caverns, and cavern collapse systems (Kerans, 2015).

The Cretaceous Buda Limestone is a resistant, massive fine-grained bioclastic limestone (limestone consisting almost entirely of fossil fragments) 14–30 m (45–100 ft) thick. It weathers dark gray to brown and contains abundant pelecypod fossils. The Buda is laterally equivalent to the upper Cuesta del Cura Formation in northeastern Mexico.

The Devils River Limestone is a hard limestone, about 210 m (700 ft) thick. It contains abundant rudists and shell fragments and is locally dolomitized, cherty, and brecciated.

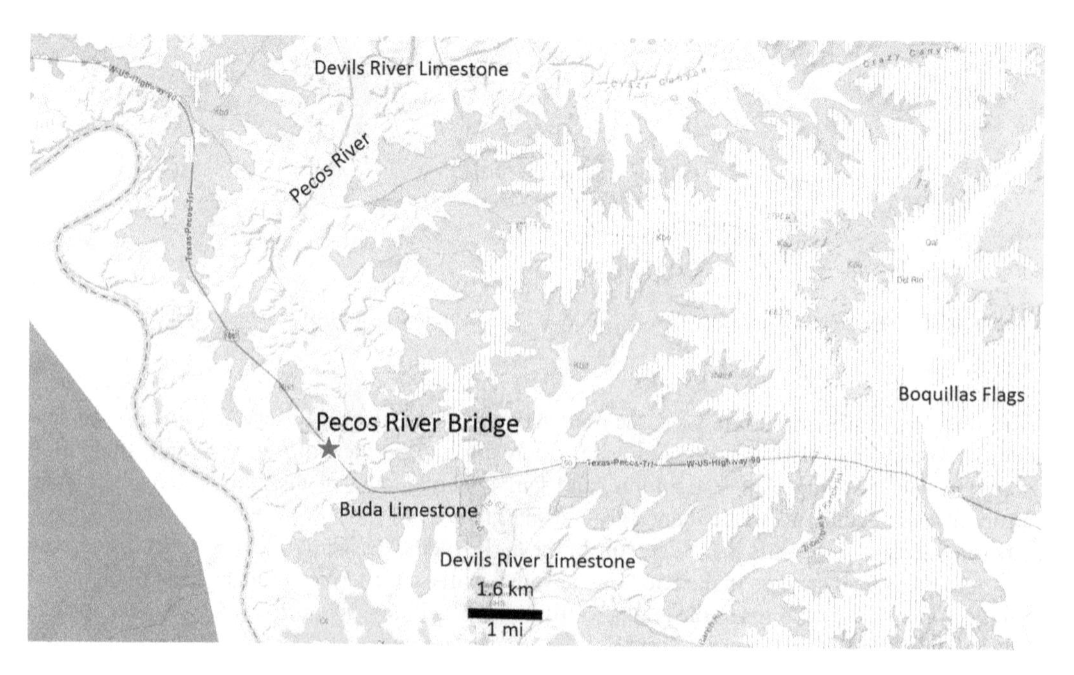

Geologic map of the Pecos River bridge area (U.S. Geological Survey and Bureau of Economic Geology, https://txpub.usgs.gov/txgeology/).

Devils River Limestone, Pecos River bridge, US-90. View north.

Pecos River Crossing to Del Rio—Devils River Contact: *Continue west on US-90 for 14.8 km (9.2 mi; 8 min). This is* ***Stop 7, Del Rio—Devils River Contact*** *(29.792826, −101.447193).*

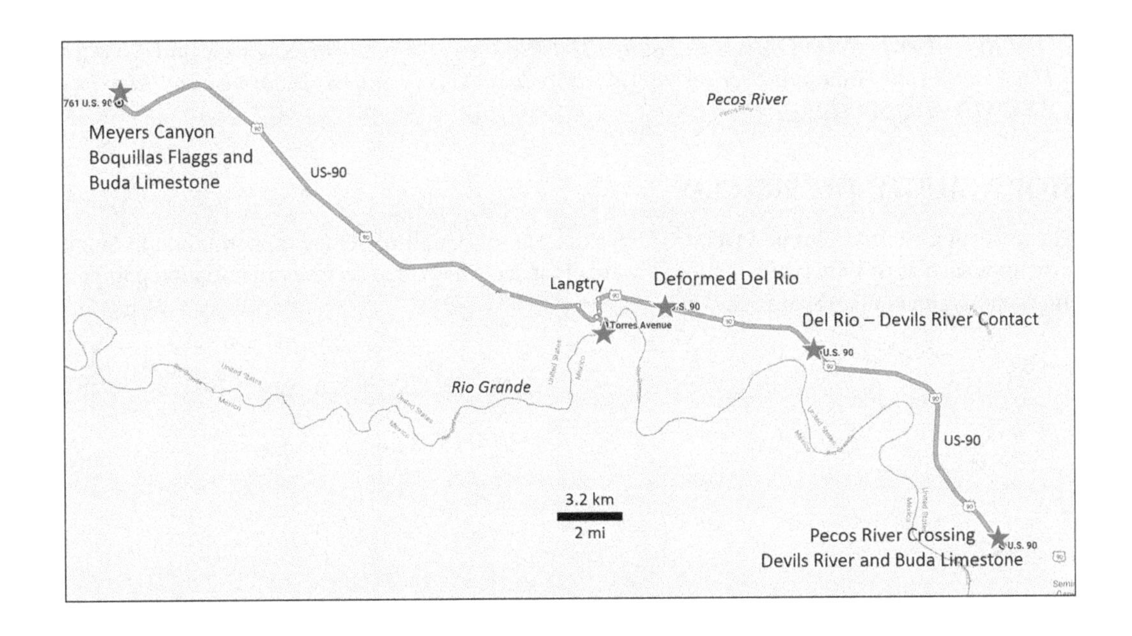

STOP 7 DEL RIO CLAY—DEVILS RIVER LIMESTONE CONTACT

The roadcut at this stop contains a striking example of the contact between the Del Rio Clay and underlying Devils River Limestone. The Del Rio Clay is laterally equivalent to the Grayson Formation along the Gulf Coast, and to the middle Cuesta del Cura Formation in northeastern Mexico. The contact indicates an abrupt change in depositional environment from clear water to muddy sedimentation.

US-90 roadcut containing the Del Rio-Devils River contact.

Del Rio—Devils River Contact to Faulted Del Rio Clay: *Continue driving west on US-90 for 8.1 km (5 mi; 4 min) and pull over on the right shoulder. This is Stop 8, Deformed Del Rio Clay (29.812783, −101.525472).*

STOP 8 FAULTED DEL RIO CLAY

This roadcut contains deformed Del Rio Clay adjacent to a small-offset (~2 m) normal fault. Small structures such as this are typical of the Edwards Plateau. They often serve as an initiation point for limestone dissolution and caverns.

Folded and faulted Del Rio Clay near Langtry, US-90 roadcut. Google Street View looking south.

Faulted Del Rio Clay to Rio Grande Valley at Langtry: *Continue driving west on US-90 to TX-25 Loop; turn left (south) on TX-25 and drive to Torres Ave; turn left (southeast) onto Torres Ave and drive to overlook for a total of 5 km (3.1 mi; 7 min). This is* ***Stop 9, Rio Grande Valley at Langtry*** *(29.805049, −101.555960).*

STOP 9 RIO GRANDE VALLEY AT LANGTRY

The wide-open, gentle Rio Grande Valley at Langtry is bracketed by cliffs of Devils River Limestone.

Eccentric "Judge" Roy Bean, the "law west of the Pecos," was actually Justice of the Peace in Langtry from 1882 till his death in 1903. The story goes that he had a crush on the actress Lilly Langtry and named the town after her.

Rio Grande valley at Langtry looking southwest. The cliffs are Devils River Limestone. The far side of the valley is Mexico.

Rio Grande Valley at Langtry to Meyers Canyon: *Return to US-90 and drive west for 28.8 km (17.9 mi; 21 min) to the east side of Meyers Canyon and pull over on the right. This is* ***Stop 10, Meyers Canyon—Boquillas Flags*** *(29.901168, −101.801441).*

STOP 10 MEYERS CANYON—BOQUILLAS FLAGS AND BUDA LIMESTONE

The Buda Limestone crops out in the lowermost part of Meyers Canyon. The overlying Boquillas Flags is divided into four units: an upper silty, medium-gray shale with some brownish-gray limestone (seen in the roadcut); below this is a silty, medium-gray shale interbedded with yellow-gray limestone; then silty dark-gray shale interbedded with siltstone; and a basal bioclastic light yellow-gray to gray-orange limestone interbedded with siltstone. Total thickness is on the order of 50–67 m (160–220 ft). The Boquillas Flags are laterally equivalent to the Eagle Ford along the Gulf Coast, and the Agua Nueva Formation in northeastern Mexico.

The Austin Chalk forms the high ground east and west of the canyon. It is a hard, ledge-forming limy mudstone to soft chalk, gray-white to white, and locally abundant fossils. Total thickness is around 177 m (580 ft). The Austin Chalk is equivalent to the San Felipe Formation in northeastern Mexico.

Boquillas Flags in roadcut above Meyers Canyon. View northeast.

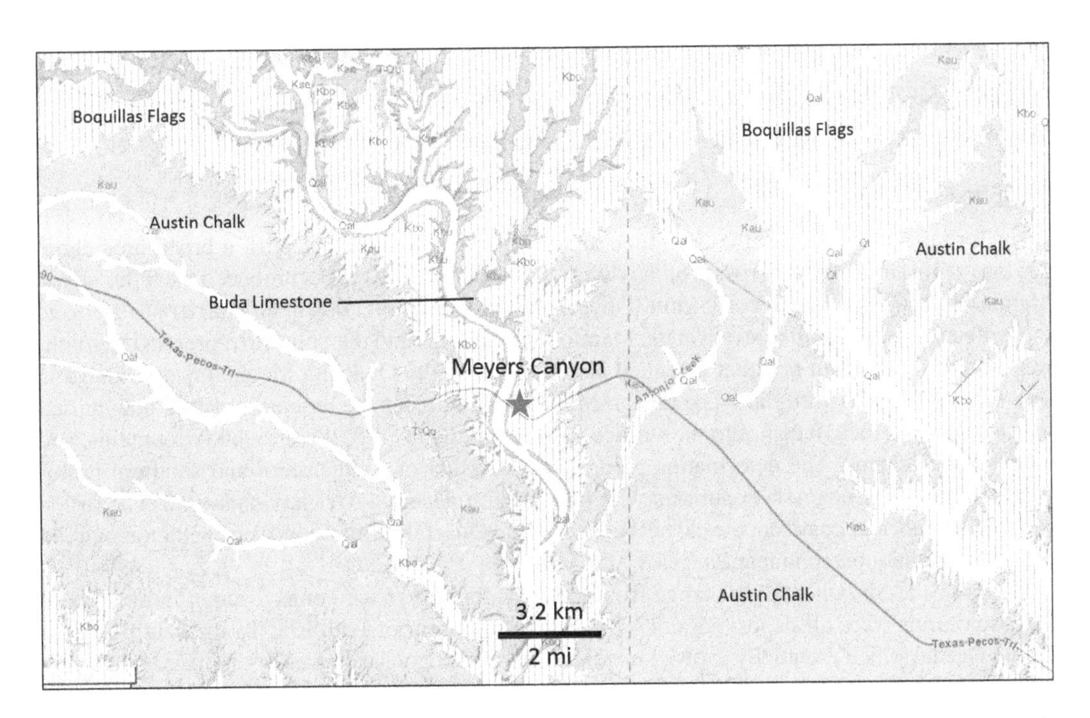

Geologic map of the Meyers Canyon area. (Courtesy of U.S. Geological Survey and Bureau of Economic Geology, https://txpub.usgs.gov/txgeology/.)

Between this stop and the next, we leave the Edwards Plateau and enter the Marathon Uplift.

Meyers Canyon to Lemons Gap East: *Continue west on US-90 for 126 km (78.6 mi; 1 h 11 min) and pull over on right. This is* ***Stop 11.1, Lemons Gap East*** *(30.202084, −102.945746).*

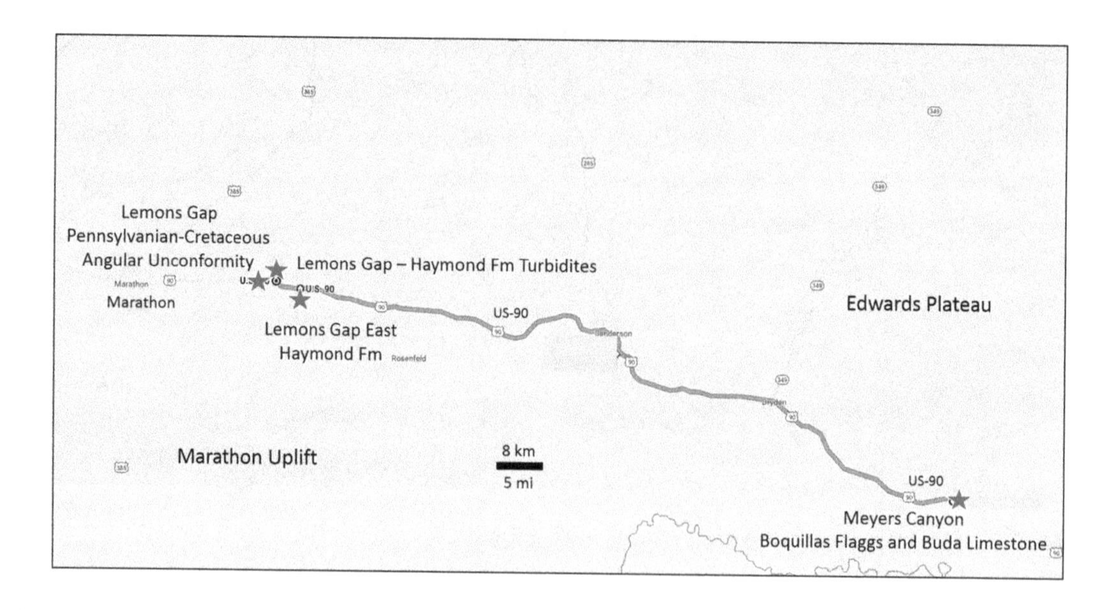

11 LEMONS GAP AREA—MARATHON UPLIFT AND THE PALEOZOIC-CRETACEOUS UNCONFORMITY

Marathon Uplift

Early Tertiary upwarping in the Marathon area created the Marathon Uplift, a broad area about 125 km (78 mi) in diameter. Erosion of overlying, mostly Cretaceous rocks produced the topographic Marathon Basin, an area 50 by 75 km (31 by 47 mi), where up to 5,000 m (16,400 ft) of Paleozoic rocks are exposed. Paleozoic strata in the Marathon Mountains and the Solitario represent the southwest end of a mountain belt that extended across Texas to the Ouachita Mountains of Oklahoma and were continuous with the Appalachian Mountains. Much of the mountain belt is now buried beneath younger rock; it comes to the surface in the Marathon Uplift, the Ouachita Mountains, and in the Appalachians. The deformation records the collision of southeastern and southern proto-North America/Europe/Asia (Laurasia) with proto-South America/Africa (Gondwana). The fusion of these two ancient continents, welded together in the Ouachita-Marathon-Appalachian mountain belt, created the supercontinent Pangea.

Strata in the Marathon Uplift range from Late Cambrian to Late Pennsylvanian in age. These Paleozoic units were all deposited along the southern margin of Laurasia. To the south was the Tethys Ocean, which eventually evolved into the Gulf of Mexico/Caribbean/Atlantic Ocean. Farther south lay Gondwana, the southern continent. The Late Cambrian to Mississippian layers record slow accumulation of 950 m (3,100 ft) of rocks over nearly 170 million years. Mississippian through

Pennsylvanian strata record rapid accumulation of 4,200 m (13,800 ft) of rock in only 60 million years, a much faster sedimentation rate. The sediments, including shale, sandstone, limestone, chert, and conglomerate, accumulated on a continental slope along the margin of ancient North America. During the Late Mississippian to Early Permian Ouachita Orogeny (when Gondwana collided with Laurasia), the sediments were pushed northwest and thrust over Pennsylvanian deep marine sediments. Shortening in this thrust belt is on the order of 80%, that is, what was originally a 100 km (60 mi) wide zone is now 20 km (12 mi) wide. Stacked imbricate faults form complex thrust zones. Progressive thrusting refolded existing thrust sheets.

The Caballos Novaculite is the most competent unit within the Marathon Uplift and controlled development of prominent detachment folds. The thin pre-tectonic sequence resulted in fault-propagation folds and detached folds, while weak horizons in the pre-tectonic sequence led to multiple detachment surfaces and the formation of duplex (stacked thrust fault) zones. The Caballos Novaculite was the strongest unit in the pre-tectonic sequence and controlled the formation of high-relief detached folds.

Deposition of the Early Pennsylvanian Dimple Limestone indicates a lull in tectonism. During Dimple deposition, a carbonate bank existed to the north, and sediments were shed southeast into a deep marine basin. The overlying Haymond Formation sandstone indicates resumption of tectonism in Middle Pennsylvanian time.

Following deposition of the Haymond Formation, a deep marine trough formed north of the developing thrust belt and began filling with coarse fluvial and deltaic sediments of the lower Gaptank Formation. As deposition continued to the north, the older Gaptank sediments were folded, thrusted, and overridden by younger thrusts. By earliest Permian time, the thrusting had overridden the Gaptank Trough and thrusting had ended.

The entire pre-Permian Paleozoic sequence was thrust at least 200 km (125 mi) to the northwest.

Following the end of thrusting in the earliest Permian, subsidence and deposition shifted north to the Delaware, Midland, and Val Verde foreland basins.

The area was again subjected to compression during the Sevier/Laramide Orogeny (75–40 Ma), when some of the old folds were refolded. The Sevier and Laramide orogenies, slightly separated in time and space farther north, appear to be overlapping in this area. The area was uplifted during Tertiary time, with pre-existing strike-slip faults reactivated as normal faults, and emplacement of numerous small igneous intrusions. The main episode of igneous activity in west Texas was between 38 and 15 Ma.

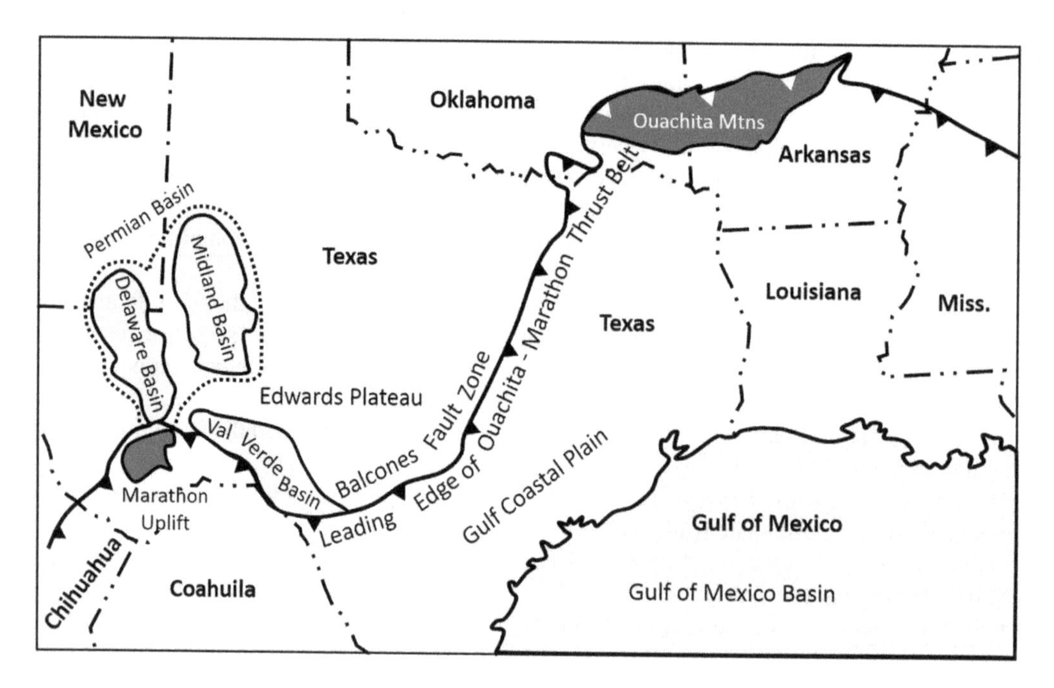

Structural setting of the Marathon Uplift. The heavy line with teeth indicates the suture zone and thrust front. Compression from the south moved rocks from the Tethys Ocean over proto-North America. (Modified after Hickman et al., 2009.)

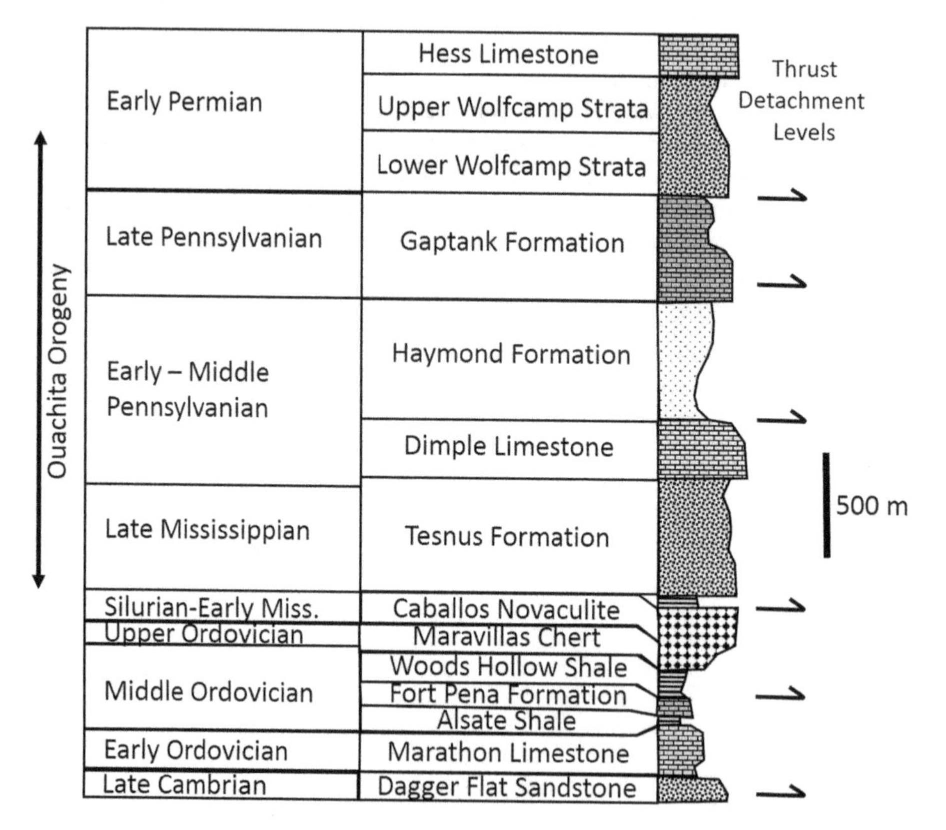

Stratigraphy of the Marathon Uplift. Major thrust horizons (detachment surfaces) are shown. (Modified after Hickman, et al., 2009.)

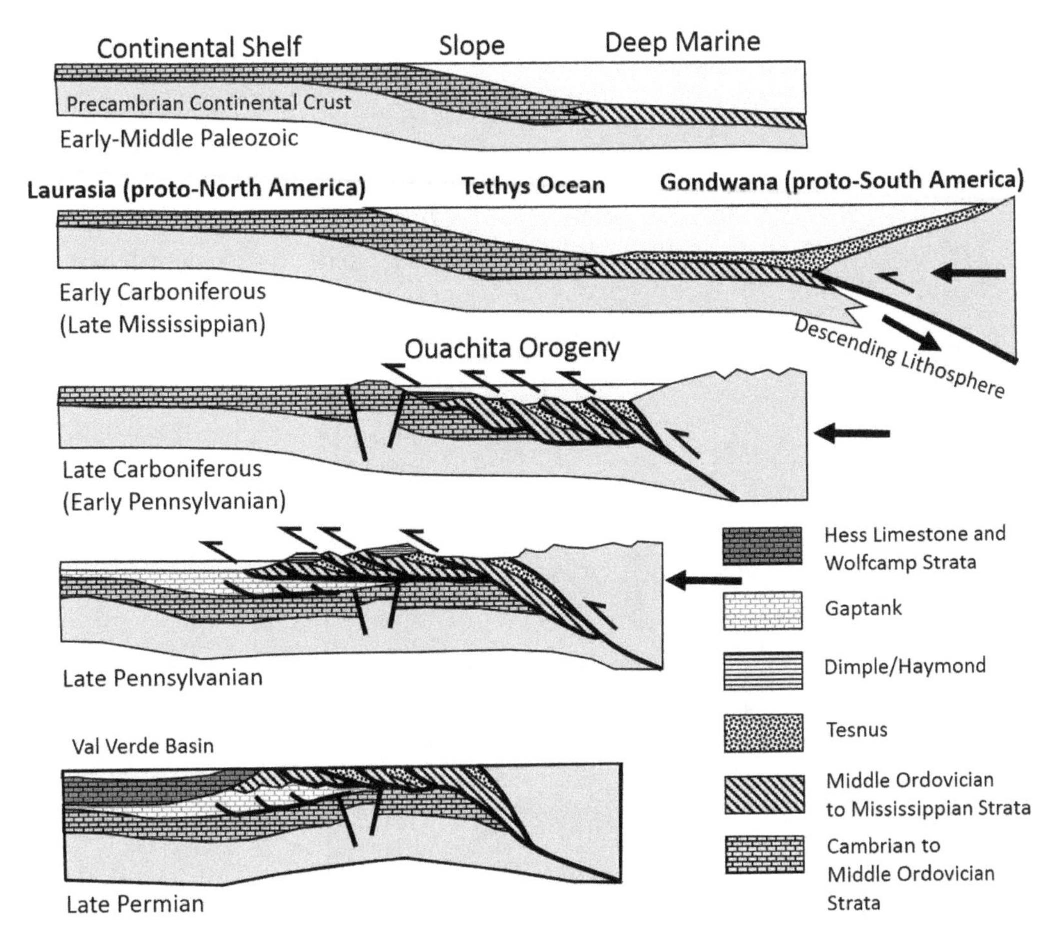

Structural development of the Marathon Fold-and-Thrust Belt and closing of the Tethys Ocean. (Modified after Hickman et al., 2009.)

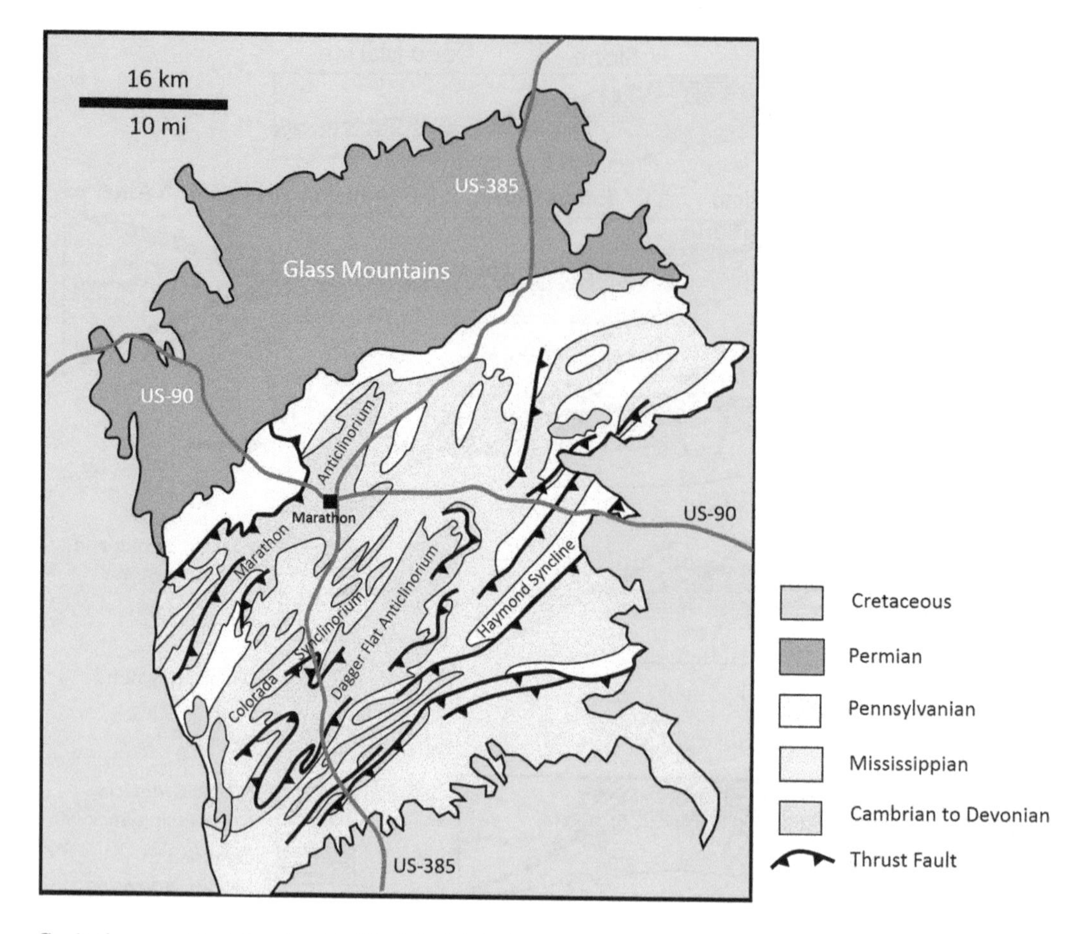

Geologic map of the Marathon Uplift. (Modified after Hickman, et al., 2009.)

Despite having the right rocks and structures for petroleum accumulation, the Marathon Uplift does not have any oil or gas fields. This may be due to breaching of traps and escape of hydrocarbons, or to excessive heat flow associated with local intrusions that over-cooked source and reservoir rocks.

Low-grade coal layers have been found in Cretaceous rocks of the region, notably lignite in the Aguja Formation. Local coal was used to fire the furnaces that produced mercury in the Terlingua Mining District.

STOP 11.1 LEMONS GAP EAST AND HAYMOND FORMATION TURBIDITES

At this spot on US-90 looking northeast, east, and south, you should be able to see that essentially flat-lying Lower Cretaceous (Fort Terrett member of the Fredricksburg Group and Glen Rose Limestone) is in angular unconformity contact over highly deformed Late Mississippian Tesnus Formation (sandstone), Early to Middle Pennsylvanian Dimple Limestone, and Haymond Formation (alternating sandstone and carbonaceous shale).

The Haymond Formation lies in the core of the northeast-trending Haymond Syncline here. This "piggyback syncline" is carried on west-northwest-directed thrusts formed during the Pennsylvanian Ouachita-Marathon Orogeny. We have just entered the Marathon Uplift.

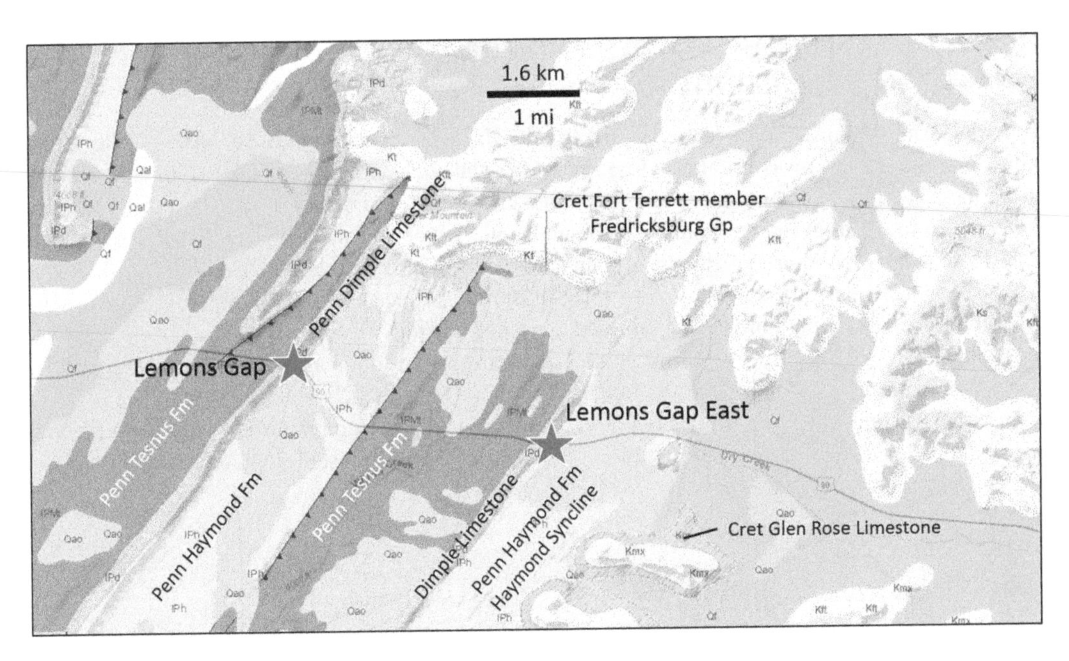

Geologic map of the Lemons Gap area, Marathon Uplift (U.S. Geological Survey and Bureau of Economic Geology, https://txpub.usgs.gov/txgeology/).

East-dipping Haymond Formation turbidites and Bouma sequences (a series of fining upward layers deposited by turbidity currents). These are outcrops of Pennsylvanian strata in the Marathon Uplift.

Lemons Gap East to Lemons Gap Turbidites: *Continue west on US-90 for 4.5 km (2.8 mi; 3 min) and pull over at historical marker pullout on the left (south side of highway). This is* ***Stop 11.2, Lemons Gap Turbidites*** *(30.2128833, −102.9884054).*

STOP 11.2 LEMONS GAP—FOLDED AND THRUSTED PENNSYLVANIAN TURBIDITES

Both the Tesnus ("Sunset" spelled backward) and the Haymond formations are continental slope turbidite deposits. They consist of alternating sandstone and shale characterized by Bouma sequencs: at the base is an erosional contact below a pebble conglomerate in a sandy matrix that grades up to coarse to medium-grained planar-bedded sandstone, which in turn grades up to a crossbedded sandstone, then ripple-marked silty sand, then tops out with laminar-bedded siltstone and shale.

The Dimple Limestone is a sandy gray, often bituminous limestone up 100–300 m (300–1,000 ft) thick that formed during a break in sedimentation coming from a southern source. These deposits are thought to represent carbonate material that was transported southeast into the Ouachita Trough, a foreland depression formed just north of the Ouachita-Marathon Fold-Thrust Belt. The Dimple Limestone was deposited in a submarine fan system and is represented by limestones, turbidite and deep-water shales, and thin black chert beds (Baker, 1985). The limestone was folded and thrust to the west during the Ouachita Orogeny.

Pennsylvanian Haymond Formation turbidites at Lemons Gap Historical Marker: "Denuded Ouachita Rock Belt."

From Lemons Gap Turbidites to Pennsylvanian-Cretaceous Unconformity: *Continue driving west on US-90 for 0.45 km (0.3 mi; 1 min) to* ***Stop 11.3, Pennsylvanian-Cretaceous Unconformity*** *(30.215180, −102.991853). Pull over on the right before the high point on the highway for a view east.*

STOP 11.3 LEMONS GAP VIEW OF THE PENNSYLVANIAN-CRETACEOUS ANGULAR UNCONFORMITY

The view east from Lemons Gap provides a stunning example of an angular unconformity. Between 180 and 200 million years of rocks are missing. After the mountain-building episode that folded Pennsylvanian turbidites to near-vertical, the Ouachita-Marathon Mountains were eroded to near sea level. The region then subsided below sea level, allowing essentially flat-lying Cretaceous limestones to be deposited over a broad area.

View east from Lemons Gap. Flat-lying Cretaceous limestone lies unconformably over steeply dipping Pennsylvanian units.

Pennsylvanian-Cretaceous Unconformity to Marathon Limestone: *Continue west on US-90 to the intersection with US-385; turn right (north) and drive 300 m; pull over on right shoulder. This is* ***Stop 12, Marathon Limestone*** *(30.209172, −103.215990), for a total of 22.5 km (14.0 mi; 12 min).*

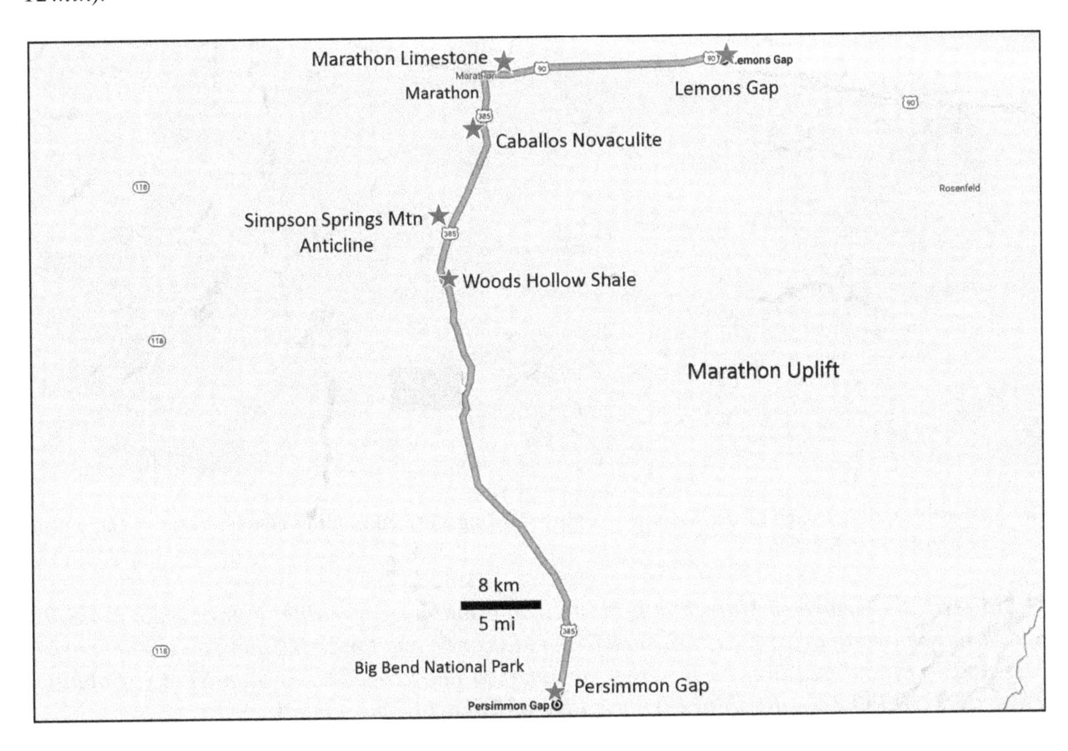

STOP 12 DEFORMED MARATHON LIMESTONE, MARATHON

The Lower Ordovician Marathon Limestone is a flaggy, dark-gray to gray-black limestone with abundant shale intervals and partings and some sandstone beds. At this roadcut, it has been highly deformed on the east flank of a faulted syncline.

Intricately folded Ordovician Marathon Limestone in roadcut near junction of US-385 and US-90.

SIDE TRIP 2 MARATHON UPLIFT, BIG BEND NATIONAL PARK, BIG BEND RANCH STATE PARK

Google Earth satellite image of the Marathon Uplift showing intricate folding characteristic of the uplift. Imagery © 2019 TerraMetrics

Side Trip 2—Deformed Marathon Limestone to Caballos Novaculite: *Return south to US-90 and turn right (west); drive to eastern outskirts of Marathon and turn left (south) on US-385; drive 5.3 km (3.3 mi) and pull over on the right shoulder on the ridge. This is* ***Stop ST 2.1, Caballos Novaculite*** *(30.158566, −103.236729), for a total of 7.6 km (4.7 mi; 6 min).*

ST2.1 CABALLOS NOVACULITE AND TESNUS FORMATION

Novaculite is a dense, fine-grained silica-rich rock that forms in a marine setting where radiolaria or diatoms (single-celled plankton with a silica shell) were especially abundant. When the diatoms die, the shells fall to the seafloor and accumulate as a siliceous ooze. It can also be derived from sponge spicules, siliceous spines that give sponges structural support. Later, during diagenesis (the transformation from sediment to rock), the silica shells or spicules convert into chert/chalcedony/flint, a microcrystalline silica. The chert is recrystallized to novaculite (microcrystalline quartz) with further diagenesis and low-grade metamorphism. Novaculite is often used to sharpen knives and other steel tools because of its hardness and fine-grained texture. Because of its resistance to erosion, the novaculite almost everywhere forms ridges.

The Caballos Novaculite is 133 m (436 ft) thick here. The Lower Chert and Novaculite are exposed on the north side of the road, and the Upper Chert and Shale outcrop on the south side. The Upper Chert member is various shades of gray, green, and brown, whereas the Shale member is red. Above the red shale is olive-green shale of the Tesnus Formation.

The Novaculite here is thought to be derived from a lime mud that contained abundant siliceous sponge spicules in addition to radiolaria. Ghosts of sponge spicules and radiolarians can be seen with a hand lens and in thin section.

An old manganese prospect pit is in the Upper Chert and Shale member north of the road.

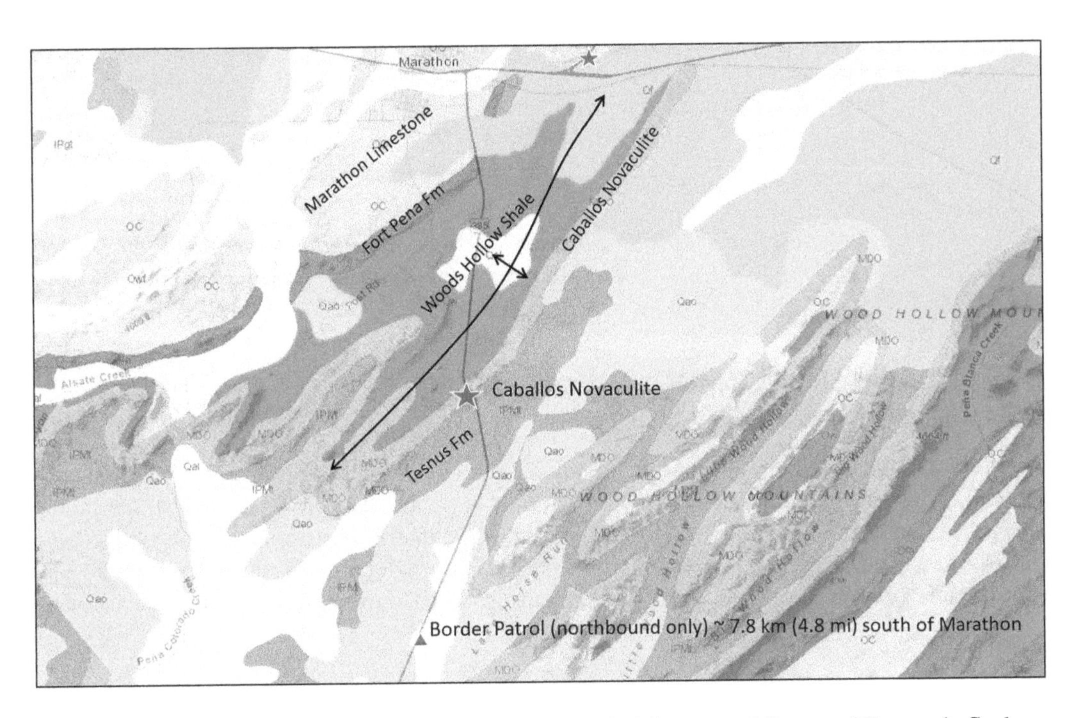

Geologic map of the Caballos Novaculite stop (U.S. Geological Survey and Bureau of Economic Geology, https://txpub.usgs.gov/txgeology/).

Caballos Novaculite outcrop 1. Layers are inclined to the southeast. Street View east.

Side Trip 2—Caballos Novaculite to Simpson Springs Mountain: *Drive south on US-385 for 11.1 km (6.9 mi; 7 min) to* ***ST 2.2, Simpson Springs Mountain Picnic Area*** *(30.066698, −103.274382) on right.*

ST2.2 SIMPSON SPRINGS MOUNTAIN PICNIC AREA—SIMPSON SPRINGS ANTICLINE

West of this pullout is Simpson Springs Mountain, a northeast and southwest-plunging breached anticline. Light-colored Caballos Novaculite forms both the flatirons (inclined strata in the foreground) and the crest of the ridge. Simpson Springs Mountain consists of complexly folded and faulted Ordovician Maravillas Formation (black chert), Silurian-Devonian Caballos Novaculite, and Pennsylvanian Tesnus Formation. It is probably a thrust-cored antiformal duplex (stacked thrusts).

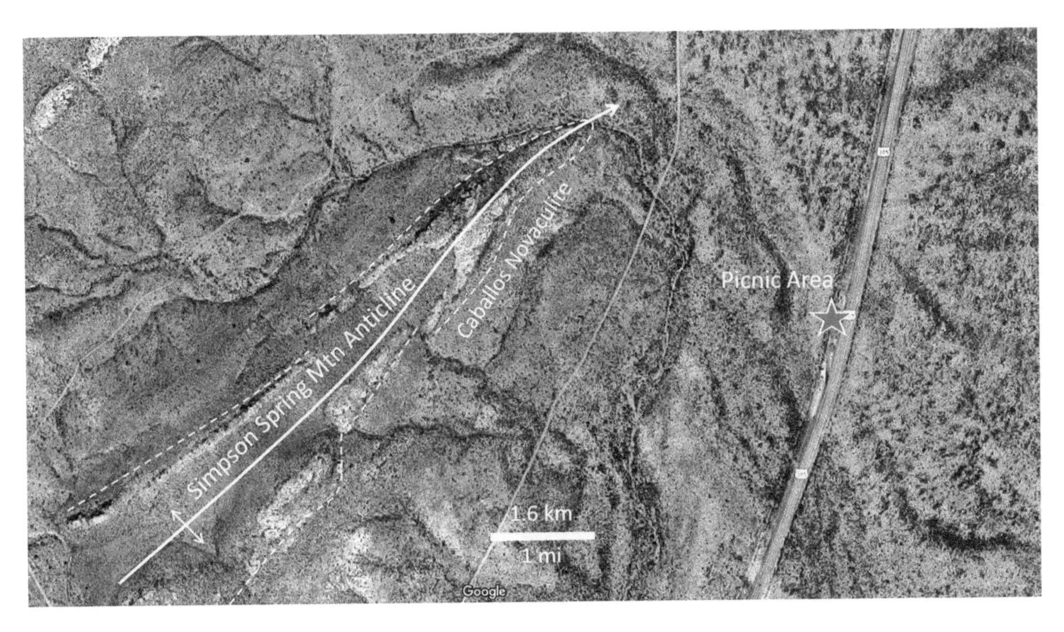

Google Earth satellite view of Simpson Springs Mountain Anticline. Imagery © 2019 CNES/Airbus; Maxar Technologies; USDA Farm Service Agency.

Simpson Springs Mountain Anticline is outlined by the resistant, white Caballos Novaculite.

Side Trip 2—Simpson Springs Mountain to Woods Hollow Shale: *Continue south on US-385 for 3.4 km (2.1 mi; 3 min) to* ***ST2.3, Woods Hollow Shale*** *(30.038230, −103.282571) and pull over on the right shoulder.*

ST2.3 WOODS HOLLOW SHALE

At this location, the road crosses a southwest-plunging anticline overturned to the northwest. The fold is cut by local, small-offset northwest-trending left-lateral strike-slip faults.

You are driving through an almost complete section of Upper Ordovician Maravillas Formation, 108 m (350 ft) of light-gray weathering black organic-rich limestone. The Maravillas was deposited in deep water as carbonate turbidity flows derived from a shelf to the northwest. Some chert beds indicate soft-sediment deformation that occurred before the rock altered to chert.

The ridge itself consists of Middle Ordovician Woods Hollow Shale, 90–120 m (300–400 ft) of green shale with interbedded gray to yellow sandy limestone and limy sandstone.

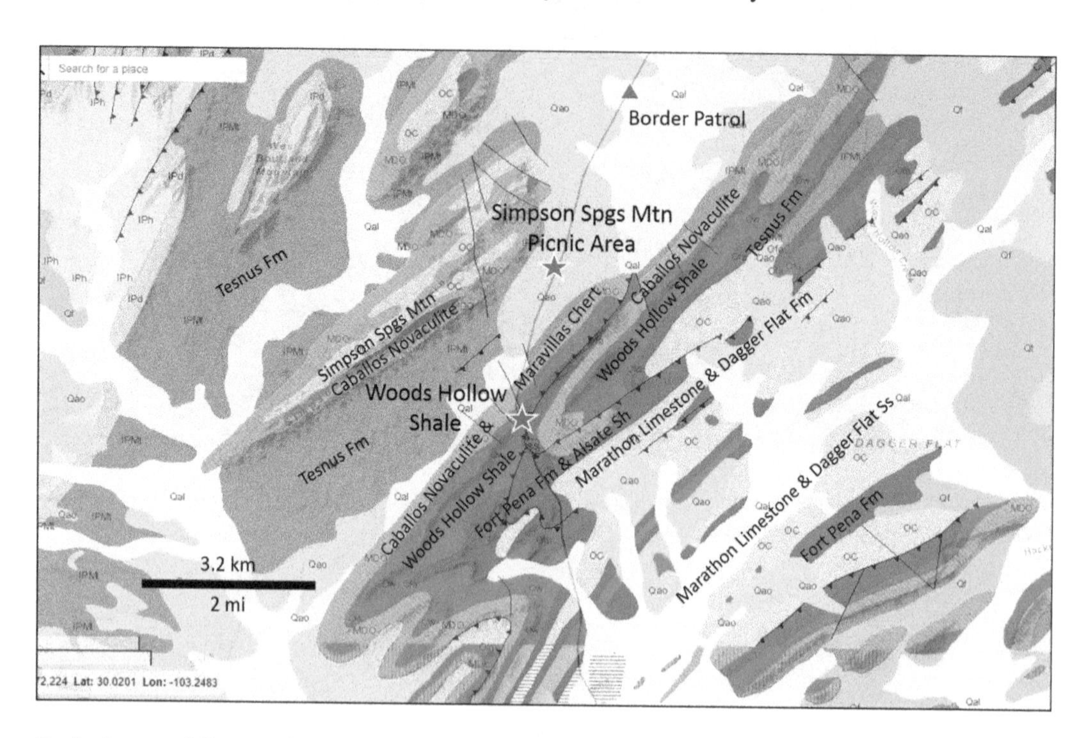

Geologic map of Simpson Springs Mountain stop and Woods Hollow Shale roadcut (U.S. Geological Survey and Bureau of Economic Geology, https://txpub.usgs.gov/txgeology/).

Roadcut in southeast-dipping Ordovician Woods Hollow Shale, shale facies, view southwest.

Roadcut in southeast-dipping Woods Hollow Shale, view northeast.

Side Trip 2—Woods Hollow Shale to Persimmon Gap: *Drive south on US-385 for 45.6 km (28.4 mi; 27 min) to* ***Stop ST2.4, Persimmon Gap*** *(29.668251, −103.170953).*

Big Bend National Park

Big Bend National Park covers 3,242 km^2 (801,163 ac). It was established in 1944 through a transfer of land from the State of Texas to the United States. The park is located along the United States–Mexico border in the Chihuahuan Desert ecosystem. The highest point occurs in the Chisos Mountains, which are mainly composed of thick Tertiary igneous rocks that reach an altitude of 2,387 m (7,832 ft).

Over 350,000 tourists visit the park annually, at least in part because of the unique geology and the resulting stunning landscape.

From Middle Mississippian to Early Permian time, some 330–285 million years ago, in an area extending from Oklahoma through west Texas to northeast Mexico, deep marine sediments were uplifted and thrust northwest as part of the Ouachita Orogeny. Only remnants of these intensely folded mountains remain here: they are the Marathon Uplift and the Solitario.

Beginning in Late Triassic time, about 200 million years ago, rifting between North and South America caused opening of the Gulf of Mexico. The Big Bend area became covered by a shallow inland sea that lasted until Late Cretaceous time, around 70 Ma. During this interval, a carbonate platform was the site of thick accumulations of limestone. As the sea began to retreat, around 100 Ma, sandstone and shale were deposited in shallow marine and shoreline environments. The remnants of these marine sediments form massive flat to gently inclined strata throughout the park.

Shallow marine layers of this time interval contain the fossil oysters, clams, ammonites, and a variety of fish and marine reptiles. Terrestrial deposits in Big Bend include petrified wood, fossil turtles and crocodiles, and the giant flying reptile, *Quetzalcoatlus northropi*, with a wingspan over 10 m (35 ft).

The Sevier/Laramide Orogeny was a time of compression related to convergence of the North American and Farallon tectonic plates along the west coast of North America. This is when the Rocky Mountains formed. Northeast-directed shortening caused northwest-southeast oriented structural features such as Mesa de Anguila in the southwest part of the park, the Sierra del Carmen-Santiago Mountains along the east edge of the park, and the Tornillo Basin between these uplifts. Mariscal Mountain, 26 km (16 mi) due south of Panther Junction Visitor Center, is considered the southernmost uplift of the Rocky Mountains in the United States.

Overlapping the Sevier/Laramide Orogeny was a period of volcanism. Volcanos, lava flows, vertical magma dikes, horizontal magma sills, laccoliths (domed sediments over magma intrusions), and ash eruptions that formed extensive air-fall tuffs occurred between 46.5 and 28 million years ago. These features can be seen in the Chisos Mountains, Rosillos Mountains, Christmas Mountains, Burro Mesa, McKinney Hills, and Grapevine Hills, as well as in Big Bend Ranch State Park west of the national park.

Basin-and-Range extension began in the Big Bend area around 25 Ma and is ongoing. This latest orogeny resulted in normal fault-bounded mountains (horsts) and valleys (grabens). The fault bounding the east side of Mesa de Anguila has on the order of 1,830 m (6,000 ft) of vertical offset. The 1995 magnitude 5.6 earthquake near Marathon, 116 km (70 mi) north of Panther Junction, proves that these stresses are still active. The area is presently being uplifted around 8 cm (3 in) every 100 years.

Today erosion is cutting away at the mountains and filling the intervening valleys. Only in the last 2 Ma has the Rio Grande reached these valleys and connected them to the Gulf of Mexico, making it the youngest major river in the United States.

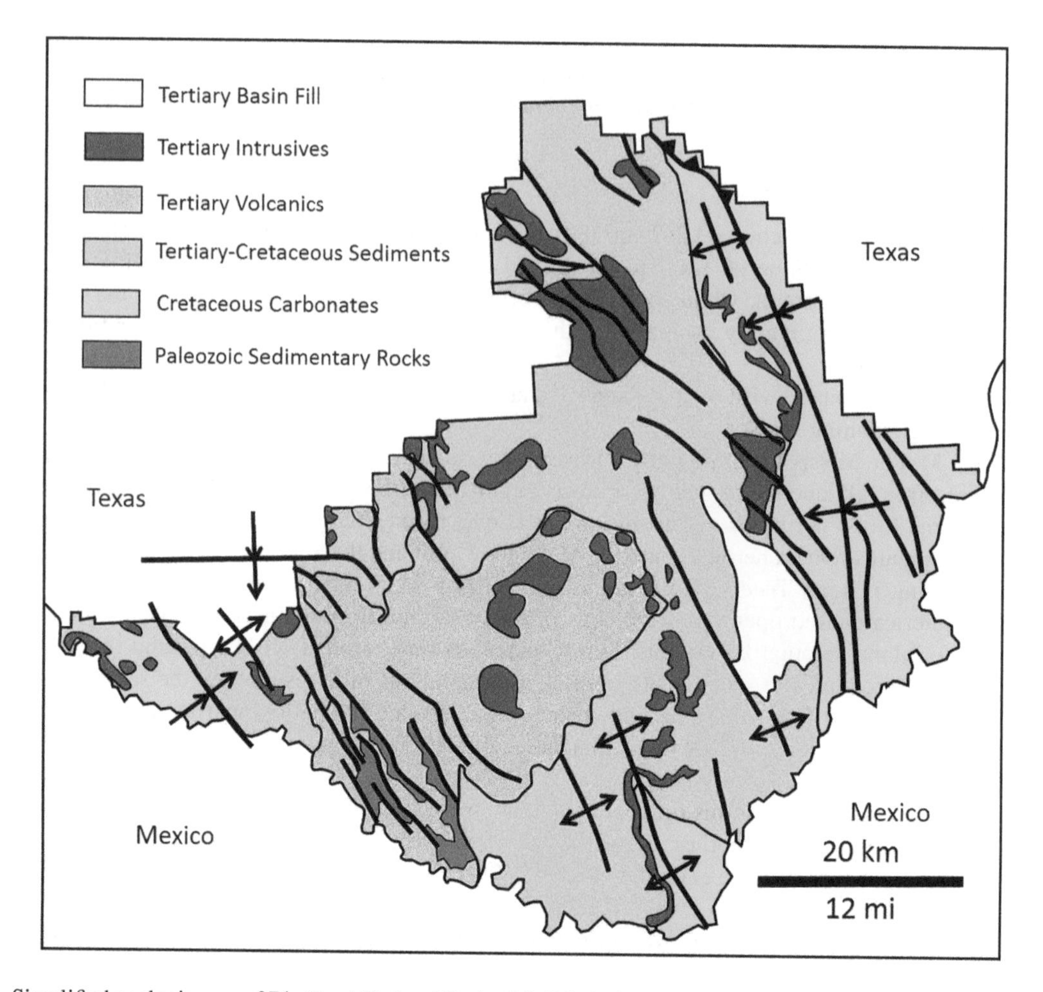

Simplified geologic map of Big Bend National Park. (Modified after Gray and Page, 2008.)

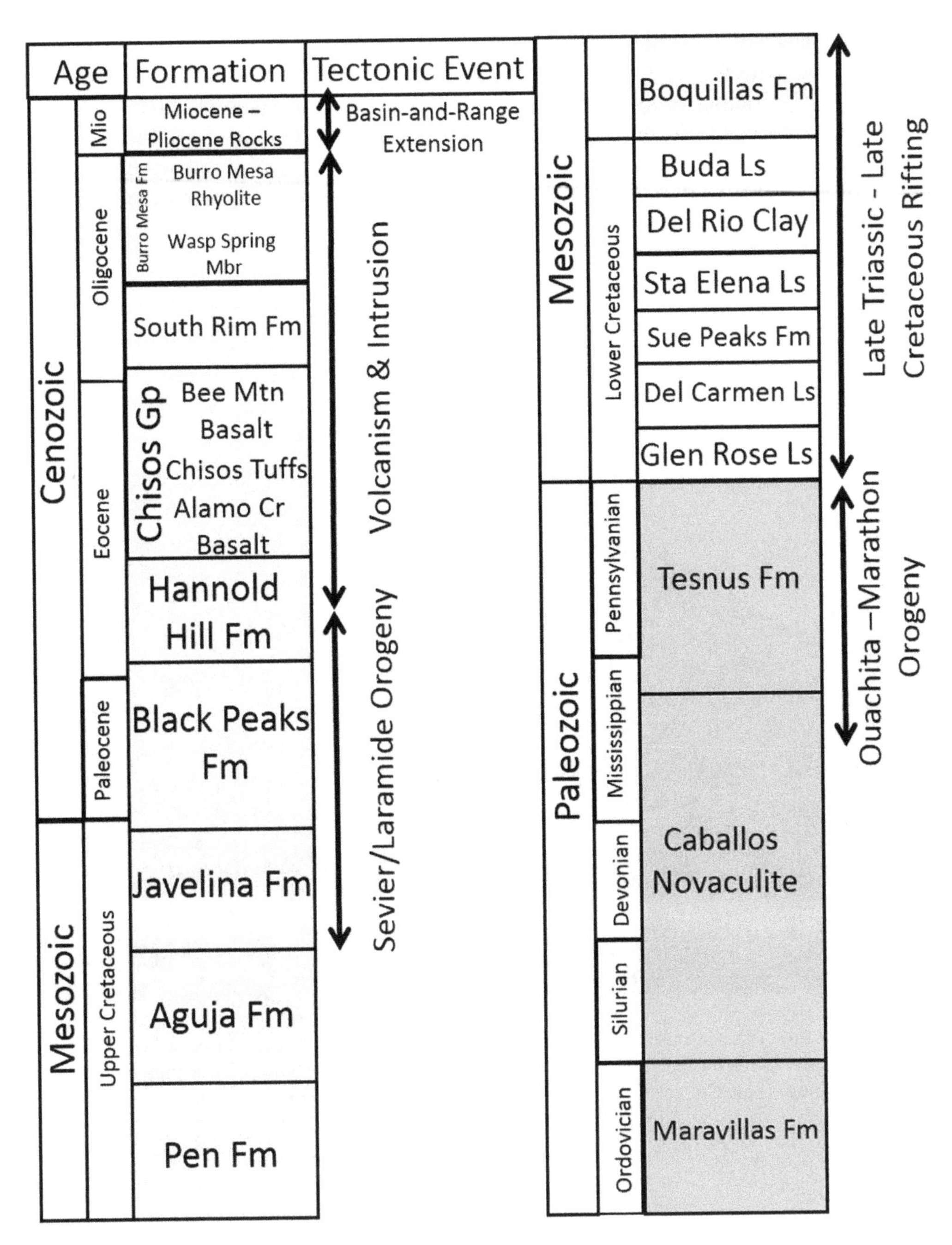

Stratigraphy and orogenic (mountain-building) episodes, Big Bend National Park. (Modified after Gray and Page, 2008.)

ST2.4 PERSIMMON GAP OVERLOOK

Persimmon Gap lies in the Santiago Mountains. West-dipping Upper Cretaceous Pen and Aguja formations are exposed along the highway below a southwest-directed Sevier/Laramide-age thrust (Lehman, 2014). The northeast-dipping Glen Rose Formation forms cliffs above the thrust. South of the Gap, across Santiago Draw, the Aguja Formation is exposed along with a mafic sill.

North of the Gap the Pennsylvanian Tesnus and Cretaceous Glen Rose and Del Carmen formations are folded by thrusting.

South of the Gap a northwest-trending, down-to-the-southwest normal fault puts Lower Cretaceous Glen Rose Limestone in fault contact with Upper Cretaceous Aguja Formation sandstone.

Persimmon Gap lies on the "Comanche Trail," a pathway used by Native Americans to move north-south across the Big Bend country. The Gap is named after native Persimmon trees that have a deep tap root and thrive in the area.

Persimmon Gap. Lower Cretaceous Glen Rose Formation is thrust over the Upper Cretaceous Aguja Formation and forms an anticline overturned to the southwest (dashed lines indicate bedding). Thrusting (dotted line) is Sevier/Laramide-age (Late Cretaceous to Eocene), not Ouachita-related (Pennsylvanian).

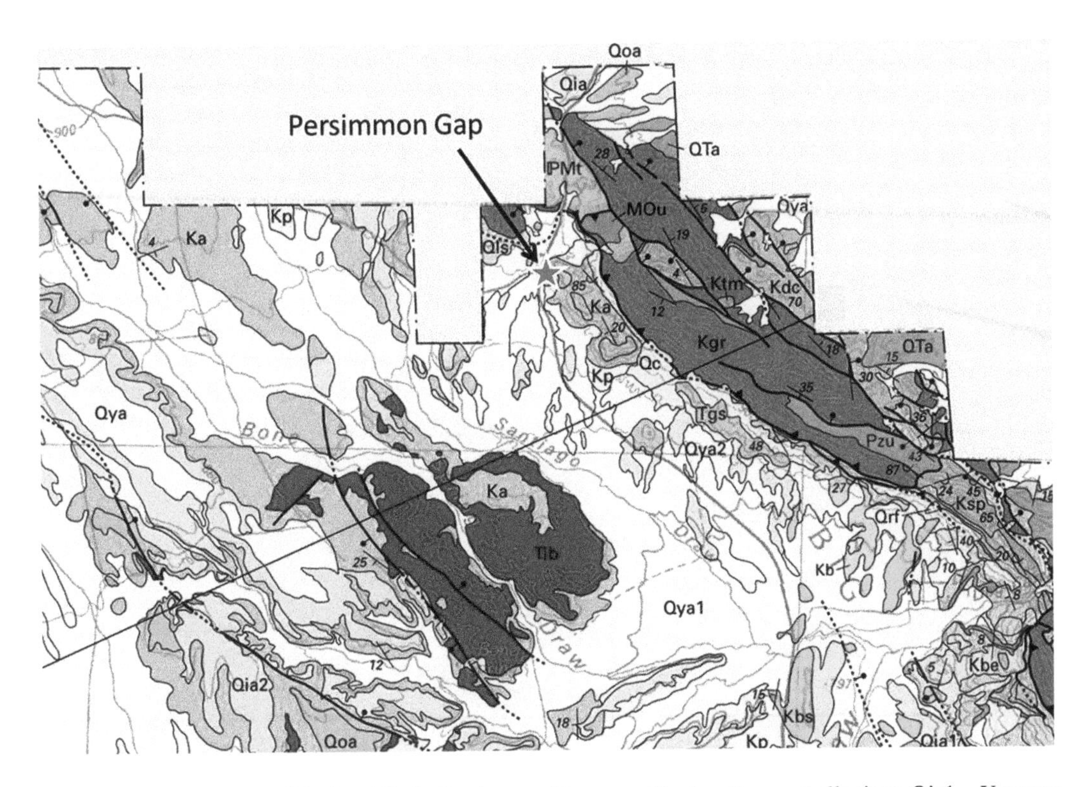

Geologic map of Big Bend National Park, Persimmon Gap area. Qya1 = Youngest alluvium; Qia1 = Younger intermediate alluvium; Qia2 = Older intermediate alluvium; Qoa = Old alluvium; QTa = Very old alluvium; Tgs = Tertiary basalt; Tib = Tertiary intrusive; Ka = Cretaceous Aguja Formation; Kp = Cretaceous Pen Formation; Kb = Cretaceous Boquillas Formation; Kbs = San Vicente member, Boquillas Formation; Kbe = Ernst member, Boquillas Formation; Ksp = Cretaceous Sue Peaks Formation; Kdc = Cretaceous Del Carmen Limestone; Ktm = Cretaceous Telephone Canyon Formation; Kgr = Cretaceous Glen Rose Limestone; PMt = Pennsylvanian Tesnus Formation; Pzu = Paleozoic rocks, undivided. (From Turner et al., 2011.)

Side Trip 2—Persimmon Gap to Fossil Trees: *Continue south on Main Park Road for 28.5 km (17.7 mi; 23 min) to* ***ST2.5, Fossil Trees/Logjam*** *(29.431413, −103.136839; between Mile Markers 9 and 10), and pull over on right shoulder. Walk about 100 m east from the highway to some low mounds.*

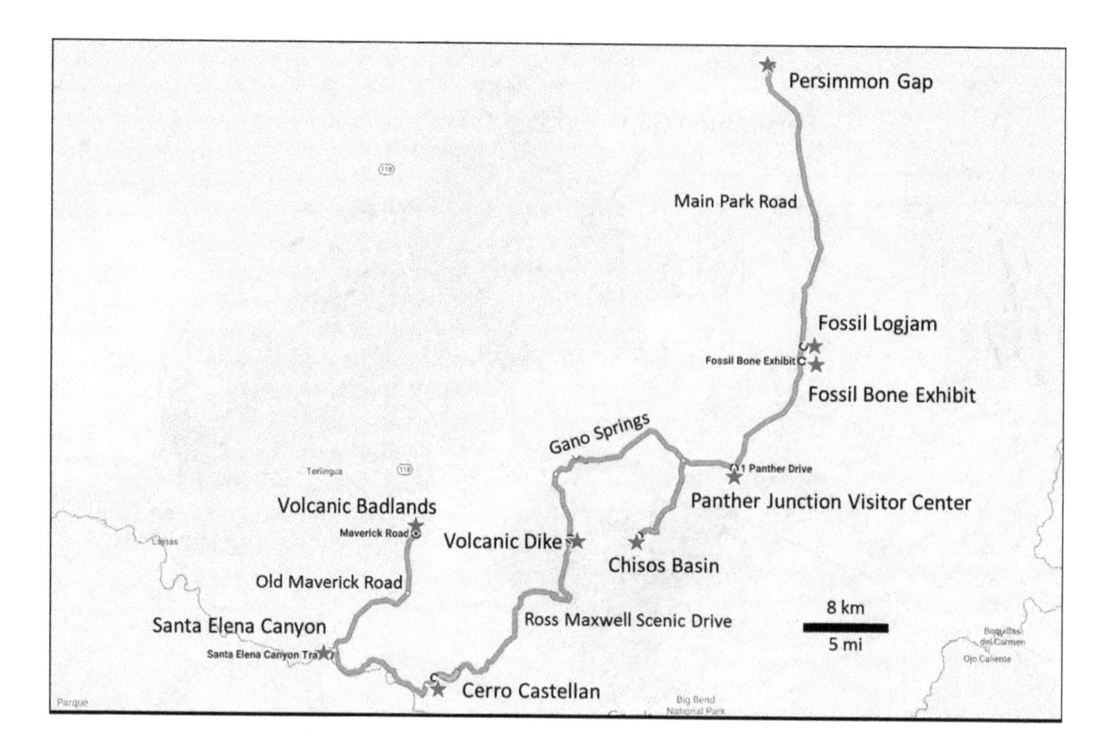

ST2.5 FOSSIL TREES/LOGJAM

About 120–150 m (400–500 ft) east of the road on a low ridge is a river channel sandstone in the Paleocene Black Peaks Formation. This is the "log jam bed," and it contains plentiful petrified logs. Wherever this sandstone is exposed, you find logs, usually one every 3 m (10 ft) or so. Some are up to 9 m (30 ft) long, with large root masses. All are *Paraphyllanthoxylon abbotti*, a Late Cretaceous-Paleocene broad-leaf tree.

The Paleocene Black Peaks Formation is alternating overbank mudstones and fluvial (river channel) sandstones, typically light-gray with black and red paleo-soil layers, often forming badlands. Badlands occur in desert areas where soft rock such as shale has been extensively eroded.

Fossil tree weathering out of the Paleocene Black Peaks Formation river channel sandstone.

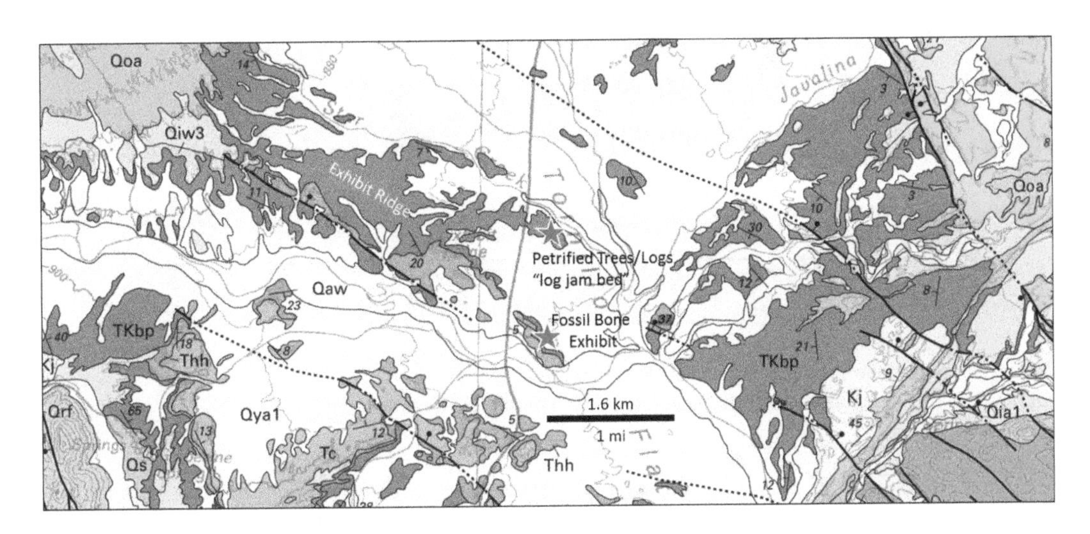

Geologic map of the petrified tree and fossil bone areas. Qaw = Active river deposits; Qya1 = Youngest alluvium; Qya2 = Older young alluvium; Qia1 = Younger intermediate alluvium; Qia2 = Older intermediate alluvium; Qoa = Old alluvium; Qiw3 = Oldest intermediate river deposits; Qs = Spring deposits; QTa = Very old alluvium; Tgs = Tertiary basalt; Tib = Tertiary intrusive; Tc = Eocene Canoe Formation; Thh = Eocene Hannold Hill Formation; TKbp = Paleocene Black Peaks Formation. (From Turner et al., 2011.)

Side Trip 2—Fossil Trees to Fossil Bone Exhibit: *Continue south on Main Park Rd for 1.6 km (1 mi; 2 min) to* ***ST2.6, Fossil Bone Exhibit*** *(29.418922, −103.138278) parking on the left.*

ST2.6 FOSSIL BONE EXHIBIT

The Exhibit Ridge Sandstone member of the Hannold Hill Formation is exposed at the Fossil Bone exhibit. It is a fluvial conglomeratic sandstone containing Eocene mammal bones.

The underlying Paleocene Black Peaks Formation consists of alternating overbank mudstones and fluvial sandstones, typically gray with black and red ancient soil (paleosol) layers, often forming badlands.

Below the Black Peak Formation is the Late Cretaceous Javelina Formation consisting of alternating shales and sandstones. The upper part of the sandstone-capped ridges contains sauropod (*Alamosaurus sanjuanensis*) and pterodactyl (*Quetzalcoatlus*) bones. The bones are found in lake deposits and abandoned stream channel deposits.

The Fossil Bone Exhibit has beautifully displayed examples of sea shells, dinosaur bones (including T-Rex, Triceratops) and mammal bones (horse, mammoth, saber tooth tiger) found throughout the park.

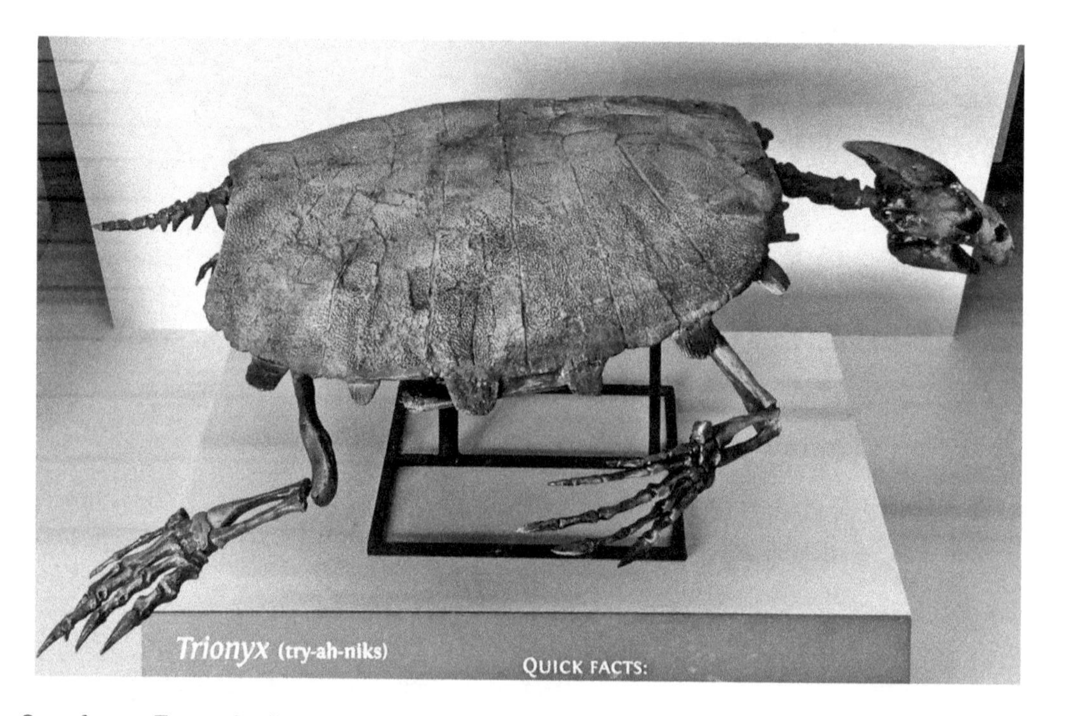

One of many Eocene fossils at the Fossil Bone Exhibit. This is *Trionyx*, a soft-shelled turtle that lived in shallow streams and ponds. Their descendants live along the Rio Grande today.

Side Trip 2—Fossil Bones to Panther Junction Visitor Center: *Continue south on Main Park Road for 13.4km (8.3 mi; 13min) to* ***ST2.7, Panther Junction Visitor Center*** *(29.328143, −103.206042).*

ST2.7 PANTHER JUNCTION VISITOR CENTER

Panther Junction contains a visitor center and park headquarters. Exhibits explain park geology, landscapes, plants, and animals as well as human history in the area. The visitor center theater shows an orientation film every 30min. A self-guided nature trail begins at the visitor center. Backcountry hiking permits and river use permits are available here. The bookstore is run by the Big Bend Natural History Association. A post office, water, and restrooms are available. A service station provides gas and groceries.

Visit

Address: 310 Alsate Dr, Big Bend National Park, TX 79834
Hours: Open all year, 8:30 am–5:00 pm. Reduced hours on Christmas Day. Closed · Opens 9 am Sunday
Phone: (432) 477-2251
Admission fees: see the website for current fees.
Website: www.nps.gov/bibe/planyourvisit/visitorcenters.htm

Side Trip 2—Visitor Center to Chisos Basin: *Drive west on Gano Springs Road to Basin Junction Road and turn left (south) drive to* ***Stop ST2.8, Chisos Basin*** *(29.269883, −103.300550), for a total of 15.5 km (9.6 mi; 19 min).*

ST2.8 CHISOS BASIN

This drive takes you through colorful Eocene and Oligocene volcanic ash, tuff, ignimbrites (tuff and pumice fragments) and lava flows of the Chisos Mountains. These were deposited between 33 and 47 million years ago on top of the existing Late Cretaceous marine and Paleogene marginal marine sediments. Vertical jointing allows these rocks to erode into spires, hoodoos, and other interesting landforms.

Eocene-Oligocene volcanics of the Chisos Basin. View to the west.

Side Trip 2—Chisos Basin Lodge to Volcanic Dike: *Return north to Gano Springs Road and turn left (west); drive to Ross Maxwell Scenic Drive and turn left (south); continue south to pull-out on the left (east) side by an interpretive plaque. This is* ***Stop ST 2.9, Volcanic Dike*** *(29.2676, −103.3687), for a total of 32.3 km (20.0 mi; 30 min).*

ST2.9 VOLCANIC DIKE

Like a wall across the landscape, this red-hued, near-vertical Oligocene rhyolitic dike was intruded into the younger parts of the Chisos Group volcanics (Turner et al., 2011).

Volcanic dike in Chisos Volcanics. View to the east.

Side Trip 2—Volcanic Dike to Cerro Castellan: *Continue southwest on Ross Maxwell Scenic Drive to Cerro Castellan pullout on the left (south) by an interpretive plaque, for a total of 27.4 km (17.0 mi; 24 min). This is* ***Stop ST 2.10, Cerro Castellan*** *(29.1486795, −103.5030757).*

ST2.10 CERRO CASTELLAN

A resistant lava flow caps this small peak. The bulk of the peak consists of volcanic ash, or tuff deposits of the Eocene-Oligocene Chisos Group volcanics and Burro Mesa Formation. The base consists of tuffs of the Chisos Group; these are overlain by the Bee Mountain basalt flow and then more Chisos tuffs. The Wasp Spring member is a 29.4 million-year-old ignimbrite. The Burro Mesa rhyolite caps the peak (Gray and Page, 2008).

Cerro Castellan consists of a lava flow over volcanic ash.

Colorful volcanic ash beds along the highway at Cerro Castellan.

Side Trip 2—Cerro Castellan to Santa Elena Canyon: *Continue west on Ross Maxwell Scenic Drive to Santa Elena Canyon Road and parking area and trail on the left. This is* ***ST2.11, Santa Elena Canyon Trailhead Parking*** *(29.167368, −103.610305), for a total of 15.6 km (9.7 mi; 17 min).*

ST2.11 SANTA ELENA CANYON

"The most spectacular result of the advent of the Rio Grande in the past 2 Ma are the canyons along its course: Mariscal Canyon, Boquillas Canyon, and the most dramatic canyon of all, Santa Elena" (Redfern, 2006–2018, online).

Santa Elena Canyon is a narrow gorge cut by the Rio Grande through the Sierra de Santa Elena/ Mesa de Anguila on the western edge of Big Bend National Park. The northwest-southeast trending uplift is tilted to the southwest and has an east-facing escarpment along the Terlingua Fault.

Strata exposed in the canyon include, from top-down, Santa Elena Limestone (upper cliffs), Sue Peaks Formation (slope-former), Del Carmen Limestone (lower massive cliff), Telephone Canyon Formation, and Glen Rose Limestone (lower slopes and ledges at the mouth of the canyon). These are all lower Cretaceous-age rocks. A short nature trail enters the canyon from the parking area at the canyon mouth. Where it enters the canyon, the walls tower 460 m (1,500 ft) above the river.

Santa Elena Canyon.

How did the Rio Grande cut canyons this deep and as narrow as 7.5 m (25 ft)? Geologists call this an antecedent river. It began eroding into flat-lying sediments high above the present-day limestone cliffs. The river became entrenched (locked-in) to its channel, and once that happened, it just kept on cutting downward by a combination of abrasion (by sand and gravel in the channel) and chemical solution (because limestone dissolves in slightly acidic water). If the river eroded the channel a mere 0.5 mm/year (0.0016 in/year), it could have cut a 500 m deep canyon in a million years. The Rio Grande has been grinding away at this canyon for around 2 million years.

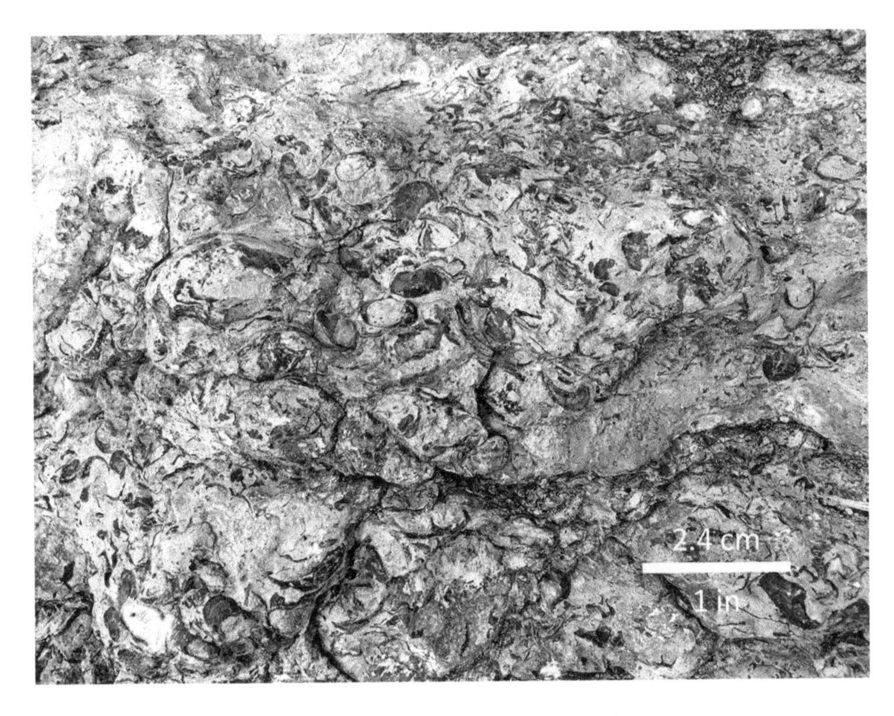

Rudist (clam) layer, Del Carmen Limestone at entrance to Santa Elena Canyon.

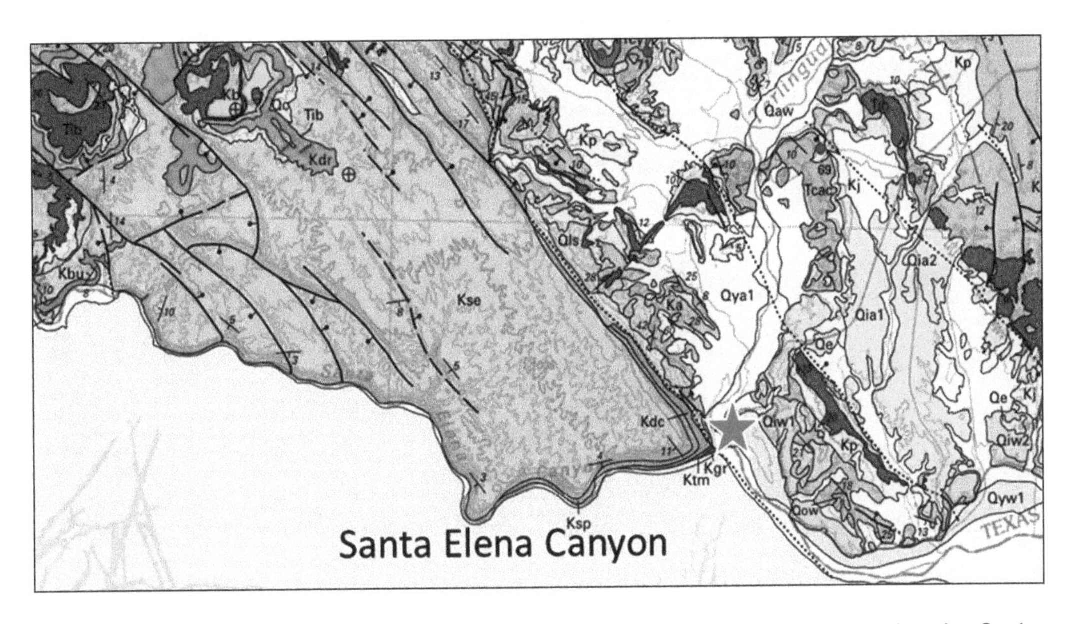

Geologic map of Big Bend National Park near Santa Elena Canyon. Qaw = Active river deposits; Qya1 = Youngest alluvium; Qya2 = Older young alluvium; Qia1 = Younger intermediate alluvium; Qia2 = Older intermediate alluvium; Qoa = Old alluvium; Qiw3 = Oldest intermediate river deposits; Qs = Spring deposits; QTa = Very old alluvium; Tgs = Tertiary basalt; Tib = Tertiary intrusive; Tc = Eocene Canoe Formation; Thh = Eocene Hannold Hill Formation; Tcac = Eocene Alamo Creek Basalt; TKbp = Paleocene Black Peaks Formation; Kj = Cretaceous Javelina Formation; Ka = Cretaceous Aguja Formation; Kp = Cretaceous Pen Formation; Kb = Cretaceous Boquillas Formation; Kbs = San Vicente member, Boquillas Formation; Kbe = Ernst member, Boquillas Formation; Ksp = Cretaceous Sue Peaks Formation; Kdc = Cretaceous Del Carmen Limestone; Ktm = Cretaceous Telephone Canyon Formation; Kgr = Cretaceous Glen Rose Limestone. (From Turner et al., 2011.)

Side Trip 2—Santa Elena Canyon to Volcanic Badlands: *Return east from parking area to Old Maverick Road and turn left (north); drive northeast on Old Maverick Road for a total of 17.6 km (10.9 mi; 36 min) and pull over on the right shoulder. This is* ***Stop ST2.12, Volcanic Badlands*** *(29.2703626, −103.5224992). NOTE: Old Maverick Road is not paved and is sometimes closed by the Park Service for safety reasons after rain or flooding. In case of closure, skip Stop ST2.12 and return northeast on Ross Maxwell Scenic Drive to Panther Junction Road; turn left (west) onto Panther Junction Road and drive to TX-118; continue straight on TX-118 to ST-2.13, Terlingua Ghost Town.*

ST2.12 VOLCANIC BADLANDS

The road traverses valley fill containing mostly volcanic material eroded off the igneous Chisos Mountains to the east. The road is on a "pediment," a low-relief gently sloping erosion surface, and crosses "bajadas," coalescing alluvial fans. The eroded "bolson" to the east, a desert valley filled by sediments, contains middle to late Pleistocene alluvial and basin fill deposits. Because these colorful sediments are unconsolidated, they are easily eroded into badlands topography.

Eroded badlands, view east from Old Maverick Road.

Side Trip 2—Volcanic Badlands to Terlingua Ghost Town: *Drive northeast on Old Maverick Road to junction with TX-118; turn left (northwest) on TX-118 and drive to FM-170 in Study Butte; turn left (west) onto FM-170 and drive to Ivey Road; bear right onto Ivey Road and drive to* ***Stop ST2.13, Terlingua Ghost Town*** *(29.320722, −103.617276), for a total of 18.7 km (11.6 mi; 23 min).*

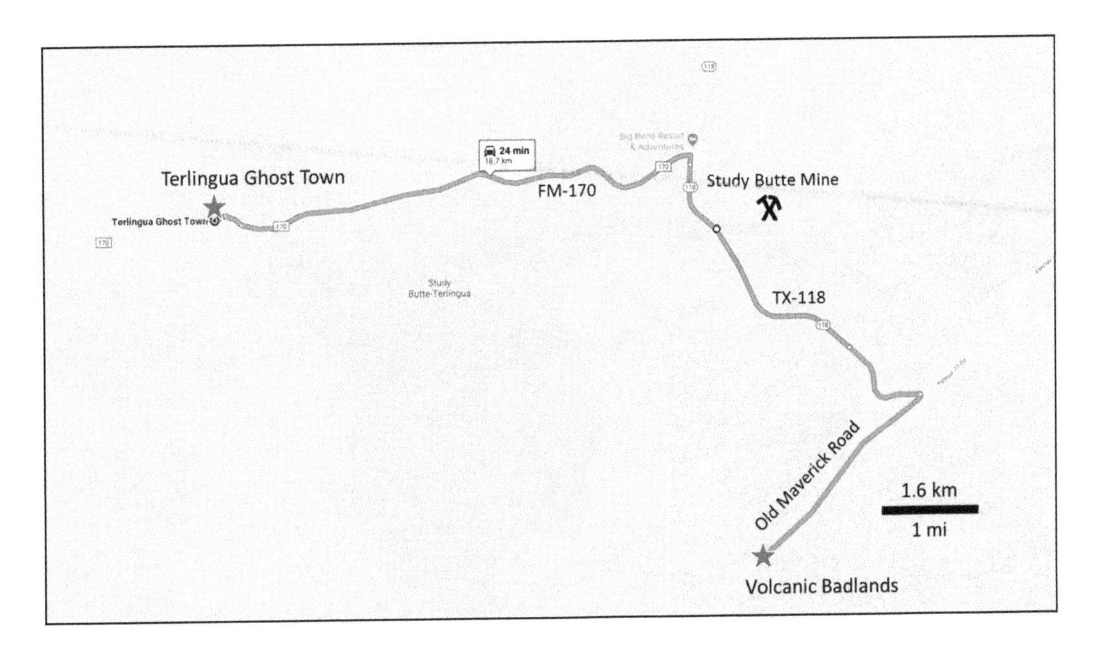

ST2.13 TERLINGUA GHOST TOWN AND MERCURY MINING DISTRICT

The Terlingua mercury district is roughly 32 by 6.5 km (20 by 4 mi). Cinnabar, a red mercury sulfide mineral, was originally used by Native Americans as a pigment. Early miners extracted the mercury by heating crushed ore to 180°C (360°F). At that temperature, the mercury becomes a vapor, which is then condensed to a liquid. The processed metal was carried 160 km (100 mi) by mule team to the nearest railroad.

The district, which includes the Study Butte area, produced over 150,000 flasks (34.5 kg, or 76 lb) of mercury between discovery in 1884 and the end of production in 1946. The district was the main producer of mercury in the United States for much of that time. World War I stimulated demand for mercury, which was used in making explosives: the price of a flask reached $115. During the Great Depression, the price dropped and mining activity tapered off.

Minerals associated with the ore include calcite, pyrite, hydrocarbons, and clays. Mineralization is found under four sets of conditions:

1. At the contact between the Santa Elena Limestone and Del Rio Formation clay.
2. Within near-vertical breccia pipes in the Buda and Ernst members of the Boquillas Formation. Breccia pipes, formed by solution collapse of limestone, contain the largest and best quality mineralization. Karst collapse breccias provided permeable and reactive pathways for deposition.
3. In calcite veins in the Boquillas Formation.
4. As veins in fractured felsic intrusions. Mineralization at the Study Butte Mine occurs in fractures in a Tertiary quartz syenite intrusion and in baked Pen Formation adjacent to the intrusion. All the deposits are related to hydrothermal fluids originating in the Tertiary intrusions.

Cretaceous rocks (Boquillas, Pen, and Aguja formations) dip gently south to southwest here. They were deformed during the Sevier/Laramide Orogeny into a north-northeast-facing monocline. Igneous rocks intruded the region between 48 and 26 million years ago. Basin-and-Range normal faulting over the past 25 Ma created northwest and northeast-trending normal faults.

Study Butte Mine and tailings, view east from pullout on right shoulder of TX-118 (29.318538, –103.530267).

The town of Study Butte, like the nearby mountain, was named for Will Study (pronounced "Stoody"), the manager of the Big Bend or Study Butte mercury mine. Mining began here around 1900 and peaked during World War I. Study Butte enjoyed a brief economic resurgence during World War II, but by the late 1940s, the mine had closed. Over the next two decades, the estimated population of Study Butte dwindled to ten or fifteen. Diamond Shamrock reopened the mine in 1970 and the population climbed to around 115. The mine closed again in 1972, but the Terlingua Ranch development began to increase the population and, along with Big Bend tourism, the population of Study Butte has grown to over 300 today.

The mine is now filled in, but you can nevertheless see the old ore roasting furnaces built in the early 1900s. The pipes that carried mercury vapor up to the condensers are still there.

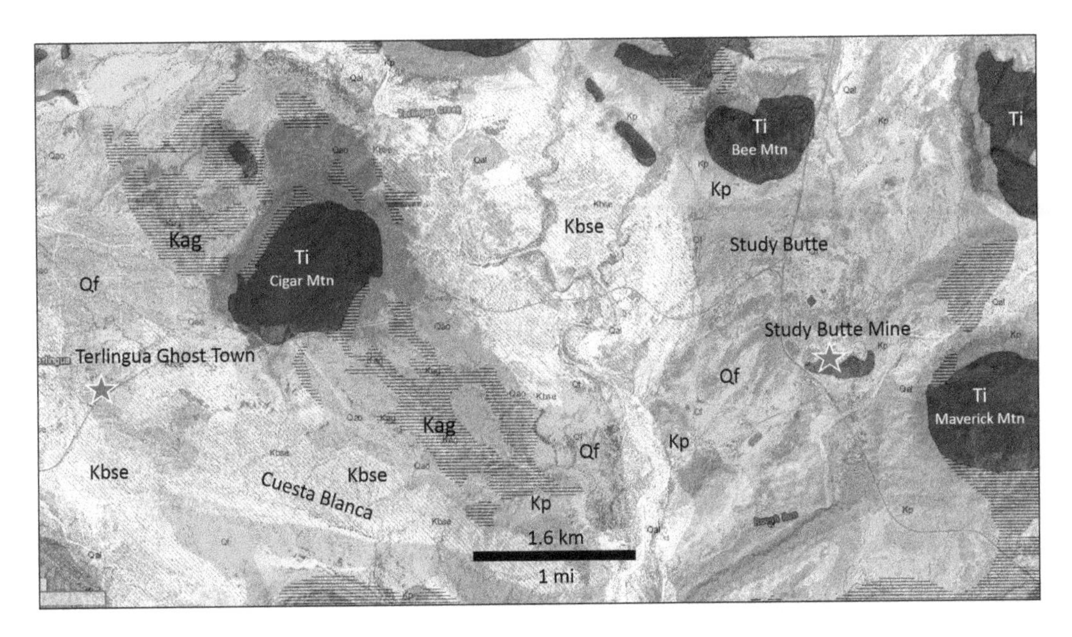

Geologic map, Terlingua and Study Butte area. Study Butte consists of the San Vicente member of the Cretaceous Boquillas Formation (Kbse), equivalent to lower Austin Chalk and Boquillas Flags. Cigar Mountain to the north is a Tertiary igneous intrusion. Kag = Aguja Formation; Kp = Pen Formation; Tc = Chisos Formation; Ti = Tertiary intrusive, Qf = Quaternary alluvial fan (U.S. Geological Survey and Bureau of Economic Geology, https://txpub.usgs.gov/txgeology/).

What is now the Terlingua ghost town was the principal producer of mercury in the United States during World War I. The Chisos Mining Company, established in 1903, was by 1918 the largest single producer of mercury in the United States. The Great Depression that followed the 1929 stock market collapse reduced demand. The deposits were mined out, and the mines went bankrupt by 1942. Although uneconomic, the mines continued working till 1946, when they were abandoned due to an influx of groundwater and depressed prices.

Since 1967, the town has become known for its world-famous Chili Cook-off that attracts as many as 10,000 visitors the first Saturday of each November.

Side Trip 2—Terlingua Ghost Town to Left-Hand Shutup Canyon, Solitario Dome: *Drive west on FM-170/Lone Star Ranch Road to Hwy 169/Bofecillos Road turnoff to Big Bend Ranch State Park; turn right (north then east) onto gravel Bofecillos Road and drive to Main Road; turn right (southeast) on Main Road/Big Bend Spring Main Road and follow the signs to the park headquarters and Sauceda Bunkhouse (29.470061, −103.958350). Ask park ranger for maps and directions to* ***Stop ST 2.14, Lefthand Shutup Canyon*** *(29.470444, −103.767125). This is a total of 145 km (90.1 mi; 3 h 32 min).*

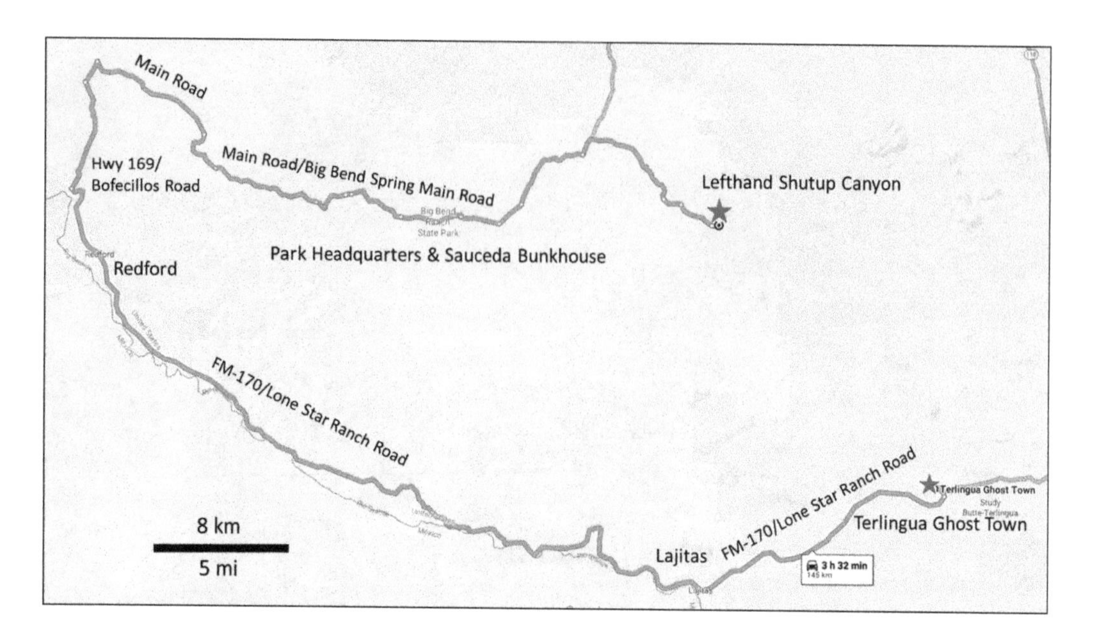

ST2.14 BIG BEND RANCH STATE PARK AND SOLITARIO DOME

Wild, undeveloped, expansive vistas, and no crowds. This is how people describe Big Bend Ranch State Park (BBRSP). Just west of Big Bend National Park, BBRSP is, at 126,000 ha (311,000 ac), the largest state park in Texas. The state park includes a number of working cattle ranches, and a herd of longhorn cattle is based there.

Part of the Chihuahuan Desert, summers are hot and winters are mild. Elevation ranges from 700 to 1,600 m (2,300 to 5,150 ft). Vegetation is sparse, and water is rare. None of the creeks flow year-round. Beware of flash floods during wet weather.

The park contains 37 km (23 mi) of frontage along the Rio Grande, and rafting is popular. Away from the river, visitors can hike, backpack, go horseback riding, and enjoy mountain biking.

Big Bend Ranch was once among the ten biggest working ranches in Texas. The Carrasco and Madrid families began sheep ranching in the area in the 1870s. In the 1910s the Bogel brothers bought and consolidated blocks of land in the area. The Bogels went bankrupt during the Great Depression, and their holdings were purchased by the Fowlkes family, who continued enlarging the ranch. Eventually Robert Anderson, chairman of Atlantic Richfield Corporation, bought the land and offered it to the state for $8 million, well under market value, but the state land commissioner could not get approval to purchase the land. On July 21, 1988, the Texas Parks and Wildlife Commission formally approved the purchase of the ranch. The park was opened to the public in 1991 as Big Bend Ranch State Natural Area. In 1995 it was renamed Big Bend Ranch State Park (Kohout, 2010).

Four mountain-building events are recorded in the rocks of Big Bend Ranch State Park. The Pennsylvanian-Permian Ouachita-Marathon Foldbelt exposes folded and thrusted Silurian and Devonian layers in the core of Solitario Dome. Erosion of the mountains formed by this event was followed by deposition of marine sediments, including limestone, shale, and sandstone. These were deformed by Late Cretaceous to Eocene Sevier/Laramide folding and faulting, seen along the Fresno-Terlingua Monocline. Mid-Tertiary intrusion and volcanism of the Trans-Pecos Volcanic Province extend throughout the park. Finally, Miocene to present-day Basin-and-Range extensional faulting and associated horsts and grabens are exposed along the Rio Grande.

Visit

Sauceda Ranger Station (Northern Unit)

Address: 1900 Sauceda Ranch Road, Presidio, TX 79845
Phone: (432) 358-4444
Hours: Open daily.
Office Hours: Sauceda Ranger Station: 8 am–6 pm daily.
The Sauceda Ranger Station is closed on December 25.
Entrance Fees: see the website for current fees.
Website: https://tpwd.texas.gov/state-parks/big-bend-ranch

Note: The Sauceda Ranger Station is 43 km (27 mi) of rugged, dirt road from FM-170. This is the only way in or out of the interior of the park. Motor homes and trailers are not recommended.

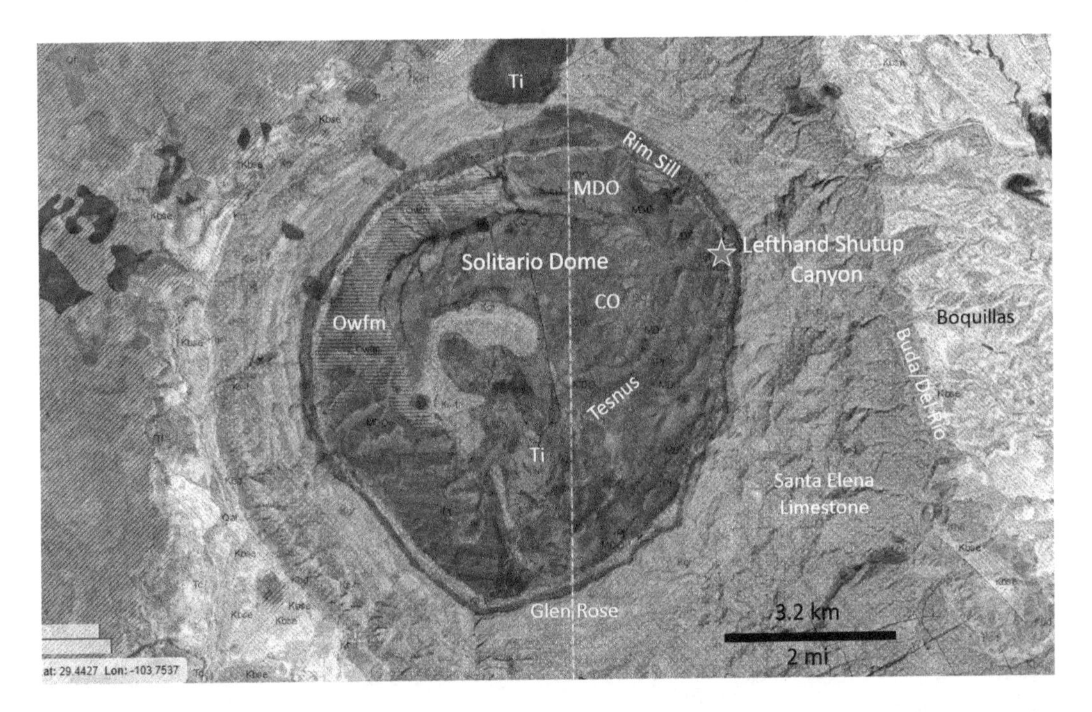

Geologic map of Solitario Dome. Ti = Tertiary Intrusive; MDO = Mississippian-Devonian-Ordovician units; Owfm = Ordovician Woods Hollow Shale, Fort Pena, and upper Marathon Limestone; CO = Cambrian Dagger Flat Sandstone and lower Marathon Limestone (U.S. Geological Survey and Bureau of Economic Geology, https://txpub.usgs.gov/txgeology/).

The Bofecillos Mountains volcanics erupted between 32 and 27 Ma from at least 8 volcanos in the park. They consist of lavas, tuffs, and volcanic sandstones and conglomerates. Solitario Peak, a quartz trachyte intrusion on the west rim of the Solitario Uplift, is part of the Bofecillos Volcanics.

El Solitario looking west. The rim consists of uplifted Cretaceous Santa Elena Limestone. (Photo courtesy of J. Fowler, https://commons.wikimedia.org/wiki/File:El_Solitario_(40189643681).jpg.)

Solitario Dome and Left-Hand Shutup Canyon

As you drive from Presidio to the Solitario, you are going through the Bofecillos Mountains. These are composed of volcanic rocks erupted from numerous volcanos in the area between 36 and 27 Ma. Upper Cretaceous rocks can be seen dipping outward around the margins of the uplift. North-northeast-oriented folds and thrusts of Cambrian to Pennsylvanian rocks, deformed during the Ouachita-Marathon Orogeny, are exposed within the crater.

Deformed Caballos Novaculite near the entrance to Lefthand Shutup Canyon, Solitario dome. View to the east.

Tesnus turbidites, north side of Lefthand Shutup Canyon.

Four main volcanic episodes affected the state park. Between 47 and 33 Ma basalt flows and ash-flow tuffs of the Chisos Group erupted in Big Bend National Park and spread west to Big Bend Ranch State Park. Volcanics of the Solitario, the oldest volcano in BBRSP, erupted between 35 and 36 Ma. The Cienega Mountains, in the northwest part of the park, formed between 32.8 and 30 Ma. The Mitchell Mesa Rhyolite buried the region under as much as 100 m (300 ft) of hot ash over a 15,500 km^2 (6,000 mi^2) area. This is 1,000 times the size of the Mount St. Helens eruption. This was followed by eruption of the Cienega Mountains Rhyolite, forming a large lava dome that is now the Cienega Mountains. Finally, the Bofecillos Mountains formed during a series of eruptions starting 32 Ma and lasting till 27 Ma.

The Solitario is a volcanic uplift. Rising magma raised the overlying rocks into a nearly circular dome. Erosion then exposed the older rocks within, providing a window into the subsurface. The first pulse of magma, around 36 Ma, barely breached the surface. It is expressed today as rhyolite dikes and sills. The second pulse, at 35.4 Ma, domed up the overlying rocks, forming the laccolith structure we see today. A Laccolith is a magma that starts as a sill, parallel to bedding, then thickens to the point where the overlying layers are pushed upward and domed outward. The solid magma is still buried about 1.5 km (1 mi) beneath the surface. As the overlying rocks stretched, fractures formed at the surface and extended down into the magma chamber. This led to a sudden, catastrophic release of pressure and resulted in explosive volcanism. Flaming hot gas and ash burst from surface vents and flowed down the slopes. So much material was ejected that the rocks over the magma chamber collapsed, forming a large caldera, or crater 1.9 × 6.4 km (1.2 × 4 mi) and up to 300 m (1,000 ft) deep. Much of the caldera is now full of debris eroded from the crater walls. The third and final pulse of volcanic activity occurred around 35 Ma, when numerous dikes intruded into the rocks of the caldera.

Tertiary rim sill with columnar jointing, Lefthand Shutup Canyon. The age of the sill is 36.6 Ma.

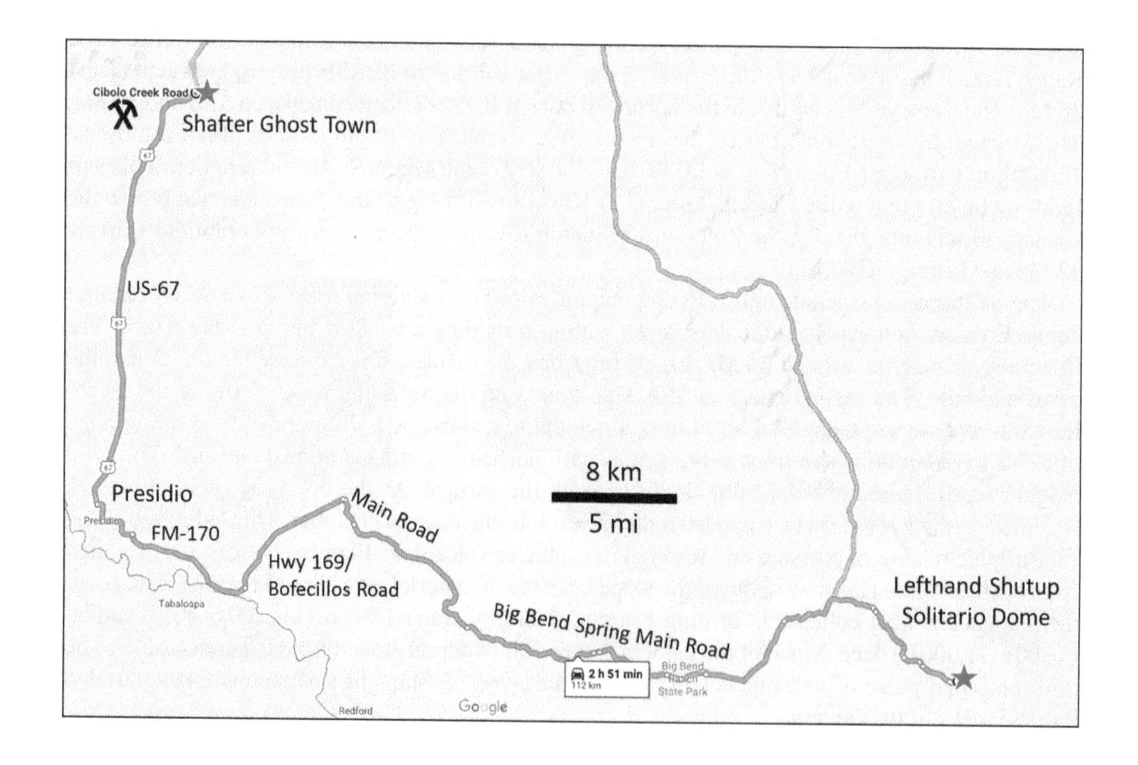

Side Trip 2—Solitario to Shafter Ghost Town: *Return west on Big Bend Spring Main Road to Bofecillos Road; turn left (southwest) and drive to FM-170; turn right (west) on FM-170 and drive to Presidio; turn right (north) onto Emma Ave/US-67 and drive to* ***ST2.15, Shafter Ghost Town*** *(29.815398, −104.306578) on the right for a total of 112 km (69.3 mi; 2 h 51 min).*

ST2.15 SHAFTER GHOST TOWN AND SILVER MINING DISTRICT

About 32 km (20 mi) north of Presidio is the town of Shafter, a silver mining ghost town. The town is named after General William Shafter, one-time commander of Fort Davis. The town is on the National Register of Historic Places: in the early 1900s, there were six silver mines operating in the area. It is always fun poking around on mine dumps looking for discarded ore rock.

Silver was discovered in the Shafter Mining District by John Spencer around 1880. The Presidio Mining Company was established to mine the silver, and the Mina Grande orebody was being mined by 1883. The ore ran to 938 g/tonne (30 oz/ton), and annual production averaged 18,100 tonnes (20,000 tons) or ore between 1890 and 1913. After a new mill was added in 1913, production averaged 76,200 tonnes (84,000 tons) from 1913 to 1926. The American Metal Company bought the mines in 1926. The mines closed between 1930 and 1934 due to the low price of silver. The mines produced again between 1934 and 1942, when they were closed due to low prices and World War II. In 1940, the ore averaged about 313 g/tonne (10 oz/ton). Between 1883 and 1942, the district produced over 992,000 kg (35 million oz) of silver.

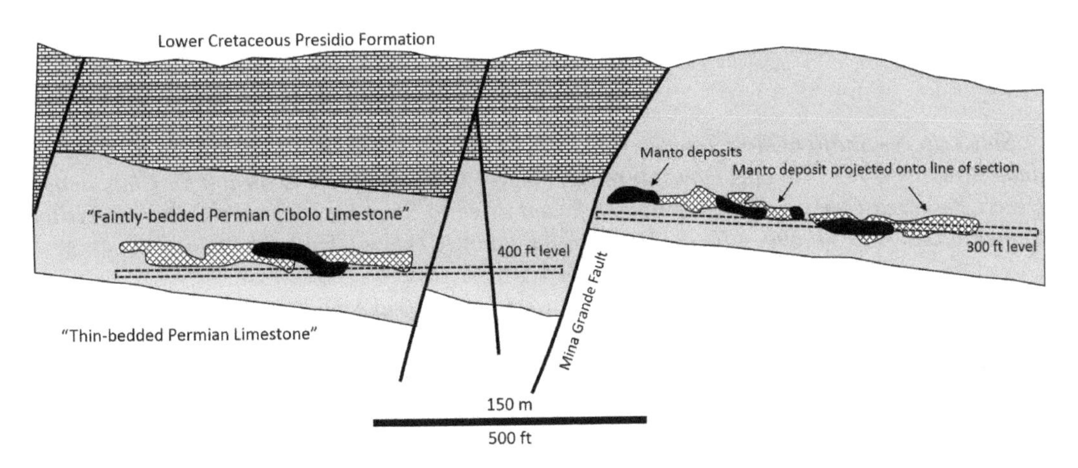

Cross section through the Presidio Mine, Shafter. Ore deposits occur in the Permian limestone and overlying Cretaceous beds. https://commons.wikimedia.org/wiki/File:Shafter_Presidio_Mine_Cross_Section.png

In 1977, the property was purchased by Gold Fields Mining, who discovered the Shafter deposit. The deposit was sold to Rio Grande Mining Company in 1994; Rio Grande was acquired by Silver Standard Resources in 2001. Aurcana bought the Shafter deposit from Silver Standard in 2008 and reopened the La Mina Grande Mine, west of US-67, in 2012. In December 2013, the project was placed on hold due to the declining price of silver and uncertainty about reserves (Aurcana Corp, www.aurcana.com/operations/shafter/overview/, Accessed April 9, 2019).

The stratigraphic section, from base to top, consists of Permian Cibolo Formation limestone overlain by the Cretaceous Presidio Formation, a massive dark-gray Cretaceous limestone. Ore occurs in the Permian Limestone as argentite (silver sulfide) and cerargyrite (silver chloride), supergene replacement minerals along bedding (manto deposits) and thrust faults. Replacement deposits and silver-bearing veins along normal faults occur in the Cretaceous limestone (Wikipedia, Accessed April 4, 2019).

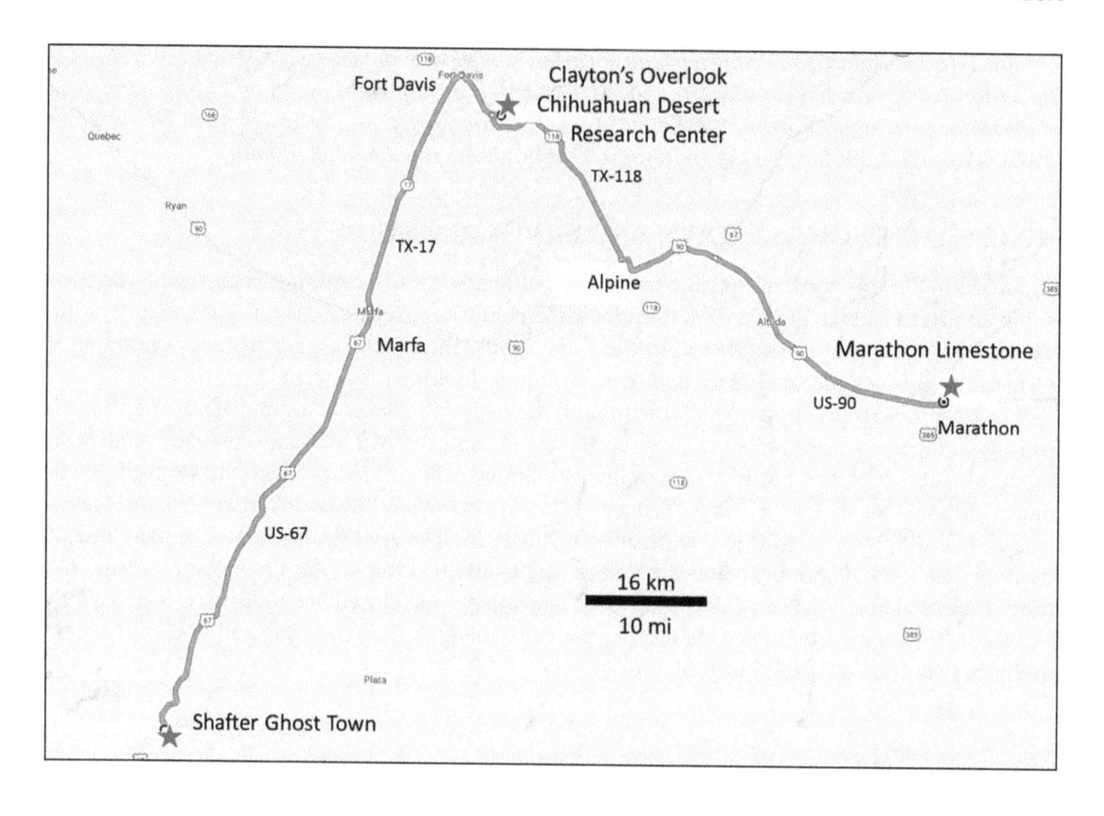

Side Trip 2—Shafter Ghost Town to Clayton's Overlook: *Drive north on US-67 to Marfa; continue straight on TX-17 to Fort Davis; turn right onto TX-118 and drive to turnoff for Chihuahuan Desert Research Center on the left; turn left and drive to* ***Stop 13, Clayton's Overlook Visitor Center*** *parking (30.540690, −103.837461), for a total of 107 km (66.6 mi; 1 h 8 min).*

Marathon to Clayton's Overlook: *Continue west on US-90 to Alpine; turn right (northwest) on TX-118 to turnoff on right for Chihuahuan Desert Research Institute; turn right and drive to visitor center parking. This is* ***Stop 13, Clayton's Overlook*** *(30.540690, −103.837461), for a total of 84.3 km (52.4 mi; 53 min).*

STOP 13 CLAYTON'S OVERLOOK, CHIHUAHUAN DESERT RESEARCH INSTITUTE NATURE CENTER AND BOTANICAL GARDENS

During the Eocene, about 35 million years ago, the Davis Mountains were formed by eruptions from two major volcanos depositing sequences of lava flows and ash layers that became the Sleeping Lion Formation. These eruptions are thought to be a result of Basin-and-Range extension and faulting that tapped into magma deep in the crust. About 2.5 million years later, magma in the subsurface pushed up the volcanic rocks at the surface, forming a large dome. The magma cooled underground and formed several laccoliths, including Musquiz Dome to the south, Pollard Dome to the east, and Barillos Dome to the southeast. Prominent Mitre Peak to the south is a 28 million-year-old quartz trachyte/rhyolite intrusion.

A 2.4 km (1.5 mi) loop trail leads from the visitor center to the top of a rise called Clayton's Overlook. There is a 60 m (200 ft) gain in elevation. From the top is a 360° panorama, and the outlook has interpretive panels explaining the geology of the Davis Mountains.

Mitre Peak looking south from Clayton's Overlook. The peak is an intrusion, not a volcano. It stands 1,887 m (6,190 ft) high.

As you climb to the overlook, you pass from the country rock, dark brown volcanic rocks of the Sleeping Lion Formation, 35.9 Ma, to the 32.8 Ma Weston Intrusion, light-gray sheets of rock that are exfoliating, or hiving off slabs of rock. Erosion has worn away the overlying volcanics and exposed the magma that uplifted this rise.

Incorporated in 1974, the Chihuahuan Desert Research Institute (CDRI) is a non-profit organization with the mission to promote public awareness of and appreciation for the Chihuahuan Desert through research and education. Located on 205 ha (507 ac) in the Davis Mountains foothills, they have interpretive trails, a nature center, and botanical gardens. CDRI was originally affiliated with Sul Ross State University in nearby Alpine. They currently work not only with Sul Ross, but also K-12 schools, colleges, and other non-profits.

Thanks to Lisa Gordon of the Chihuahuan Desert Research Institute and Blaine Hall for providing geologic information and a tour of the overlook.

Visit

Hours: The Nature Center is open Monday through Saturday from 9:00 am to 5:00 pm. Sundays (March 17–December 1) 12:30 pm to 5:30 pm. Closed Thanksgiving, Christmas, and New Years Day.
Address: 43869 Hwy 118, PO Box 905, Fort Davis, TX 79734
Phone: 432.364.2499
Entrance fee: see the website for current fees.
Website: www.cdri.org/

Clayton's Overlook to Carlsbad Caverns: *Drive north on TX-118 to Davis; turn left (south) onto TX-17 and drive to TX-166; turn right (west) on TX-166 and drive to US-90; turn right (northwest) on US-90 and drive to Van Horn; continue straight (north) on TX-54 to US-62; turn right (northeast) on US-62 and drive to Whites City; turn left (north) on TX-7 to visitor center and cavern entrance. This is* ***Stop 14, Carlsbad Caverns Visitor Center*** *(32.176832, −104.442147), for a total of 306 km (190 mi; 3 h 4 min).*

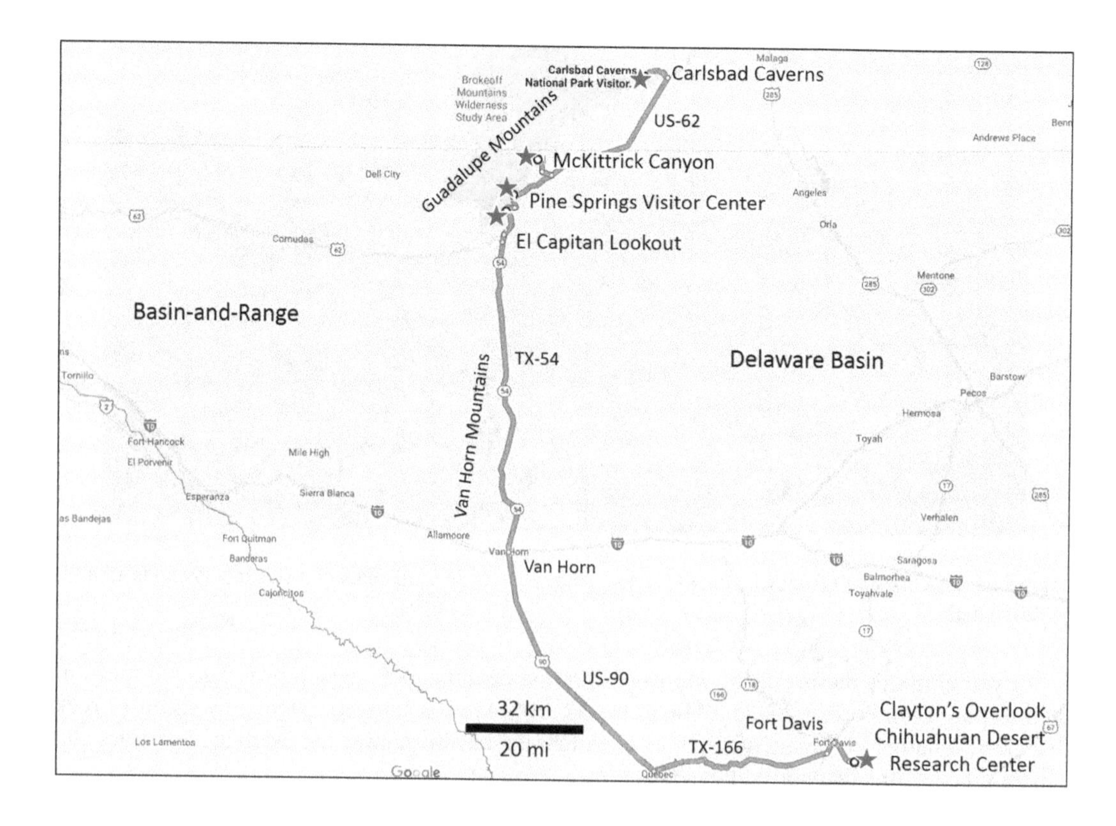

STOP 14 CARLSBAD CAVERNS NATIONAL PARK AND THE PERMIAN BASIN

Driving north from Marathon to Carlsbad Caverns, you enter a new geologic province, the Basin-and-Range, as well as the world-famous oil-producing Permian Basin and its western sub-basin, the Delaware Basin. The Van Horn Mountains, north of Van Horn, are bounded on the east by a nearly north-south series of Basin-and-Range normal faults. The west flank of the Guadalupe Mountains is also defined by a large Basin-and-Range fault that tilts the mountain range to the east (Kerans and Janson, 2012; Bebout et al., 2007). Basin-and-Range faulting dies out near the western margin of the Permian Basin.

The Permian Basin

The Permian Basin in west Texas and southeast New Mexico is one of the hottest areas in the U.S. oilpatch at the time of this writing. The basin covers about 223,000 km^2 (86,000 mi^2) and extends about 400 km (250 mi) east-west and 480 km (300 mi) north-south. The basin contains the world's thickest section of Permian-age rocks.

From late Precambrian through Mississippian time (850–310 Ma), the ancestral Permian Basin was a broad marine passive margin. A continental margin sequence is present throughout the southwestern United States and is up to 1.50 km (5,000 ft) thick. Passive margins are the transition zone between oceanic and continental crust that are *not* active plate margins. They usually contain a more-or-less undisturbed pile of sediments.

Between late Mississippian and Permian time, the collision of Laurasia and Gondwana, known in Europe as the Hercynian Orogeny and in North America as the Ouachita-Marathon and Appalachian (Alleghanian) orogenies, created the Ouachita–Marathon Thrust Belt and the associated foreland basins. Down-warping of the foreland basin (the region in front of the thrust belt) formed the Delaware Basin to the west, the Midland Basin to the east, and the Central Basin Platform between them. The mountain-building event to the south resulted in rapid and voluminous erosion and sedimentation into these basins. Siliciclastic (quartz-rich sand and shale) deposition in Pennsylvanian time was followed by the formation of limestone platforms and reefs along the basin flanks in the Early Permian. The Delaware Basin continued to fill until the late Permian. Sandstones and deep-water, organic-rich shales were deposited within the basins while reef carbonates were deposited on the Central Basin Platform and along the margins of the basins. The extensive reef deposits fringing the Delaware Basin comprise the Capitan Limestone. Eventually, the Permian sea was filled in, and the basins were capped with evaporite deposits, including salts and gypsum, when the sea slowly evaporated.

The Permian Basin is one of the largest petroleum-producing basins in the United States. The first commercial well in the Permian was the Santa Rita No. 1, drilled between 1921 and 1923. Since that time, the basin has produced a cumulative 4.5 billion m^3 (28.9 billion barrels of oil) and 2.1 trillion m^3 (75 TCF of gas; Texas Railroad Commission). More than 40,000 exploration wells and 200,000 development wells have been drilled in the Permian Basin.

Active drilling in the Permian Basin near Monahans, Texas.

The Permian section produces mainly oil, with over 10.3 billion m^3 (65 billion barrels, BBL) of oil-in-place in 2,188 fields. Gas comes primarily from pre-Mississippian strata. The basin contains about 1.38 trillion m^3 (48.5 trillion ft^3) of non-associated gas-in-place.

Period	Series	Group	Delaware Basin Formation
Permian	Guadalupian	Delaware Group	Lamar Bell Canyon
			Cherry Canyon
			Brushy Canyon
	Leonardian		Upper Avalon Shale
			Lower Avalon Shale
			1st Bone Spring
			2nd Bone Spring
			3rd Bone Spring
Wolfcampian			Wolfcamp
Penn.			

Period	Series	Group	Central Basin Platform Formation	
Permian	Guadalupian	White-horse	Tansill	
			Yates	
			7 Rivers	
			Queen	
			Grayburg	
		Ward	San Andres	
			Glorieta	
	Leonardian	Yeso	Paddock	
			Blinebry	
			Tubb	
			Drinkard	
		Abo		
		Wolfcamp	Hueco	
			Bursum	
Penn.				

Period	Series	Group	Midland Basin Formation
Permian	Guadalupian	White-horse	Tansill
			Yates
			7 Rivers
			Queen
			Grayburg
		Ward	San Andres
			Glorieta
	Leonardian	Clear Fork	Upper Leonard
		Upper Spraberry	
		Lower Spraberry	
		Dean	
		Wolfcamp	
Penn.			

Permian Basin stratigraphy. (Modified after Shale Experts, Manti Petroleum.)

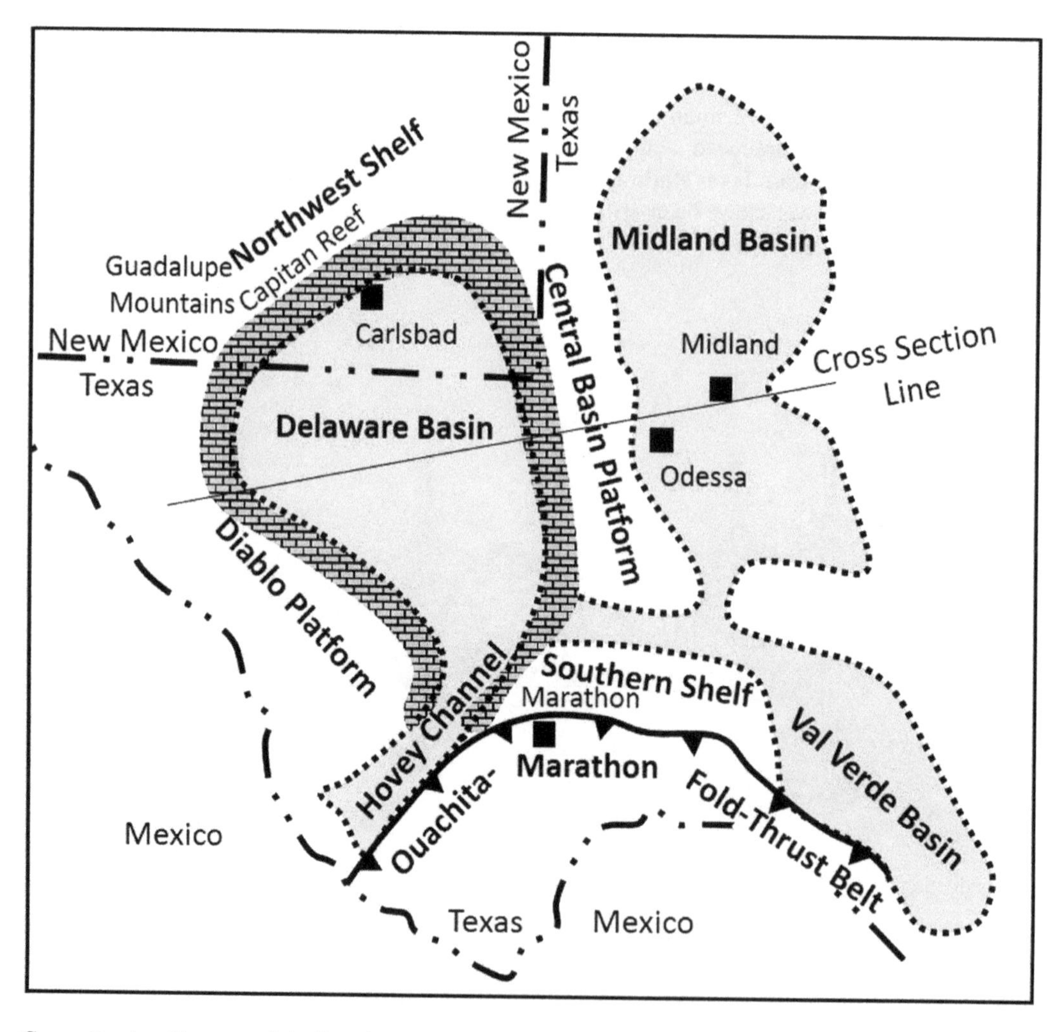

Generalized outline map of the Permian Basin. Cross section line refers to next figure.

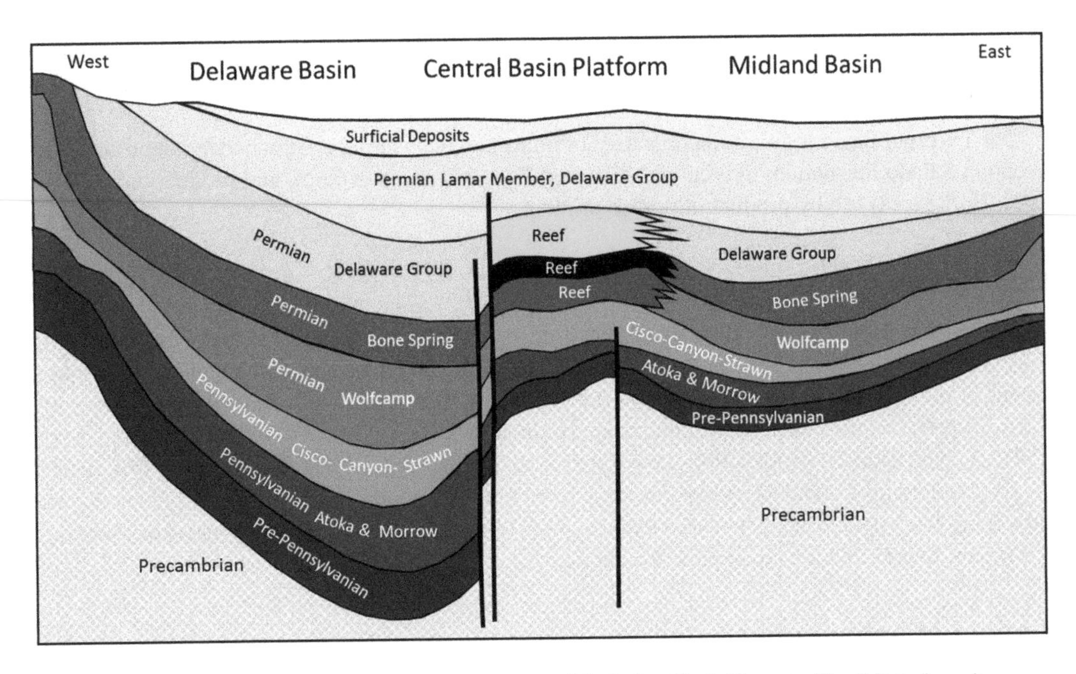

West-east cross section through the Permian Basin. (Modified after Shale Experts, Manti Petroleum.)

Hydrocarbon traps are usually a combination of stratigraphic and structural, although each type does occur alone. The sealing mechanisms are porosity and permeability barriers of carbonate, evaporite or shale, especially at updip transitions from porous limestones to anhydrite plugged dolomites. The best production is from near-back-reef or grainstone margin facies that did not undergo the marine cementation of reefs or the evaporite plugging of the farther back-reef layers. Porosities range from 1.5% to 25% and reservoir permeabilities from 0.02 to 200 millidarcies.

More than 90% of production is from primary or early secondary porosity in back-reef dolomites, limestones, and sandstones of the Tansill, Yates, Seven Rivers, Queen, and Grayburg formations or the open-shelf facies of the San Andres Limestone. Reef and fore-reef facies are totally non-productive.

The source for most Permian oil is presumably from the oxygen-starved, organic-rich basin sediments such as the Bone Spring Limestone and parts of the Delaware Mountain Group.

Horizontal drilling and hydraulic fracturing have led to a production boom that began around 2008. Between January 2007 and August 2015, crude oil production in the Permian Basin grew from 134,000 m^3 (843,000 BBL) of oil per day to 312,000 m^3 (1,961,000 BBL) per day. Monthly production for the entire Permian Basin during October 2015 was just over 318,000 m^3 (2 million BBL) of oil a day and increasing (Jarvie et al., 2017). Natural gas production rose from 100 million m^3 (3,529 million ft^3) per day to 119 million m^3 (4,201 million ft^3) per day between 2008 and 2014 in the Texas part of the basin. By September 2015, gas production averaged 131 million m^3 (4,636 million ft^3/day).

In 1995, the U.S. Geological Survey (USGS) estimated that remaining reserves in the basin are 15 billion m^3 (100 billion BBL) of oil-in-place. The EIA (US Energy Information Administration) estimated there is 510 billion m^3 (18 trillion ft^3) of gas remaining.

The major hydrocarbon reservoirs are high porosity limestone, dolomite, and sandstone. However, advances in horizontal drilling and hydraulic fracturing have expanded production to unconventional reservoirs such as oil shales (e.g. the Wolfcamp Shale).

Petroleum source rocks in the Permian Basin include Ordovician, Devonian, Mississippian, Pennsylvanian, and Permian (Wolfcampian [299–280 Ma], Leonardian [280–271 Ma], and

Guadalupian [271–260 Ma]) age rocks. Most of these systems contain oil-prone organic material. While most source rock is organic-rich shale, the Bone Spring is a limestone to marly shale (Jarvie et al., 2017).

The Permian Basin is also a major source of potash, which is mined from bedded deposits in the Permian Salado Formation. Sylvite was discovered in drill cores in 1925, and production began in 1931. Halite (salt) is a by-product of potash mining.

Delaware Basin

As previously mentioned, the Delaware Basin occupies the western third of the Permian Basin. We will be driving along the northern flank of the Delaware Basin, the larger of the two major basins in the foreland of the Ouachita–Marathon Thrust Belt. The Delaware Basin covers roughly 2.6 million ha (6.4 million ac) and contains approximately 7,600 m (25,000 ft) of siltstone, sandstone, and carbonate deposits (Cook, 1966; Jarvie et al., 2017; Keller et al., 1980; Sutton, 2014; Wikipedia).

As you drive to Carlsbad Caverns from the town of Carlsbad, you pass through oil fields that are producing from the buried, basinward extension of the Capitan Reef carbonates that are exposed in Carlsbad Caverns and the Guadalupe Mountains.

Geology of the Caverns

Carlsbad Caverns National Park is in the Guadalupe Mountains of southeastern New Mexico. The park contains part of the Capitan Reef, one of the best-preserved Permian reefs in the world.

During the Permian, between 265 and 250 Ma, the area was located along the coast of a shallow sea. Fossils indicate the reef consisted mainly of sponges, bryozoa, and algae, not coral as in modern reefs. The coastline was subsiding slowly, and the reefs grew upward and outward, keeping pace with the subsidence. The reefs occur as limestone layers up to 549 m (1,800 ft) thick, 3–5 km (2–3 mi) wide, and over 640 km (400 mi) long. Behind the reefs was a lagoon and tidal flats; in front of the reefs was a talus slope formed when storms broke off pieces of the reef, which then tumbled down into the depths. The upper levels of the cavern are developed in the lagoon and tidal flat zone; the large caverns and lower levels are developed in the reef and reef talus.

As the sea slowly evaporated, salt and gypsum were deposited. The entire area was then buried under thousands of meters of new sediments. During the Sevier/Laramide Orogeny, the Guadalupe Mountains were uplifted at least 3,000 m (10,000 ft). Erosion began exposing the old reef. Uplift and faulting created fractures in the limestone. Rain is naturally slightly acidic. As it percolated through fractures in the limestone, it gradually dissolved the rock. As well, hydrogen sulfide gas emanating from oil deposits deep in the subsurface worked upward to the near surface. The hydrogen sulfide combined with oxygen in groundwater to form sulfuric acid. The sulfuric acid aggressively dissolved the limestone deposits to form a labyrinth of caverns extending from the surface to at least 490 m (1,600 ft) deep, making this the second deepest limestone cave system in the United States. The Big Room, the main cavern, is 1,220 m (4,000 ft) long, 191 m (625 ft) wide, and up to 78 m (255 ft) high. The Big Room is the fifth largest chamber in North America and the 28th largest in the world. The presence of gypsum in the caverns confirms this process, as it is a by-product (calcium sulfate) of the reaction between sulfuric acid and limestone. Blocks of gypsum can be seen in the floor of the Big Room. Most of the 119 known caves had been formed by 4–6 million years ago.

Over the past 2–4 million years, the mountains were uplifted again, raising the caverns above the water table. This allowed two things to happen. First, without the support of groundwater, the caves began to collapse. Eventually, this led to an opening to the outside and air filling the caverns. Once this happened, water from rain and snowmelt that made its way to the caverns was able to deposit dissolved calcite as it dripped from the ceilings. This led to precipitation of stalactites and stalagmites. This process would have been more active during the last ice age, when the climate was wetter. Due to the arid climate, few speleothems (cave deposits) are actively growing today.

History

In 1898, a teenager named Jim White discovered the cavern while looking for the source of bats emerging from the cave. He explored and named many of the rooms, including the Big Room, New Mexico Room, Kings Palace, and others. He also named many of the cave's more prominent formations, such as the Totem Pole, Witch's Finger, Giant Dome, and Bottomless Pit.

In October 1923, President Coolidge signed a proclamation establishing Carlsbad Cave National Monument to preserve "… a limestone cavern known as the Carlsbad Cave, of extraordinary proportions and of unusual beauty and variety of natural decoration…."

On May 14, 1930, the United States Congress established Carlsbad Caverns National Park. In 1932, the national park opened a new visitor center building with two elevators for visitors to access the caves. The new center included a cafeteria, museum, ranger stations, book store, and first-aid area.

In November 1978, the Carlsbad Caverns Wilderness southwest of the national park was established by President Carter.

Entrance to Carlsbad Caverns.

Visit

Carlsbad Caverns has two entries on the National Register of Historic Places: The Caverns Historic District and the Rattlesnake Springs Historic District. The developed portion around the cave entrance is The Caverns Historic District. A detached part of the park, Rattlesnake Springs Picnic Area, is a natural oasis with landscaping, picnic tables, and wildlife habitats. Roughly two-thirds of the park has been set aside as wilderness.

About 410,000 people visit the caverns annually. The park contains over 119 caves, of which three are open to the public. Carlsbad Caverns is the most famous and is fully developed with lights, paved trails, and elevators. Slaughter Canyon Cave and Spider Cave are undeveloped, except for designated paths for the guided "adventure" caving tours. Over 190 km (120 mi) of cave passage has been explored and mapped.

Three hiking trails and unpaved roads provide access to the backcountry. Camping is permitted in the backcountry of the park if you have a permit from the visitor center. In addition to exploring the caves, you can watch bats emerge from the caves. Optimal viewing normally occurs in July and August when the current year bat pups first join the flight of adult bats. Seventeen species of bats live in the park, including a large number of Mexican free-tailed bats.

The natural entrance to the caverns is also known for its colony of cave swallows, possibly the world's largest. Rangers host star parties at night throughout the year.

Travertine formations, Carlsbad Caverns.

Entrance Fee: see the website for current fees.
Website: https://www.nps.gov/cave/index.htm

Ranger-Guided Tours: Reservations are highly recommended for all ranger-guided tours. To make reservations, call 877-444-6777 or go to Recreation.gov. Children under the age of 4 are not permitted on any ranger-guided tours. Other age limits apply depending on the tour.

Operating Hours and Seasons:

The visitor center and cavern are closed on Thanksgiving, Christmas, and New Year's Days.

September 4, 2018–May 24, 2019

Carlsbad Caverns Visitor Center Hours: 8 am–5 pm
Last entrance ticket sold: 3:15 pm

Hike INTO the cavern: 8:30 am–2:30 pm
Last time to hike OUT of the cavern: 3:30 pm (complete hike out by 4:30 pm)
Elevator service INTO the cavern: 8:30 am–3:30 pm
Last elevator OUT of the cavern: 4:45 pm

May 25, 2019–September 2, 2019

Carlsbad Visitor Center Hours: 8 am–7 pm
Last entrance ticket sold: 4:45 pm
Hike INTO the cavern: 8:30 am–3:30 pm
Last time to hike OUT of the cavern: 5 pm (complete hike out by 6 pm)
Elevator service INTO the cavern: 8:30 am–5 pm
Last elevator OUT of the cavern: 6:45 pm

Mailing Address: 3225 National Parks Highway, Carlsbad, NM 88220
Physical Address: Carlsbad Caverns National Park, 727 Carlsbad Caverns Highway, Carlsbad, New Mexico 88220
Phone: (575) 785-2232

Carlsbad Caverns to McKittrick Canyon Visitor Center: *Return to White City and turn right (southwest) on US-62/US-180; drive to McKittrick Road; turn right and drive to* ***Stop 15.1, McKittrick Canyon Visitor Center and trail*** *(31.977489, −104.752108), for a total of 61.6 km (38.3 mi; 44 min).*

15 GUADALUPE MOUNTAINS NATIONAL PARK

The Guadalupe and Delaware Mountains, in particular, contain some of the finest outcrops of reef and reef-related rocks in the world. The Western Escarpment provides outstanding exposures of the transition from shallow-water deposits to deep-water deposits. Some 3.2 km (2 mi) of Permian strata are exposed in the Guadalupe Mountains.

Depositional environments from far-back-reef to deep basin can be seen in outcrops with little or no structural deformation. The reef complex is cut by a series of deep canyons, especially McKittrick Canyon, that provide outstanding cross-sectional views of the lateral and vertical sequence of depositional environments through time. Geologists from around the world come here to admire and study the ancient reef complex. These exposed reefs are a direct analog to oil-producing Permian reefs here and elsewhere in the world.

Geology

During the Permian (299–252 Ma), the landmass of Pangea had not yet split apart and New Mexico and Texas occupied the southwestern margin of this continent about 10° north of the paleo-equator. The Delaware Basin was located in the foreland (in front) of the Ouachita-Marathon Fold-Thrust Belt, previously formed when the southern continent of Gondwana had collided with Laurasia during Pennsylvanian time. By earliest Permian time, the subsiding Delaware Basin extended over 26,000 km^2 (10,000 mi^2). The Hovey Channel connected the Permian Ocean with the isolated Permian/Delaware Basin. The Delaware Sea (think of the Black Sea today) occupied a basin about 240 by 120 km (150 mi by 75 mi). About 3.5 km (12,000 ft) of Permian strata have been measured in outcrops from the Guadalupe Mountains, on the western side of the Permian Basin (Scholle, 2000; Bebout et al., 2007, 2009; Standen et al., 2009; Stieb, 2015; National Park Service, 2015).

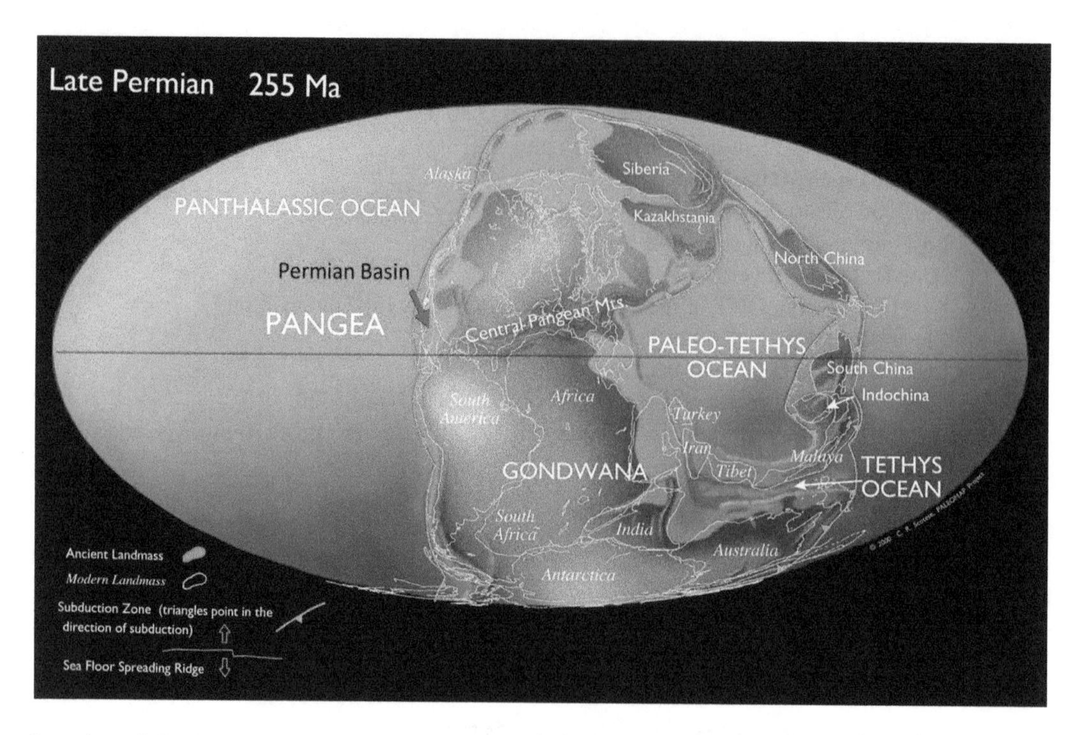

Location of the Permian Basin during Permian time (arrow). The basin was a shallow sea north of the suture zone between Gondwana and Laurasia. (From Scotese, C.R. 2001. *Atlas of Earth History*, Volume 1, Paleogeography, PALEOMAP Project, Arlington, TX, 52 pp., with permission; www.scotese.com/earth.htm.)

Rock exposures in Guadalupe Mountains National Park comprise reef, back-reef, fore-reef, and basin sediments. The reef, a partially submerged mound formed by accumulation of plant and animal skeletons, is now the Capitan Limestone. The Capitan reef is a massive, fine-grained limestone containing fossils of algae, sponges, and bryozoans. There was relatively low wave and current activity in the back reef, an area between the reef and shoreline. Only muds were deposited this area, and the water was often stagnant and turbid, and had high salinity. Brachiopods, crinoids, and fusulinids (a Pennsylvanian-Permian foraminifera [plankton] about the size and shape of rice grains) were common.

Waves battered the Capitan Reef front, causing large fragments of the reef to slide downslope. Thus, the fore-reef consists of debris that extended into the basin. In addition to debris, the fore-reef also contained lime mud and fossils, such as trilobites, brachiopods, sea urchins, algae, and bryozoans.

The basin in front of the reef was nearly a kilometer (half mile) deep. The sediments that washed into the basin during the building of the Capitan Reef later became thin black limestones separated by thicker beds of fine-grained sandstone and siltstone. The black limestone contains the remains of the dead plants and animals that settled to the bottom of the sea. Most of the organic matter was buried and preserved. Heat and pressure changed the organic matter to oil and gas over millions of years.

The Capitan Reef extended almost 640 km (400 mi) along the margins of the Delaware Sea. The basin temporarily stopped subsiding in the Leonardian epoch, near the start of mid-Permian time. Small carbonate shoals developed along its margin, along with small discontinuous patch reefs just offshore. The first formation that resulted was the Yeso, which consists of alternating beds of dolomite, limestone, gypsum, and sandstone. Thin beds of Bone Spring Limestone accumulated as limy ooze in the stagnant, deepest part of the basin.

Subsidence began again in mid-Permian time, and by the Guadalupian epoch, the patch reefs had grown larger. Sediments deposited close to the shore included dolomites of the San Andres Formation and patch reefs of the Brushy Canyon Formation. Rapid subsidence of the basin commenced in the middle Guadalupian, resulting in the Goat Seep Reef. Subsidence of the basin stopped for good in late Guadalupian time.

Sea level dropped as sedimentation continued to fill the Delaware Basin into the Ochoan epoch (a division of the Permian between 260–251 Ma). The basin was sporadically cut off from the sea, and the resulting brine became the deep-water evaporites (gypsum, calcite, and salt) of the Castile Formation. As the basin filled, rivers deposited the red silt and sand of the Rustler and Dewey Lake Formations.

After filling in and being buried, the Permian Basin region was part of a stable, largely non-depositional region during most of Mesozoic time.

Beginning about 75 million years ago, during the Sevier/Laramide Orogeny, compression along the western margin of North America caused west Texas and southern New Mexico to slowly elevate.

Near-vertical Basin-and-Range normal faulting began in this area about 26 million years ago. The western edge of the Guadalupe Mountains has been lifted more than 3 km (2 mi) from its original position below sea level. The uplift caused an eastward dip of 5°–10° in the elevated block. The area down-dropped to the west is now under Salt Flats.

History

Artifacts, such as spear tips, knife blades, basket work, and potshards, indicate humans have been in the Guadalupe Mountains at least 12,000 years. People first came at the end of the last ice age, when the climate was wetter. Mescalero Apaches roamed the area when the Spanish appeared around 1550. Apache and Spanish legends about gold and silver hidden in the mountains eventually drew American prospectors to the area.

In 1858, Pinery Station was constructed for the Butterfield Overland Mail near Pine Springs. The Overland Mail traveled over Guadalupe Pass, 1,687 m (5,534 ft) above sea level. During the winter of 1869, Lt. Cushing of the U.S. Army led his troops into the Guadalupe Mountains to stop Indian raids: he destroyed two Mescalero Apache camps. The Mescalero made their last stand in the Guadalupe Mountains, but by 1890, they had either been killed or forced onto a reservation.

Felix McKittrick was one of the first Americans to settle in the Guadalupe Mountains; he worked cattle during the 1870s. In 1921, Wallace Pratt, a geologist for Humble Oil, was impressed by the beauty of McKittrick Canyon and bought the land. He later donated about 24 km^2 (6,000 ac) of McKittrick Canyon to Guadalupe Mountains National Park.

Guadalupe Mountains National Park was formally opened to the public in September 1972.

Visit

Things to do in the park include camping, hiking the 138 km (86 mi) of trails, horseback riding, driving off road, and picnicking.

Entrance Fee: See the website for current fees.
Website: www.nps.gov/gumo/index.htm
Hours: 8:00 am to 4:30 pm
Address: 400 Pine Canyon, Salt Flat, TX 79847
Phone: (915) 828-3251

STOP 15.1 MCKITTRICK CANYON TRAIL

McKittrick Canyon is a day-use area; the entrance gate is open from 8:00 am to 4:30 pm, and open until 6:00 pm during daylight savings time. Much of the following discussion is from trail guides by Brown and Loucks (2007) and Bentley (2014).

Most of the reef that rims the Delaware Basin is buried; faulting, however, uplifted and exposed the reef in the Guadalupe, Apache, and Glass Mountains. The west flank of the Guadalupe Mountains is a fault scarp that exposes the reef and its basin equivalents in cross-sectional view. The east side of the Guadalupe Mountains reveals a shelf-to-basin depositional profile. McKittrick Canyon cuts into the carbonate platform and provides cross-sectional views. McKittrick Canyon exposes nearly the complete platform section.

The Permian Reef Geology Trail in the mouth of McKittrick Canyon traverses 610 vertical meters (2,000 ft) of Permian (upper Guadalupian stage) facies through one of the world's best exposures of a carbonate platform margin. Present-day topography is roughly equivalent to that formed by the ancient Capitan Reef along the edge of the Delaware Basin.

The Capitan Reef and its associated carbonate platform define the margin of the Delaware Basin. Back-reef equivalents of the Capitan Formation include the sandstones, carbonates, and evaporites of the Seven Rivers (oldest), Yates, and Tansill (youngest) formations. Basinward equivalents are the sandstones of the Bell Canyon Formation, and along the edge of the basin, the Hegler (oldest), Pinery, Rader, McCombs, and Lamar (youngest) carbonate members.

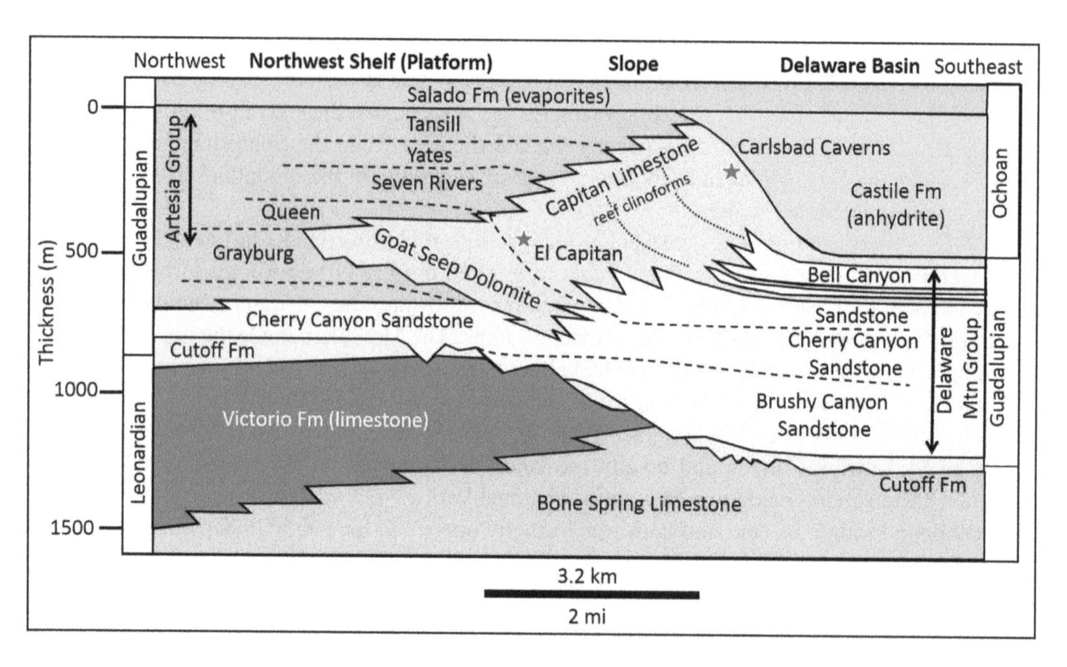

Relationship of Permian strata in the Guadalupe Mountains. Stars indicate the stratigraphic position of our stops. (Modified after SEPM, 2013.)

The Capitan Reef is the youngest of a series of shelf-margin complexes developed around the Delaware Basin. The Capitan Reef separates shallow-water deposits to the northwest from deep-water deposits to the southeast. The 16–24 km (10–15 mi) wide reef tract is basinward of a broad, restricted lagoon evaporite and beach (shoreface) sandstone-siltstone-shale tract. Seaward of the reef tract exposed along the trail are units deposited in shelf crest and outer shelf, reef, slope, and toe of slope environments. Deep-water deposits southeast of the reef tract are siltstone to fine-grained sandstone with interbedded, basinward-thinning limestones of the Bell Canyon Formation (Bebout and Kerans, 1993).

View northeast to the Capitan Reef limestone showing bedding that prograded south (deposition moved south) into the Permian Basin, McKittrick Canyon, Guadalupe Mountains National Park. View northeast.

McKittrick Canyon to Pine Springs Canyon: *Return to US-62 and turn right (southwest) and drive to Pine Canyon Drive; turn right and drive to* ***Stop 15.2, Pine Springs Visitor Center*** *(31.893654, −104.822209), for a total of 19.2 km (12 mi; 17 min).*

STOP 15.2 PINE SPRINGS VISITOR CENTER

The Pine Springs Visitor Center provides information regarding park geology, hiking trails, park maps, park newspaper, and backcountry permits. A bookstore and shop is also located in the visitor center where you can find books, hiking essentials, water bottles, and collectibles.

This is a great little visitor center. It contains displays of plants and animals, a geological timeline, and an electronic ranger touch-screen display for you to use to identify plants and animals. The theater offers a 12-min film about the Guadalupe Mountains that highlights the geology, history, and importance of this park.

Starting in the parking lot behind the visitor center, a trail leads up Pine Springs Canyon. You start in the Permian Cherry Canyon Formation (sandstone, siltstone, limestone), pass through the Bell Canyon Formation (mostly sandstone, some limestone), and end in reef limestone of the Capitan Formation. A 6.4 km (4 mi) round-trip hike takes you to a scenic slot canyon in the Capitan Limestone at Devils Hall.

Visit

Address: The Pine Springs Visitor Center is located at Pine Springs and can be accessed via U.S. Highway 62/180 between Carlsbad, New Mexico, and El Paso, Texas.

Hours: The Pine Springs Visitor Center is open daily except Christmas; hours are 8:00 am to 4:30 pm.

Phone: (915) 828-3251

Pine Springs Canyon and Permian reef strata. (Photo courtesy of Fredlyfish4, https://commons.wikimedia.org/wiki/File:Pine_Spring_Canyon_Fall.JPG.)

Pine Springs Canyon to El Capitan Lookout: *Return to US-62 and turn right; drive 7.4 km (4.6 mi; 5 min) to* ***Stop 15.3, El Capitan Lookout*** *on the right (31.854017, −104.844558).*

STOP 15.3 EL CAPITAN LOOKOUT

El Capitan Peak (2,464 m, or 8,085 ft) and adjacent Guadalupe Peak (2,667 m, or 8,749 ft) are the two tallest mountains in Texas. The massive rock faces are composed of the Capitan Limestone, ancient reefs. Below the cliffs of Guadalupe Peak and El Capitan are Cherry Canyon Formation siltstones and sandstones and the Brushy Canyon Formation sandstones. These sandstones and siltstones were deposited in submarine channels in the Delaware Basin.

El Capitan in Guadalupe Mountains National Park, Texas.

Once we cross the Border Fault Zone along the west side of the Guadalupe Mountains, we are once again in the Basin-and-Range Province. Driving west from the Guadalupe/Delaware Mountains, we enter Salt Flat Basin, a topographic depression. As we approach the Hueco Mountains, we enter the Orogrande Basin, a former arm of the sea that contains Mississippian to Permian-age rocks.

El Capitan to Southern Outlier: *Drive west on US-62/US-180 to Desert Storm Rd in Butterfield; turn left (south) and drive 3.6 km (2.2 mi) on gravel road and turn right (west); drive 0.48 km (0.3 mi) on unnamed road and turn left (south); drive 0.6 km (0.4 mi) and turn right (west); drive 0.15 km (0.1 mi); turn left (SW) and drive 0.63 km (0.4 mi) to* ***Stop 16.1, Southern Outlier View Point*** *(31.787805, −106.108258), for a total of 136 km (84.5 mi; 1 h 20 min). BE EXTREMELY CAREFUL AS YOU APPROACH THIS STOP: IT IS VERY EASY TO GET STUCK IN THE SAND HERE. 4-WHEEL DRIVE AND A SHOVEL ARE RECOMMENDED.*

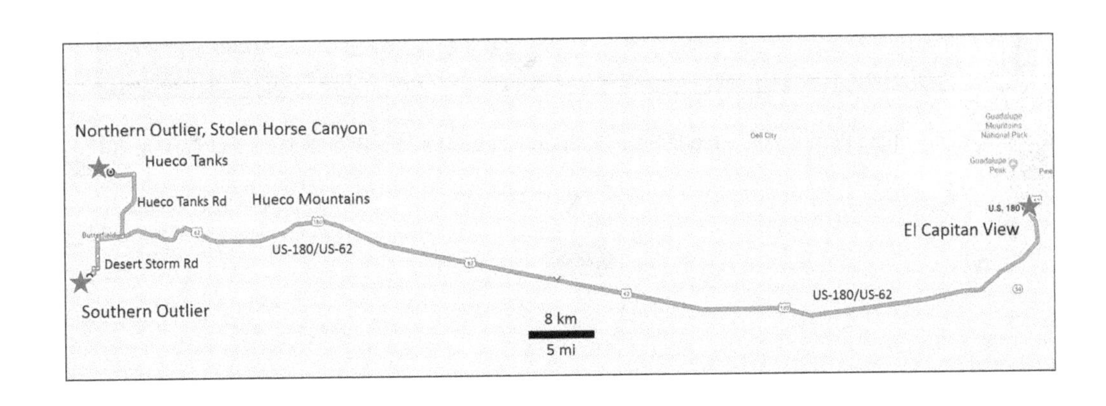

16 HUECO GROUP, OROGRANDE BASIN

The contemporary Hueco Bolson is a Quaternary Basin-and-Range fault-bounded valley (graben) imposed on the Mid-Mississippian-Early Permian Orogrande Basin. The Orogrande Basin was a north-northeast-south-southwest-oriented relatively shallow continental margin basin developed on the southern edge of ancestral North America. The basin is filled with up to 2,000 m (6,000 ft) of sediments mainly from the Pedernal Uplift to the northeast. Starting with Mississippian limestone, the section changes upward to sandstone, siltstone, and shale deposition during the Pennsylvanian. As the sediment source was lowered by erosion, the basin became sediment-starved, with interbedded clastics and limestones. Carbonates increased upward in the section until ultimately the basin was flooded by sandstone and mudstone red beds of the early Permian Abo Formation and buried beneath gypsum of the mid-Permian Yeso Formation.

These stops are western outliers to the main Hueco Mountains. The rocks are almost entirely early Permian Hueco Limestone, with the exception of Hueco Tanks, which is a Tertiary syenite porphyry intrusive.

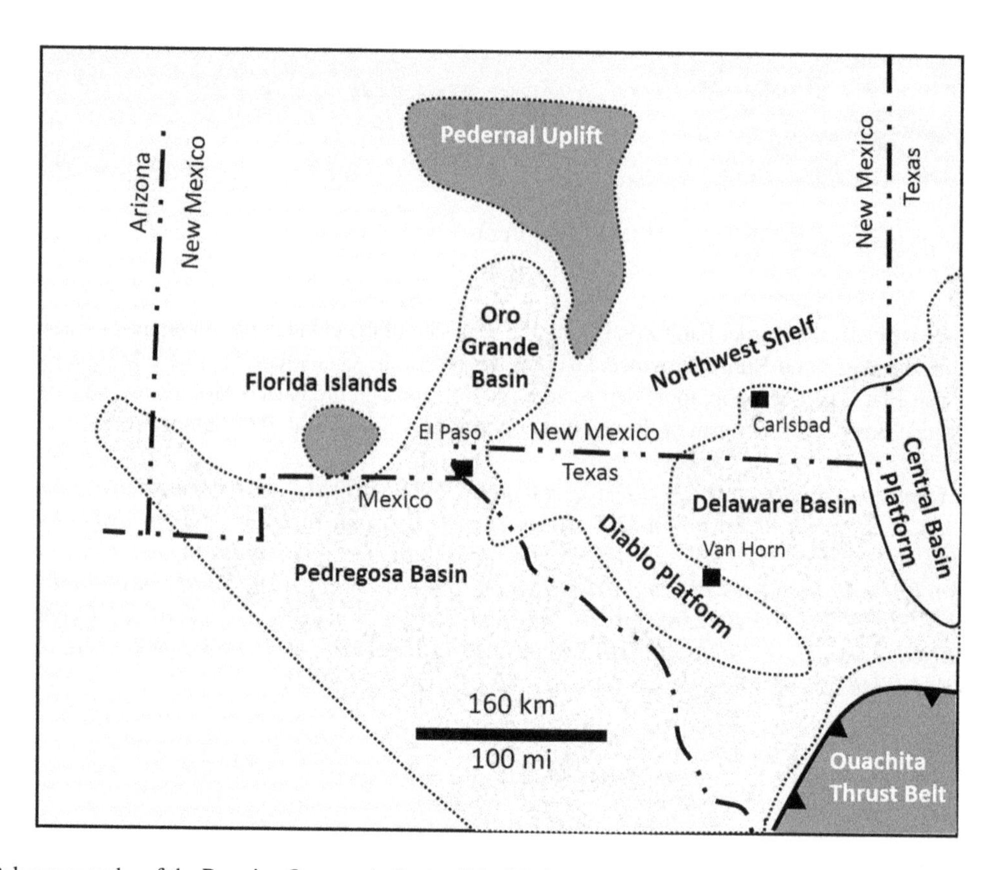

Paleogeography of the Permian Orogrande Basin. (Modified after Jordan, 1975; Kerans and Janson, 2012.)

Age	Group/Fm		Thickness
Quaternary	Alluvium		
Permian	Hueco Group	Alacran Mtn Fm Ls	190 m
		Deer Mtn red shale mbr	
		limestone	
		Cerro Alto Limestone	140 m
		Hueco Canyon Fm (limestone)	200 m
		Powwow mbr	
	Magdalena Limestone	Upper Division (limestone & marl)	150 m
Penn-sylvanian		Middle Division	90 m
		Lower Division (limestone)	150 m

Stratigraphy of the Hueco Basin. (Modified after Kerans and Janson, 2012.)

STOP 16.1 SOUTHERN OUTLIER

From this stop, you look directly west at the ridge. This 1.6km (1 mi) long exposure of the east side reveals 60m (200ft) of Hueco Group, including upper slope to lower shelf-margin phylloid algal mound facies, and mid-slope to toe-of-slope turbidites and debris flows. A phylloid algal-mound is a local thickening of limestone due to the presence of leaflike or phylloid algae. As mentioned previously, these algae were a primary component of Permian reefs. From base to top of the slope, you see (1) the thin-bedded, low-angle, southwest dipping cherty mudstone that represents toe-of-slope deposits; (2) lens-shaped laterally discontinuous carbonates; (3) massive cliff-former at the south end of the ridge consisting of high-angle clinoforms (basinward-inclined strata characteristic of a basin margin) dipping southwest; and (4) at and just below the ridgeline, a massive phylloid algal mound and time-equivalent toe-of-slope debris flows with transport direction to the west. These are an excellent analog for Early Permian (Wolfcampian) reservoirs in the Permian Basin (Kerans and Janson, 2012).

Permian Hueco Reef, Southern Outlier. View west.

Southern Outlier to Northern Outlier: *Return on Desert Storm Road to US-62/US-180 and turn right (east); drive3.2 km (1.9 mi) to Hueco Tanks Road/TX-2775 and turn left (north); drive north on Hueco Tanks Road to Hueco Mountain Road and turn left (west); take Hueco Mountain Road to Old Butterfield Trail and turn left (west); take Old Butterfield Trail to Tunisia Street and turn right (north); drive to Bass Court and turn right (east); drive to* ***Stop 16.2, Northern Outlier*** *(31.904321, −106.080153) for a total of 20.8 km (12.9 mi; 27 min).*

STOP 16.2 NORTHERN OUTLIER

From this stop, we look west toward Stolen Horse Canyon and see a massive, offlapping clinoform package prograding to the southwest (a sequence of inclined strata advancing basinward). This is the margin of a Lower Permian carbonate platform.

Near-horizontal strata in the upper part of the ridge are shelf-top deposits that are laterally equivalent to the clinoforms. Near the base of the ridge, at the mouth of the canyon, are massive phylloid algal mounds. The slope-forming parts of the ridge consist of thin-bedded muddy toe-of-slope deposits (Kerans and Janson, 2012).

Similar strata form important oil and gas reservoirs in the Permian Basin.

Phylloid Algal Mounds, Permian Hueco Formation, Northern Outlier. View west.

Just east of this stop is Hueco Tanks, "the best bouldering in the world." This is an Oligocene granitic intrusion. The number of people that enter in a day is limited, and you need a permit from Texas State Parks Department. Two options are available: guided and self-guided tours.

- Guided tour: You must book tours a minimum of 1 week in advance by calling (915) 849-6684.
- Self-guided visit: Permits are issued for 70 people to access the North Mountain area each day. To make sure you get in, reserve permits up to 90 days before your visit by calling (512) 389-8911.
  Website: https://tpwd.texas.gov/state-parks/hueco-tanks

Hueco Tanks, view northeast.

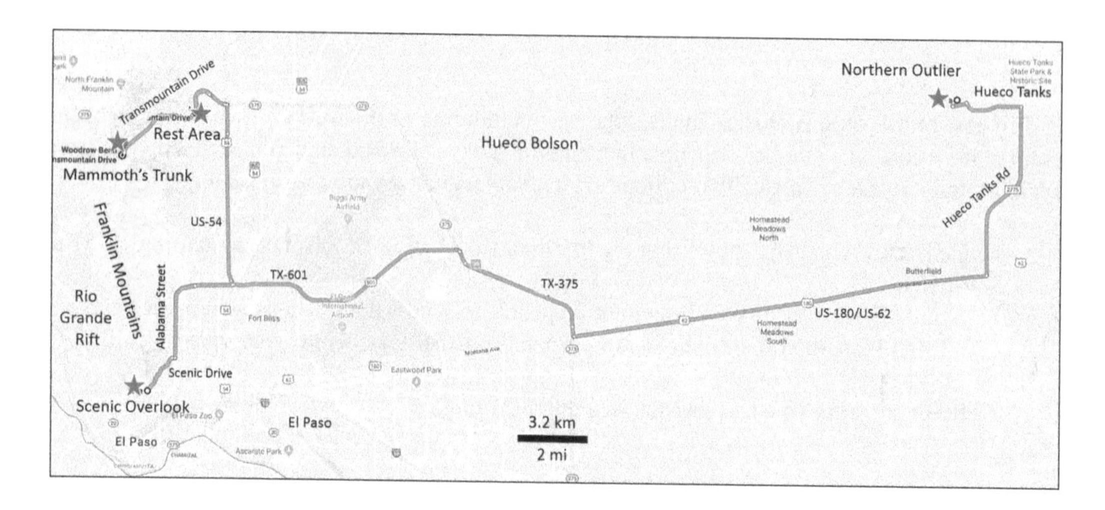

Northern Outlier to Scenic Drive, Franklin Mountains: *Return to US-62/US-180 and turn right (west); drive to TX-601 Spur (TX-375/Purple Heart Memorial Highway) and turn right (northwest); merge left (west) onto TX-601 and drive to Fred Wilson Ave; continue straight on Fred Wilson Avenue to Alabama Street; bear left onto Alabama Street and continue to Scenic Drive; turn right (west) on Scenic Drive and drive to* ***Stop 17.1, Murchison Rogers Park Scenic Drive Overlook*** *(31.782592, −106.479727), for a total of 58.4 km (36.3 mi; 50 min).*

17 FRANKLIN MOUNTAINS AND RIO GRANDE RIFT

As you approach El Paso from the east, you are entering the Rio Grande Rift, a north-south zone extending from just south of here to Leadville, Colorado, along which the continent is tearing itself apart. The rift is part of the east-west crustal extension happening in the western United States, part of the same spreading that caused the Basin-and-Range Province in Nevada, Utah, Arizona, and New Mexico over the past 20–35 million years and that continues today. The process of rifting involves uplift and spreading. This causes the center to drop along faults, leaving mountains on either side. Rifting caused the Hueco Mountains, the Hueco Bolson (or enclosed basin), the Franklin Mountains, and the Mesilla Bolson west of El Paso (Cornell, online). The Franklin Mountains are a west-tilted, north-trending uplifted fault block (horst) along the east side of the Rio Grande Rift. Mountain building may be ongoing, as shown by recent fault scarps along both sides of the Franklin Mountains. About 9,000 m (30,000 ft) of vertical offset occurs along the boundary fault zone east of North Franklin Mountain. More than 2,750 m (9,000 ft) of recent valley fill sediment lies beneath El Paso International Airport, and 3,660 m (12,000 ft) of basin fill has been measured under the Mesilla Bolson (Cornell, online).

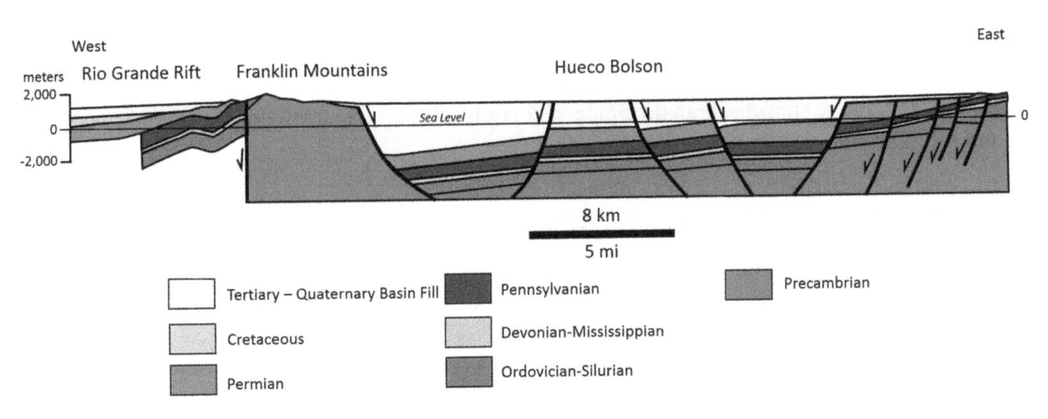

Cross section from the Rio Grande Rift to the Hueco Bolson. (Modified from Collins and Raney, 2000.)

The oldest rocks in the El Paso area were deposited along the shore of a shallow tropical sea between 1.2 and 1.4 Ga. The oldest layers are in the Castner Formation, lime-rich muds that were later metamorphosed to marble. Overlying the Castner Marble is a thin basalt flow known as the Mundy Breccia. It is, in turn, overlain by thick quartz sands that were metamorphosed to the Lanoria Quartzite. Around 1.1 Ga granitic rocks, the Red Bluff Granite, intruded the area along with eruption of ash-flow tuffs, the Thunderbird Group (Cornell, online).

These Precambrian rocks were then eroded until about 500 Ma (the 'Great Unconformity'), when a rising sea flooded the region. The first sediments deposited above the erosion surface are the Cambrian Bliss Sandstone. For the next 250 million years, from Ordovician to Permian time, this region was part of the proto-North American continental shelf, a low-lying region very close to sea level. During submerged episodes, mainly limestones and dolomites were deposited (Cornell, online); during emergent periods, the area was subject to erosion.

The region was above sea level during most of the Mesozoic. In mid-Cretaceous time, the El Paso area was part of the Chihuahuan Embayment, an arm of the Gulf of Mexico. Shallow marine carbonates were deposited. These sedimentary rocks are exposed along the western slope of the Franklin Mountains and in the Sierra de Juarez just south of here. During the Sevier/Laramide Orogeny, these carbonates were thrust to the northeast, forming the Sierra de Juarez (Cornell, online).

The Campus Andesite and Cristo Rey pluton were emplaced in early Cenozoic time (45–50 Ma; Cornell, online).

History

El Paso del Norte, or "the pass of the north," was where the Spanish Camino Real trail crossed the Rio Grande and passed through a gap between the Sierra de Juarez and the Franklin Mountains. The valley was settled in 1680, mainly by refugees fleeing the Pueblo Indian Revolt in New Mexico.

El Paso was originally on the south side of the river. In 1888, the city south of the border was renamed Ciudad Juarez in honor of the Mexican leader Benito Juarez. The city on the Texas-side formed around Fort Bliss after the Mexican-American War.

Spanish and Mexican settlers avoided the mountains because Apache and Comanche marauders used them as bases to raid the river settlement. After the Indians were removed from the area, outlaws hid there. U.S. Deputy Marshall Charles Fusselman was killed in a shootout with rustlers at Fusselman Canyon in 1890.

Small amounts of tin were produced in the Franklin Mountains. A mine opened in 1909 by the El Paso Tin Mining and Smelting Company ultimately proved unsuccessful, and mining ended in 1915. A total of 160 ingots, each weighing 45 kg (100 lb), of tin were produced.

The legendary Lost Padre Mine is said to be hidden somewhere in the Franklin Mountains. The legend claims that some 300 burro loads of silver were left by Jesuits, who filled in the mine before leaving the area. In other versions, 5,000 silver bars, 4,336 gold ingots, and nine burro loads of jewels were hidden in the shaft by Juan de Oñate, a conquistador and Spanish governor of New Mexico around the year 1600 (Miles).

STOP 17.1 MURCHISON ROGERS PARK SCENIC DRIVE OVERLOOK, EL PASO

As you approach the Scenic Drive Overlook, you are driving through roadcuts in the lower Ordovician El Paso Group carbonates (around 460 Ma). These strata are laterally equivalent to the Ellenburger Formation in central Texas and were part of a carbonate platform that extended along the southern margin of Laurasia all the way to New York. This was part of the "Great American Carbonate Bank" mentioned in our discussion of the Appalachians, Tour 1 of this volume. The El Paso Group contains carbonate mounds, fossil-bearing shallow-water limestone, and dolomite.

The roadcut on the ridge crest to the north contains the contact between the El Paso Group carbonate (Scenic Drive Formation dolomite and limestone) and the overlying Ordovician Montoya Group (Upham Formation dolomite with a basal sandstone).

From Scenic Point, you get a panoramic view of El Paso and Juarez, the Hueco Valley to the east, and the Mesilla Valley to the west. These valleys are normal fault-bounded grabens at the south end of the Rio Grande Rift. They are filled with thousands of meters of sand, gravel, and playa (dry lake) deposits.

The low hills southwest of this stop (by the smelter stack) are Campus Andesite, actually a range of igneous intrusive rocks. Due west is Crazy Cat Mountain, a jumble of Ordovician to Cretaceous-age rocks that is thought to be a gravity slide block.

Ordovician El Paso Group dolomite and limestone, and Montoya Group basal sandstone and overlying dolomite. Scenic Drive parking area, view to the north.

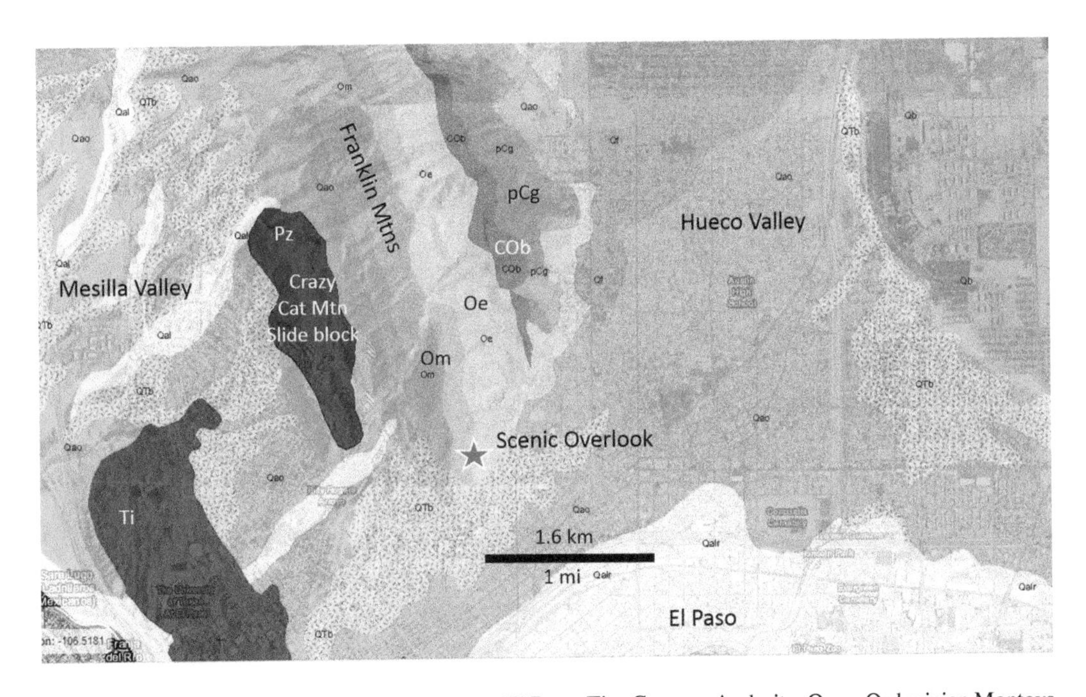

Geologic map of southern Franklin Mountains at El Paso. Ti = Campus Andesite; Om = Ordovician Montoya Dolomite; Oe = Ordovician El Paso Gp (ls, dol, ss); Cob = Upper Cambrian Bliss Sandstone; pCg = Precambrian granite (U.S. Geological Survey and Bureau of Economic Geology, online. https://txpub.usgs.gov/txgeology/).

Scenic Drive to Mammoths Trunk: *Return northeast on Scenic Drive to Alabama Street; turn left (north) and drive straight on Fred Wilson Ave; use left lane to turn left (north) on Gateway Blvd and merge onto US-54; use left 2 lanes to turn left onto Woodrow Bean Transmountain/TX-375; drive to pullout on the left,* ***Stop 17.2, Mammoths Trunk Trailhead*** *(31.877919, −106.492863), for a total of 23.2 km (14.4 mi; 25 min). You have to pass the parking area driving west and make a U-turn on this divided highway to reach the trailhead.*

STOP 17.2 MAMMOTHS TRUNK TRAIL ROADCUT, CENTRAL FRANKLIN MOUNTAINS

Measuring from the intersection with US-54 and TX-375, at 4 km (2.5 mi; 31.894116, −106.470328), dark gray-brown Castner marble is exposed. The outcrop includes stromatolites (layered rocks secreted by cyanobacteria) and black mafic igneous sills. Garnets occur near the top of the unit.

At 4.5–4.8 km (2.8–3.0 mi), a long roadcut exposes granite, but you cannot stop—there is a guardrail.

A poor roadcut in Precambrian Lanoria Quartzite is at 5.3 km (3.3 mi), after the end of the guardrail (31.887787, −106.481231).

The Mammoths Trunk Trailhead parking area is at 7.1 km (4.3 mi). The parking area is in Ordovician El Paso Formation dolomite. The trail to Mammoths Trunk ends in west-dipping Silurian Fusselman Dolomite.

View south toward Mammoths Trunk from near the parking area.

At Smugglers Pass there is a large roadcut in the Precambrian Thunderbird Group. The Thunderbird Group consists of a basal conglomerate overlain by trachyte tuff and rhyolite ignimbrite.

Smugglers Pass roadcut in Precambrian Thunderbird Group volcanic rocks. View west from Mammoths Trunk parking area.

Mammoths Trunk to Rest Area: *Drive east on Woodrow Bean Transmountain Drive/TX-375 for 4.5 km (2.8 mi; 3 min) to rest/picnic area pullout on the right. This is* ***Stop 17.3, Rest Area*** *(31.895347, −106.456886).*

STOP 17.3 REST AREA

In the roadcut across the highway to the north, you see Precambrian Castner Marble intruded by Precambrian granite. The light-colored granite contains xenoliths (literally "strange rocks") of dark Castner Formation, blocks of an original limestone that were engulfed by the granite and metamorphosed to marble. At the west end, the Castner Formation bedding is near-vertical.

Precambrian Castner Marble (dark) and granite (light) along TX-375 looking north from Rest Area.

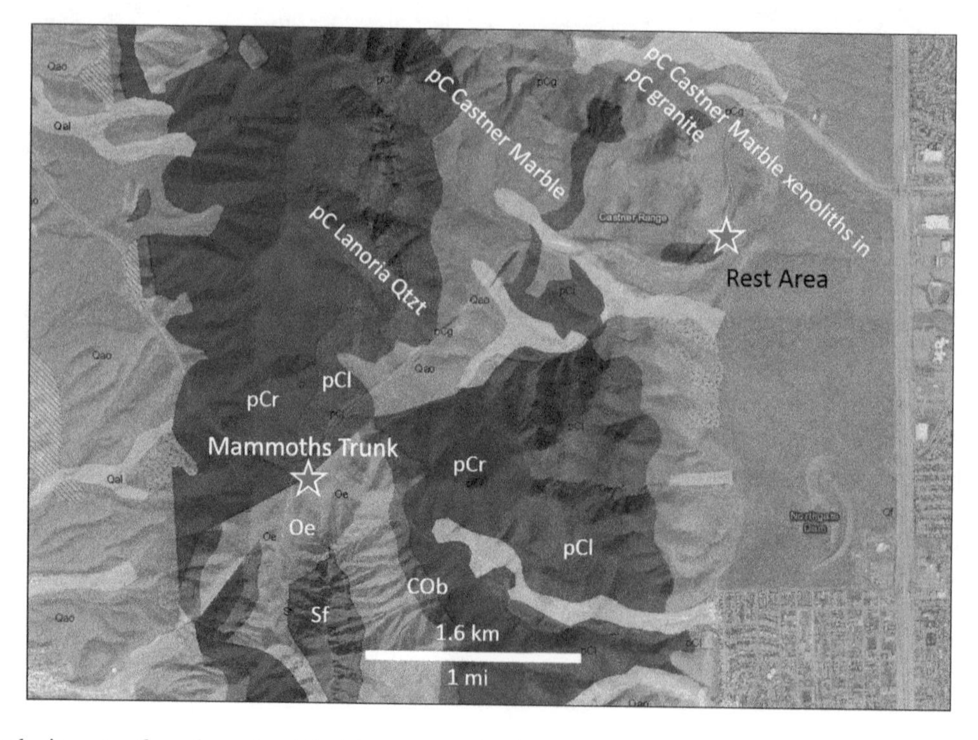

Geologic map of southern Franklin Mountains at El Paso. Sf = Silurian Fusselman Dolomite; Om = Ordovician Montoya Dolomite; Oe = Ordovician El Paso Formation (limestone, dolomite, sandstone); Cob = Upper Cambrian Bliss Sandstone; pCr = Precambrian Rhyolite; pCl = Precambrian Lanoria Quartzite (U.S. Geological Survey and Bureau of Economic Geology, https://txpub.usgs.gov/txgeology/).

Rest Area to White Sands: *Drive east on TX-375 to Gateway Blvd/US-54; turn left (north) on Gateway/US-54 and drive northeast to the outskirts of Alamogordo; turn left (west) on US-70 and drive to Dunes Drive/White Sands Gate; turn right (north) and drive to* ***Stop 18.1, Playa Trail*** *parking (32.793869, −106.212566) for a total of 149 km (92.8 mi; 1 h 31 min).*

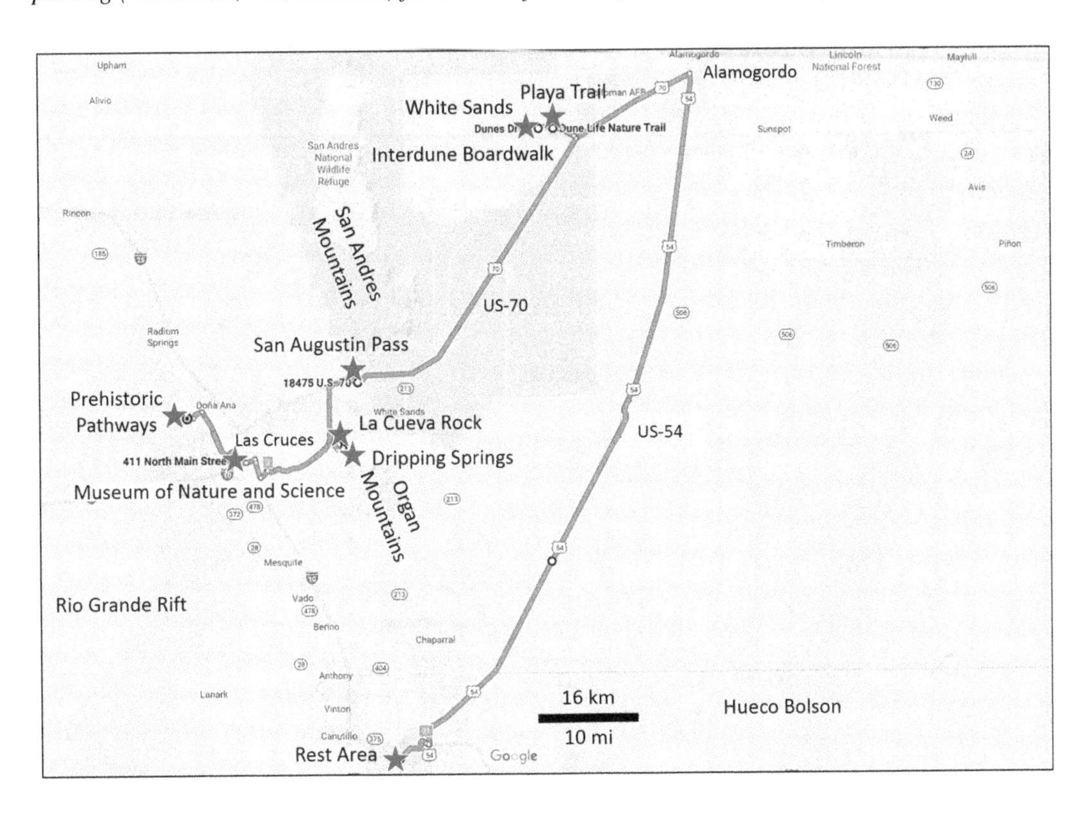

18 WHITE SANDS NATIONAL MONUMENT

Geology

White Sands National Monument contains a portion of the largest gypsum dune field in the world. The entire dune field covers 712 km² (275 mi²). Most of the dune field, about 60%, is located within White Sands Missile Range, which surrounds White Sands National Monument. The national monument hosts about 200 km² (115 mi²) of the dune field.

White Sands lies in the Tularosa Basin, a valley bounded by Basin-and-Range normal faults. The gypsum that forms the white sands was deposited during evaporation of a shallow Permian sea. These evaporite deposits later became the Yeso and San Andres formations. The Sevier/Laramide Orogeny caused the area to be uplifted into a broad arch. Rain and snowmelt dissolved the soluble evaporites and carried them to nearby basins. Around 10 million years ago, crustal rifting began in an east-west direction, forming north-south faults. The Tularosa Basin was dropped down, and the Sacramento Mountains to the east and San Andres Mountains to the west were uplifted along these faults. The Tularosa Basin is a "bolson" (pouch in Spanish) because it has no outlet. Occasionally playa Lake Lucero fills due to rain or runoff from the San Andres Mountains. As the playa lake evaporates, gypsum crystallizes and is deposited on the lake floor. It is later picked up by the wind and carried east to the dune fields.

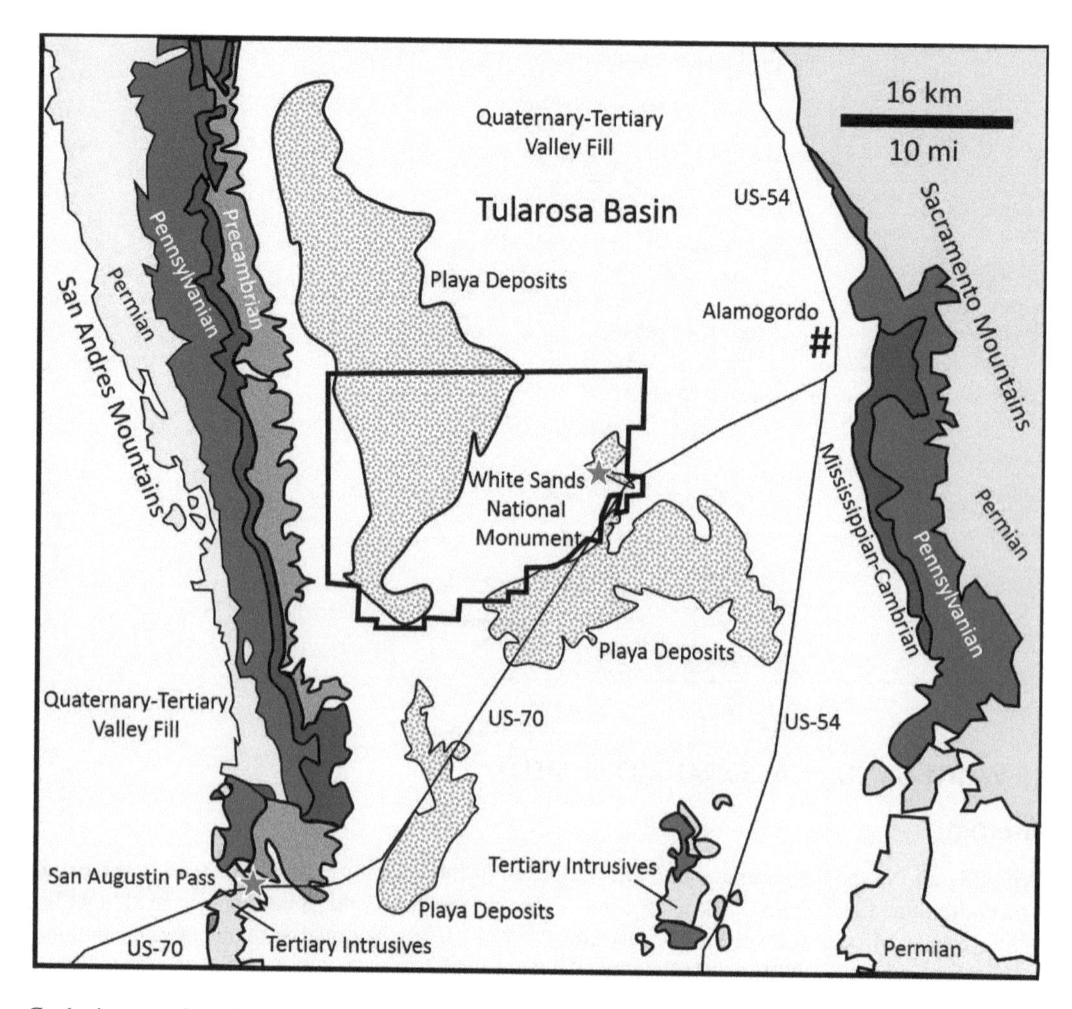

Geologic map of the Tularosa Basin and White Sands National Monument. The range west of the Tularosa Basin is the San Andres Mountains, which become the Organ Mountains south of San Augustin Pass. (Modified after New Mexico Bureau of Geology and Mineral Resources, 2007–2019, https://geoinfo.nmt.edu/tour/federal/monuments/whitesands/geology.html.)

Era	Period	Ma	Events
Cenozoic	Quaternary		Region becomes arid; playa lakes form. Large glacial lakes.
		2.6	
	Neogene		Rio Grande Rift forms. Basins fill with sediments.
		23.0	
	Paleogene		Basin-and-Range extension begins. Laramide mountain building episode ends.
		65.5	
Mesozoic	Cretaceous		Western Interior Seaway extends across area. Laramide Orogeny begins.
		145.5	
	Jurassic		Sevier Orogeny.
		199.6	
	Triassic		Breakup of the supercontinent Pangaea.
		251	
Paleozoic	Permian		Deposition of Permian Yeso Formation rich in gypsum, ultimate source of sand grains at White Sands.
		299	
	Pennsyl-vanian		Formation of Ancestral Rocky Mountains, Ouachita-Marathon Orogeny, and the Orogrande Basin.
		318	

Geologic history of the White Sands area. (Modified after KellerLynn, 2012.)

History

In January 1933, President Hoover proclaimed 57,867 ha (142,987 ac) of gypsum sand in south-central New Mexico as "White Sands National Monument." In the early 1940s, the U.S. government deemed this part of New Mexico to be ideal for military operations. Just after the attack on Pearl Harbor in 1942, the Alamogordo Bombing and Gunnery Range was established. The Alamogordo Army Air Base, also established in 1942, was used for aircrew training during World War II. The first atomic-bomb test detonation was made at the Trinity Site at White Sands. In 1945, White Sands Proving Ground was set aside. In 1958, the Gunnery Range and proving ground were consolidated into the White Sands Missile Range.

Visit

Mailing Address: PO Box 1086, Holloman AFB, NM 88330
Phone: (575) 479-6124
Entrance Fees: see the website for most recent fee information:
Website: www.nps.gov/whsa/planyourvisit/fees.htm
Hours: White Sands National Monument is open daily year-round except for Christmas Day.
Dunes Drive opens at 7:00 am daily except on Christmas Day. Depending on the season, you must exit before 6:00 pm (winter) or 9:00 pm (summer).
The visitor center is generally open from 9:00 am to 5:00 pm (winter) or 7:00 pm (summer).

Note: Due to missile testing in White Sands Missile Range, it is occasionally necessary to close the road into the monument for up to three hours. U.S. Highway 70 between Alamogordo and Las Cruces is also closed during missile testing. Visitors are encouraged to call the monument the day before arrival to confirm hours of operation.

STOP 18.1 PLAYA TRAIL

Length: 800 m (0.5 mi) round-trip.
Completion Time: 30 min
Difficulty: Easy
Restroom: No restroom facility is available

A playa is a dry lake that occasionally fills with rainwater or snowmelt. The playa changes throughout the season. Hike out and see what it looks like, but stay on the trail.

This is not the only playa at White Sands National Monument. Intermittent Lake Lucero, the largest playa at the monument, is the source of the gypsum sand. It covers approximately 26 km^2 (10 mi^2). Lake Lucero is 25 km (15.5 mi) southwest of the visitor center.

This short, level trail leads to a small playa. Interpretive panels explaining local plants and animals are located along the trail. Learn about this landscape that changes with the seasons.

Playa, dunes, and Sierra Blanca in background. View east.

Playa Trail to Interdune Boardwalk: *Continue north on Dunes Drive for 3.1 km (2 mi; 4 min) to parking on the right at* ***Stop 18.2, Interdune Boardwalk*** *(32.793539, −106.239334).*

STOP 18.2 INTERDUNE BOARDWALK

Length: 650 m (0.4 mi) round-trip
Completion Time: 20 min
Difficulty: Easy walk

The Interdune Boardwalk is an elevated boardwalk that leads through the fragile interdune area to a scenic view of the dune field and the Sacramento Mountains. Exhibits along the boardwalk explain the environment, wildlife, and geology.

The boardwalk is fully accessible for wheelchairs and strollers. A shade structure with seating is located midway along the walk. You must stay on the boardwalk. To play in the sand, drive 4.8 km (3 mi) down the road into the heart of the dunes.

Gypsum sand dunes.

White Sands to San Augustin Pass: *Return to US-70 and turn right (southwest); drive 64.2 km (39.9 mi; 40 min) to pullout on right shoulder before a guardrail and across from the San Augustin Pass Overlook. This is* ***Stop 19.1 San Augustin Pass and the Organ Batholith*** *(32.425616, −106.564352).*

Side Trip 3—White Sands to Valley of Fires: *Return to US-70 and turn left (northeast); drive to US-54 and turn right and immediately left to merge onto US-54 north; use left lanes to continue on US-54/US-70 north and drive to Carrizozo; turn left (north) onto East Ave; turn left (northwest) onto US-380 and drive to* ***Side Trip 3, Valley of Fires*** *parking area (33.684608, −105.920457) on the left for a total of 131 km (81.7 mi; 1 h 29 min).*

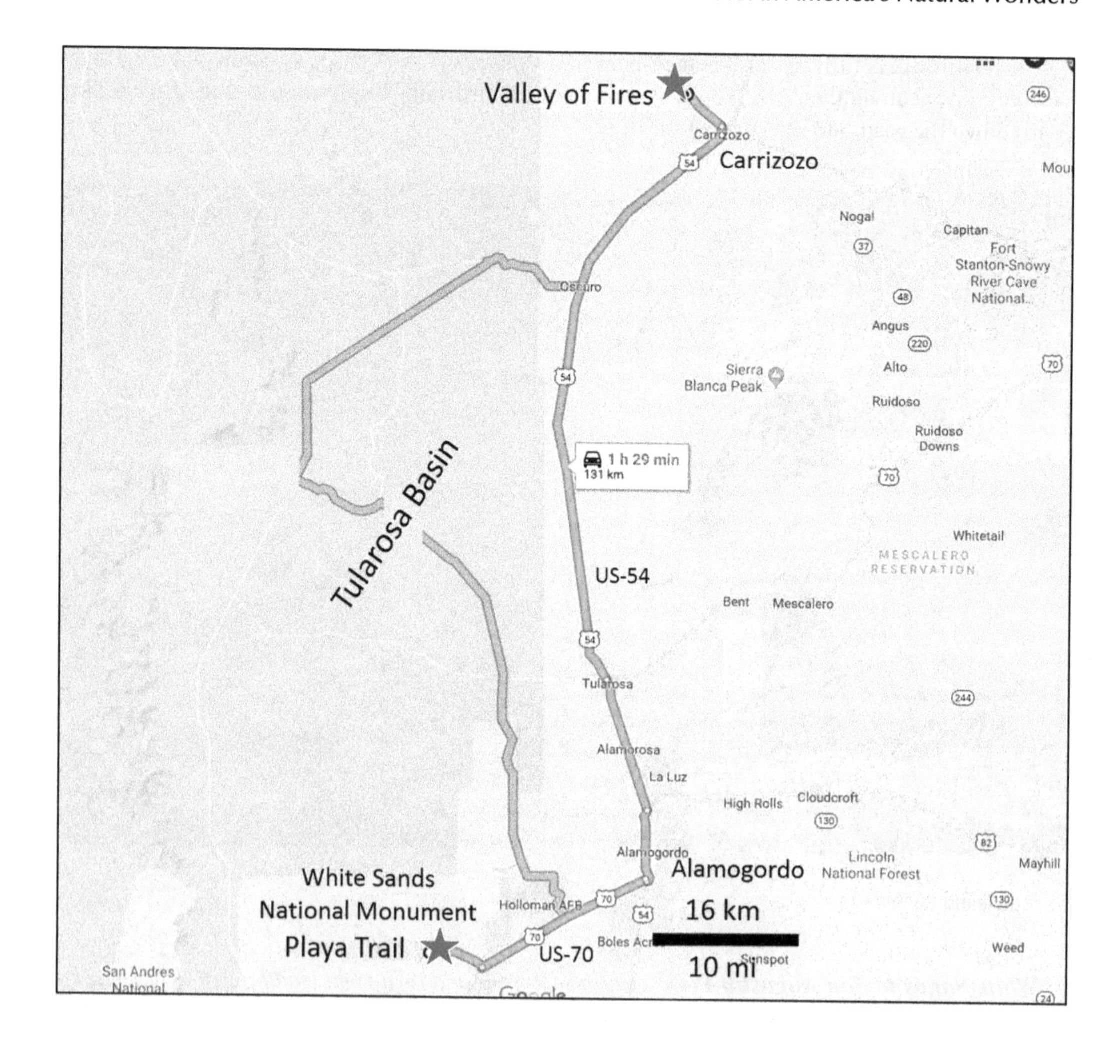

SIDE TRIP 3 MALPAIS AND VALLEY OF FIRES NATIONAL RECREATION AREA

This is bad country. Literally. Valley of Fires National Recreation Area is located adjacent to and within the Carrizozo Malpais (Spanish for 'bad country') Lava Flow in the Tularosa Basin.

Little Black Peak erupted approximately 5,200 years ago and flowed up to 70 km (44 mi) south in the basin. The lava flow is 6.5–10 km (4–6 mi) wide, 49 m (160 ft) thick, and covers 324 km^2 (125 mi^2). It is thought to be one of the youngest lava flows in the continental United States: in fact, it is so recent that plants have hardly had a chance to establish themselves on the black, forbidding rock.

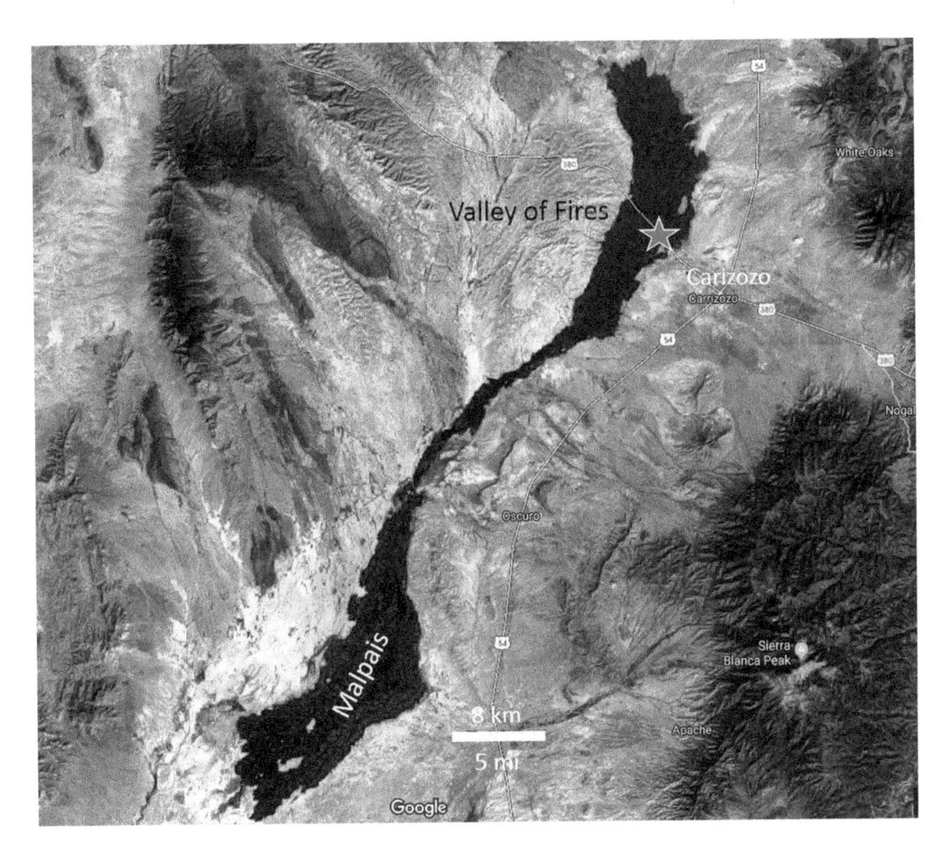

Google Earth image of the Carrizozo Malpais and Valley of Fires. The black slash crossing the image is the malpais, recent basalt lava flows. Imagery © 2019 TerraMetrics.

Valley of Fires. (Photo courtesy of Bureau of Land Management, www.flickr.com/photos/blmnewmexico/7349471372/in/album-72157630075953674/.)

Visit

The recreation area is run by the Bureau of Land Management (BLM). You can hike along an interpretive trail and you can camp, but there is no off-roading. The visitor center has an information center, gift shop, and bookstore.

Address: Roswell Field Office, Valley of Fires Recreation Area, Carrizozo, NM 88301
Phone: 575-648-2241, 575-840-6243
See the website for current day use and camping fees.
Website: www.blm.gov/visit/valley-of-fires

Side Trip 3—Valley of Fires to San Augustin Pass: *Return SE on US-380 to Carrizozo; turn right (south) onto US-54 and drive to Alamogordo; turn right (west) onto US-70 and drive to a pullout on right just before the guardrail and opposite San Augustin Pass Overlook, for a total of 180 km (112 mi; 1 h 46min). This is* ***Stop 19.1 San Augustin Pass and the Organ Batholith*** *(32.425616, −106.564352).*

19 ORGAN MOUNTAINS-DESERT PEAKS NATIONAL MONUMENT

These mountains appear on old Spanish maps as "la Sierra de la Soledad"—the Mountains of Solitude. The current name, Organ Mountains, is thought to refer to the resemblance between the jagged spires of The Needles and the pipes of an organ. The mountains rise nearly 1.5 km (1 mi) above the Mesilla Valley to the west and Tularosa Basin to the east. The Needles, at 2,740 m (8,990 ft), are the range's iconic feature. Climbers come from around the world to ascend the smooth granite walls.

History

Abundant archeological sites indicate that Native Americans have lived and hunted in these mountains for thousands of years. As with the Franklin Mountains to the south, they were largely bypassed by Spaniards and Mexicans on their way to Santa Fe, mainly because the Apache and others use the mountains as a base for raids on settlements.

In the 1870s, Colonel Eugene Van Patten built Van Patten's Mountain Camp and Resort at Dripping Springs. The two-story 14-room hotel attracted a few famous guests, including the lawman Pat Garrett, the Mexican revolutionary Pancho Villa, and possibly the outlaw Billy the Kid. By 1916, the resort was having financial problems and was sold to Dr. Nathan Boyd, who converted it to a sanatorium for the treatment of tuberculosis. Conventional wisdom of the time was that dry, clean air would relieve the disease. The sanatorium ran into difficulties during the 1920s and shut down. Soon after it was sold to rancher A.B. Cox, who used the water for his livestock.

On May 21, 2014, President Obama designated the Organ Mountains-Desert Peaks National Monument. The purpose was to permanently protect more than 201,000 ha (496,000 ac) of unique desert landscape, geology, wildlife, flora, and cultural and historical sites. The Organ Mountains form only one part of the many parcels included in this National Monument.

Granite spires of the Organ Mountains. View east.

Geology

The San Andres, Organ, and Franklin Mountains are part of a 240km (150 mi) long north-south-trending series of west-tilted ranges along the east edge of the Rio Grande rift. The northern half of the Organ Mountains consists of Precambrian granite, Paleozoic sedimentary rocks, Tertiary intrusions and volcanic rocks, and Tertiary-to-Recent gravel and alluvium. Quartz monzonite intrusions of the Organ Mountain Batholith are the dominant rock type. The Organ Batholith is the upper part of a magma chamber that erupted as pyroclastic flows (hot gas and volcanic fragments) and lavas about 33 million years ago. The pyroclastic rocks are up to 3.2km (2 mi) thick. The southern half of the Organ Mountains are part of a middle Tertiary eruptive center. Cueva Rhyolite and Soledad Rhyolite fill the caldera and dip toward the center of the structure.

Granite and metamorphic basement rocks of the Organ Mountains are the roots of an ancient uplift formed 1.3–1.4 Ga. Later in the Precambrian, the region underwent extension and injection of diorite-diabase dikes. During late Precambrian time, the mountain ranges were eroded to a broad, low-relief erosion surface.

Paleozoic and Mesozoic sedimentary rocks at least 2,600m (8,500ft) thick were deposited on Precambrian granite and metamorphic basement in the Organ Mountains. Starting in Late Cambrian time, seas moved over the top-Precambrian erosion surface. The Bliss Sandstone formed in a near-shore environment. Tidal to shallow-water shelf dolomites of the El Paso Group accumulated above the Bliss Formation. In Middle Ordovician time, southern New Mexico was gently uplifted, tilted, and eroded. Shallow seas in Middle to Late Ordovician time spread across the erosion surface and deposited the Cable Canyon Sandstone followed by Montoya carbonates. Dolomitization of the carbonates suggests a warm, shallow, possibly restricted saline marine environment. In Middle Silurian time, the Fusselman Dolomite was deposited in a shallow marine setting. Tilting and erosion from Late Silurian to Early Devonian was followed in Middle and Late Devonian time by deposition of the Devonian Percha Shale. The black, organic- and pyrite-rich shale indicates anaerobic bottom conditions. The Percha Shale grades upward into the Mississippian Caballero Limestone. Above the Caballero Formation, much of the Lake Valley, Las Cruces, and Rancheria formations contain crinoid debris separated by stretches of calcareous mud with abundant brachiopods, bryozoans, corals, and other fossils.

Withdrawal of the Rancheria Sea in latest Mississippian time was followed by nearshore sandstone, siltstone, oolitic limestone (limestone made up of small spherical grains), and fossiliferous rocks of the Helms Formation. An Early Pennsylvanian return of marine conditions led to deposition of the Lead Camp Limestone and La Tuna and Berino formations in a shallow, warm marine setting.

In Late Pennsylvanian time, the Orogrande Basin began to form by subsidence of a large area east of the Organ Mountains. Initial deposition in the basin was mostly sandstone, siltstone, and shale of the Panther Seep Formation. Deposition occurred at or above sea level, as indicated by mud cracks, ripple marks, stromatolites, and local caliche (soil cement) and gypsum beds.

The Orogrande Basin was deepest during the Permian, when normal marine conditions led to flourishing of algae, fusulinids, gastropods, and echinoids. Carbonates that would later become the Hueco Formation were deposited over southern New Mexico. Uplift of the Ancestral Rocky Mountains to the north during Early Permian time shed large amounts of sand, gravel, silt, and mud that was deposited in the Hueco Basin as the Abo Formation. In middle Permian, shallow marine waters spread across the region. Gypsiferous sand and silt, deposited in lagoons or in shallow, restricted basins, formed the Yeso Formation. These were buried in turn by shallow marine carbonate mud of the Yeso and the San Andres formations.

Most of the Organ Mountains-Las Cruces area was uplifted and was subject to erosion during the Mesozoic. At the end of Early Cretaceous time, the area was on the western flank of the Cretaceous Interior Seaway that extended from the Gulf of Mexico to the Arctic Ocean. Beach and nearshore sandstones of the Sarten and Dakota formations were deposited, followed by deeper marine mudstone of the Mancos Shale. The sea retreated for good in Late Cretaceous time, leaving behind beach deposits of the Gallup Sandstone.

The Late Cretaceous-Eocene Sevier/Laramide Orogeny is represented by conglomerate beds of the Love Ranch Formation that were shed off the growing Rocky Mountains. The orogeny is characterized by northwest-trending and northeast-verging basement-involved thrust-faulted uplifts and associated asymmetric and overturned folds. Basement-cored uplifts are flanked by monoclines and steeply dipping faults. The main structural features associated with uplift in the Organ Mountains and southern San Andres Mountains are the Bear Peak Fold-Thrust Belt and the Torpedo-Bennett Fault Zone. These structures form the northern and western boundaries of the Sevier/Laramide-age uplift in this area, respectively.

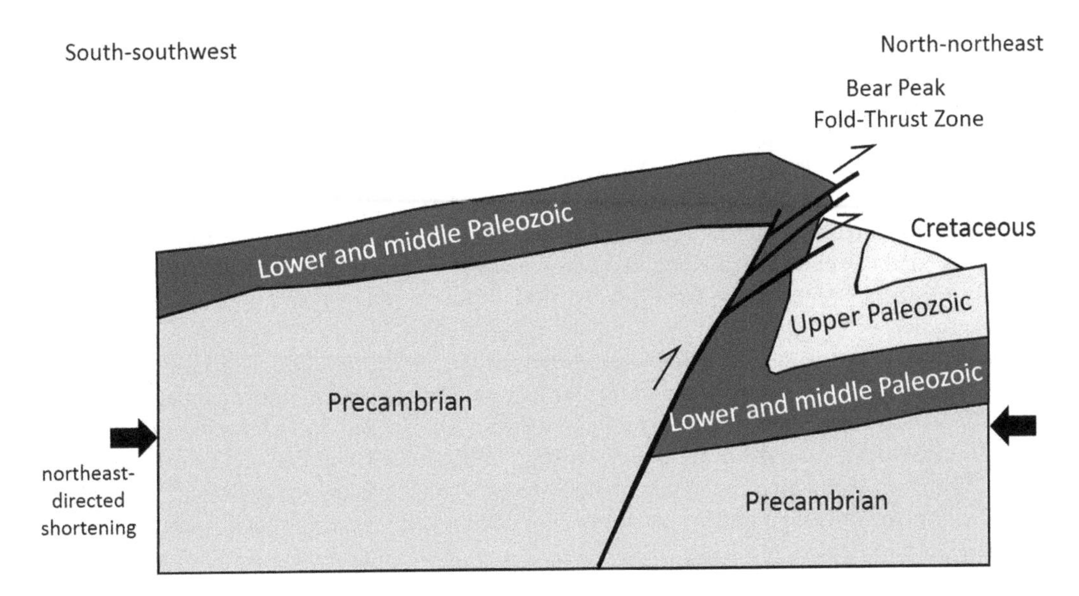

Sevier/Laramide uplift and shortening across the Southern San Andres Mountains. (Modified from Seager, 1981 with permission from the New Mexico Bureau of Geology and Mineral Resources.)

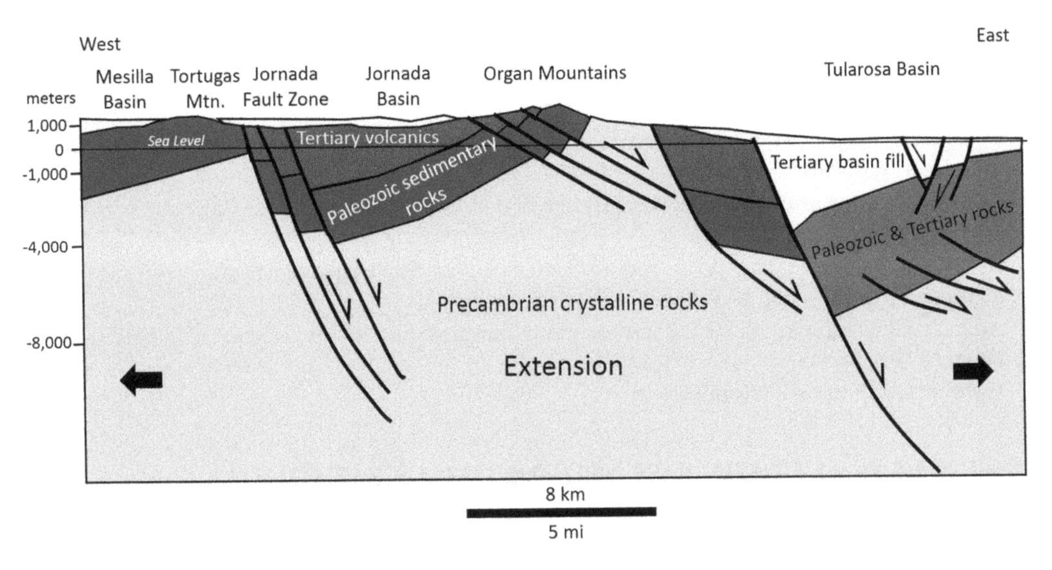

Tertiary Basin-and-Range extension across the Organ Mountains. (Modified from Seager, 1981 with permission from the New Mexico Bureau of Geology and Mineral Resources.)

Eocene to Miocene lava flows and tuffs buried the Sevier/Laramide-age uplifts in the Organ Mountains area. The Organ Batholith, source of most of the volcanic rocks, later intruded its own volcanic cover. Igneous activity is thought to be ultimately related to subduction of the Farallon plate beneath western North America. Volcanic rocks are over 3.3 km (2 mi) thick in parts of the Organ Mountains. The lower 610 m (2,000 ft) of the volcanic section is mainly andesitic; the upper 2,750 m (9,000 ft) is mostly rhyolitic.

The Eocene-Oligocene quartz monzonite of the Organ Batholith forms the rugged backbone of the range and is well exposed along the eastern slopes. The similar composition and age of the volcanics (33.0–33.7 Ma) and the batholith (32.5–34.4 Ma) strongly suggest they are derived from the same source.

Late Tertiary east-west extension formed the narrow, north-trending Rio Grande rift that extends from northern Mexico to around Leadville, Colorado. Extension in the rift began between 32 and 26 Ma and continues to the present day (Chapin and Seager, 1975; Chapin, 1979; Seager, 1981). Vertical offset is on the order of 300–900 m (1,000–3,000 ft) on the major range-bounding faults. The Organ Mountains Fault and Artillery Range Fault Zone are part of the rift that extends from El Paso to Mockingbird Gap along the eastern base of the Franklin-Organ-San Andres Mountains. These faults are identified by nearly continuous modern-day scarps in alluvial fans along the eastern base of the mountains. Gravity modeling suggests 2,100–3,000 m (7,000–10,000 ft) of Tertiary-Quaternary valley fill and total displacement across the fault zone of at least 3,650–4,600 m (12,000–15,000 ft).

Visit

This National Monument, run by the BLM, contains hiking trails, archeological sites, volcanic geology, pioneer sites, and climbing routes. Horseback riding, mountain biking, camping, backpacking, and landscape photography are popular pastimes.

Dripping Springs Visitor Center has interpretive displays, 22 picnic sites, and restrooms. This natural desert oasis also serves as the La Cueva Trailhead, where a short, 1.6 km (1 mi) hike takes visitors to a 5,000-year-old rock shelter, just one of 243 known archeological sites.

Hours: The entrance gate is open from 8:00 am to sunset. The Dripping Springs Visitor Center is open from 8:00 am to 5:00 pm daily, year-round.
Address: BLM Las Cruces District Office, 1800 Marquess Street, Las Cruces, NM 88005-3370
Phone: 575-525-4300
Website: www.blm.gov/visit/omdp

STOP 19.1 SAN AUGUSTIN PASS AND THE ORGAN BATHOLITH

This outcrop is an exposure of the Organ Batholith, the Oligocene-Eocene quartz monzonite intrusive that forms the backbone of the Organ Mountains. The prominent peak to the northwest is San Augustin Peak, part of the intrusive.

Organ Mountains pullout on US-70. The roadcut exposes Tertiary intrusives (Oligocene-Eocene quartz monzonite) of the Organ Batholith.

San Augustin Pass to La Cueva Rock: *Continue west on US-70 to Nasa Rd/Baylor Canyon Rd exit; turn left (south) on Baylor Canyon Rd and drive to Dripping Springs Rd; turn left (east) and drive to* ***Stop 19.2, La Cueva Rock*** *parking area and trailhead (32.334872, −106.599256), for a total of 19.3 km (12.0 mi; 19 min).*

STOP 19.2 LA CUEVA ROCK

From the La Cueva parking area, a 1.5 km (1 mi) trail leads to a rock shelter occupied by Native Americans for millennia. La Cueva rock shelter is an archeological site associated with the Jornada branch of the prehistoric Mogollon culture. In the mid-1970s, the Centennial Museum of the University of Texas at El Paso excavated the area and found 100,000 artifacts. This work suggested that the shelter had been occupied from about 5000 BC. During the 1700 and 1800s, the rock shelter was likely used by Apaches living in the area.

During the late 1860s, it was the home of an eccentric hermit, Giovanni Maria de Agostini (Rosales, 2017). Agostini was an Italian noble born in the year 1800. He spent many years walking through Europe, South America, Mexico, and Cuba. In 1869, he visited the Barela family on the plaza in Old Mesilla, New Mexico. He told them of his plans to live at La Cueva. They warned him of the dangers of staying there alone, but he replied "I shall make a fire in front of my cave every Friday evening while I shall be alive. If the fire fails to appear, it will be because I have been killed." One Friday night in 1869, the fire failed to appear at La Cueva. Antonio Garcia led a group up the mountain and found the Hermit lying face down on his crucifix with a knife in his back. The murder was never solved.

Cueva Rock consists of 35–36 Ma Cueva Tuff (McIntosh et al., 1991). The tuff is distinguished by its light color, abundant pumice fragments, fragments of rhyolite, andesite, granite, and gneiss, and high silica content (Seager, 1981).

Cueva Rock (buff colored, in foreground) consists of Lower Tertiary Cueva Tuff. View northeast.

La Cueva Rock to Dripping Springs: *Return to Dripping Springs Road and turn left (east); drive to* ***Stop 19.3, Dripping Springs Natural Area Visitor Center*** *parking and trailhead (32.329616, −106.590741), for a total of 1.8 km (1.1 mi; 4 min).*

STOP 19.3 DRIPPING SPRINGS

A remarkable vertical-walled, slot canyon descends from the west side of Baldy Peak. At the mouth of the narrow canyon is Dripping Springs, confined by walls of rhyolite tuff. The Oligocene Squaw Mountain Tuff erodes to dramatic cliffs nearly 200 m (600 ft) high (Seager, 1981).

Take the Dripping Springs Trail through rock spires and desert scrub to the ruins of the resort buildings and spring issuing from a granite wall. The Dripping Springs Trail is a 4.3 km (2.7 mi) moderate difficulty round-trip that gains 126 m (413 ft) in elevation.

Oligocene Squaw Mountain Tuff at Dripping Springs. (Photo courtesy of Bureau of Land Management.)

The Dripping Springs Visitor Center offers interpretive displays of the Organ Mountains. Camping is not allowed, but there are 12 picnic sites. Pets are allowed only on designated trails.

Visit

Hours: Open year-round, except closed on Thanksgiving Day, Christmas Day, New Year's Eve, and New Year's Day.
April through September, the entrance gate is open 7:00 am until sunset.
October through March, the entrance gate is open 8:00 am until sunset.
The Visitors Center is open from 8:00 am until 5:00 pm
Phone: (575) 522-1219
Fee: see the website for current entry fee.
Website: www.blm.gov/visit/dripping-springs-natural-area

Dripping Springs Natural Area to Las Cruces Museum of Nature and Science: *Drive west on Dripping Springs Rd; continue straight onto E University Ave; turn right (north) onto I-25 and drive to Exit 3; turn left (west) onto Lohman Ave; continue straight on Amador Ave; turn right on Main St and drive to 411 North Main Street, Las Cruces. This is* ***Stop 20.1, Museum of Nature and Science*** *on the left for a total of 22.7 km (14.1 mi; 26 min).*

20 LAS CRUCES AREA

Las Cruces lies in the heart of the southern Rio Grande Rift. The Mesilla Basin here contains up to 840 m (2,750 ft) of Tertiary and Quaternary valley fill sediments above Oligocene-Eocene volcanics in the deepest part of the basin beneath the Rio Grande. The Rio Grande, moving water from southern Colorado to the Gulf of Mexico, became established as a through-going river system in this area around 2 Ma.

STOP 20.1 LAS CRUCES MUSEUM OF NATURE AND SCIENCE

The Las Cruces Museum of Nature and Science is focused on the natural environment of the Chihuahuan Desert and southern New Mexico. Three permanent exhibits feature Permian Trackways, flora and fauna of the desert, and light and space. The Permian Trackways exhibit displays fine examples of dinosaur, insect, and other tracks found at Prehistoric Trackways National Monument.

Dimetrodon skeleton over his tracks.

Visit

Address: 411 North Main Street, Las Cruces
Phone: 575-522-3120
Email: museum.monas@las-cruces.org
Hours: Tuesday through Friday 10:00 am to 4:30 pm
Saturday 9:00 am to 4:30 pm
Closed on all statutory holidays
Admission: Free

Museum of Nature and Science to Prehistoric Pathways National Monument: *Continue north on Main Street to US-70/Picacho Ave; turn left (west) on Picacho and drive to NM-185/Valley Dr and turn right (north); drive to Shalem Colony Trail and turn left (west); drive to Rocky Acres*

Trail/County Road DO-13 and turn right (west); drive west approximately 400 m (¼ mi) to Permian Tracks Road on the left. Bear left on Permian Tracks Road, cross over a cattleguard and continue west to the second parking area. After about 1.5 km (1 mi), this road is for high-clearance vehicles only. There are no signs directing you to the Monument. This is ***Stop 20.2, Prehistoric Pathways National Monument*** *(32.373208, −106.866896), for a total of 13.9 km (8.7 mi; 18 min).*

STOP 20.2 PREHISTORIC TRACKWAYS NATIONAL MONUMENT

Prehistoric Trackways National Monument is located in the Robledo Mountains just northwest of Las Cruces. The monument encompasses 2,127 ha (5,255 ac) and is administered by the BLM.

Prehistoric Trackways lies on the west side of the Rio Grande rift in sedimentary rocks that were originally deposited along the coast of the Hueco Seaway. Fossil footprints of land animals, sea creatures, and insects are preserved, as well as fossil plants and petrified wood and plenty of marine invertebrate fossils such as brachiopods, gastropods, cephalopods, bivalves, and echinoderms are found in the Permian Hueco Group, mainly the Abo Formation redbeds.

Tracks were discovered on June 6, 1987 by Jerry Paul MacDonald. Prehistoric Trackways was designated a National Monument on March 30, 2009.

Age	Robledo Mountains		Franklin Mountains		San Andres Mountains	
					Yeso Group	
Early Permian	Hueco Group	Apache Dam Formation	Hueco Group	Alacran Mountain Formation	Hueco Group	Abo Formation
		Robledo Mountains Formation				Robledo Mountains Formation (upper)
		Community Pit Formation		Cerro Alto Formation		Community Pit Formation
		Shalem Colony Formation		Hueco Canyon Formation		Robledo Mountains Formation (lower)

Formations that comprise the Early Permian Hueco Group vary slightly from area to area. (Modified after Lucas et al., 2011.)

Permian tracks, probably from a Dimetrodon, in Prehistoric Trackways National Monument. Circles added to highlight the tracks.

Visit

There are no marked trails, nor are there directional signs, so you'll need to explore to see anything. The nearest fossils (32.380981, −106.881607) are about 2.5 km (1.5 mi) up a trail that follows the bottom of the ravine just to the right (north) of the second parking area. The tracks can be difficult to find, as the monument is largely undeveloped.

Guided hikes are a good idea; they are periodically offered by BLM staff (address and phone number are provided below). As a fallback, if a guided hike is not available and you can't find the trail, there is an excellent Prehistoric Trackways exhibit at the Las Cruces Museum of Nature and Science.

Season/Hours: Generally open year-round, all hours.
Fees: None
Address: BLM Las Cruces District Office, 1800 Marquess Street, Las Cruces, NM 88005
Phone: 575-525-4300
Website: www.blm.gov/visit/ptnm

Our transect ends in the heart of the Rio Grande Rift. Look around at the mountains on either side of the valley. One day a few million years from now, these will most likely be the coastlines of eastern and western North America.

ACKNOWLEDGMENTS

Assembling this geological tour was the result of a collaboration with Jon Blickwede, Ralph Baker, and John Berry, long-time associates of the author. Their participation is gratefully acknowledged.

REFERENCES

Aurcana Corp. Shafter silver project overview. www.aurcana.com/operations/shafter/overview/. Accessed 9 April 2019.

Baker, L.M. 1985. Sedimentology and diagenesis of the basinal facies of the Dimple Limestone, Marathon Basin, West Texas. *MSc Thesis*, Texas Tech University, 107 p.

Bebout, D.G., and C. Kerans (eds.) 1993. *Guide to the Permian Reef Geology Trail, McKittrick Canyon, Guadalupe Mountains National Park, West Texas.* Univrsity of Texas Bureau of Economic Geology, Austin, TX, Guidebook 26, 48 p.

Bebout, D.G., C. Kerans, and P.M. Harris. 2007. *Introduction, Guide to the Permian Reef Geology Trail, McKittrick Canyon, Guadalupe Mountains National Park, Texas.* Texas Bureau of Economic Geology, Austin, TX, 8 p.

Bebout, D.G., C. Kerans, and P.M. Harris. 2009. *Introduction: Guide to the Permian Reef Geology Trail, McKittrick Canyon, Guadalupe Mountains National Park, Texas.* American Association of Petroleum Geologists Search and Discovery Article #60033. Texas Bureau of Economic Geology, Austin, TX, 6 p.

Bentley, C. 2014. McKittrick Canyon, Guadalupe Mountains National Park, Texas. Blog, https://blogs.agu.org/mountainbeltway/2014/05/06/mckittrick-canyon-guadalupe-mountains-national-park-texas/

Brown, A., and R.G. Loucks. 2007. Toe of slope. In Bebout, D.G., and C. Kerans (eds.), *Guide to the Permian Reef Geology Trail, McKittrick Canyon, Guadalupe Mountains National Park, Texas.* Texas Bureau of Economic Geology, Austin, TX, 15 p.

Chapin, C.E. 1979. Evolution of the Rio Grande rift—A summary. In R. Riecker (ed.), *Rio Grande rift—Tectonics and Magmatism.* American Geophysical Union, Monograph, Washington, D.C., pp. 1–5

Chapin, C.E., and W.R. Seager. 1975. Evolution of the Rio Grande rift in the Socorro and Las Cruces areas: New Mexico Geological Society. *Guidebook 26th Field Conference*, Socorro, NM., p. 297

Collins, E.W., and J.A. Raney. 2000. Geologic map of West Hueco Bolson, El Paso Region, Texas. The University of Texas at Austin, Bureau of Economic Geology, Miscellaneous Map No. 40, Map scale 1:100,000, 25 p.

Condon, S.M., and T.S. Dyman. 2006. Chapter 2: 2003 Geologic assessment of undiscovered conventional oil and gas resources in the Upper Cretaceous Navarro and Taylor Groups, Western Gulf Province, Texas. U.S. Geological Survey Digital Data Series DDS-69-H, 47 p.

Cook, V.O. 1966. Regional geology of the Delaware Basin. *Journal of Petroleum Geology*, v. 18, no. 10, 7 p.

Cornell, W.C. A brief geological history of the El Paso-Juarez (ELP-J) Region. www.geo.utep.edu/loca/fieldtrip.html. Accessed 6 May 2019.

Eckhardt, G. 1995–2018 online. Introduction to the Edwards aquifer. www.edwardsaquifer.net/intro.html. Accessed 6 May 2019.

Fails, T.G. 1990. *The Northern Gulf Coast Basin: A Classic Petroleum Province.* Geological Society of London, London, Special Publications, 50, pp. 221–248. doi: 10.1144/GSL.SP.1990.050.01.11.

Galloway, W.E. 2009. Giant fields of North America: Gulf of Mexico. *GeoExpro*, v. 6, no. 3, 10 p.

Gray, J.E., and W.R. Page. 2008. *Geological, Geochemical, and Geophysical Studies by the U.S. Geological Survey in Big Bend National Park, Texas.* U.S. Geological Survey, Reston, VA, Circular 1327, 104 p.

Hickman, R.G., R.J. Varga, and R.M. Altany. 2009. Structural style of the Marathon Thrust Belt, West Texas. *Journal of Structural Geology*, v. 31, pp. 900–909.

Jarvie, D.M., D. Prose, B.M. Jarvie, R. Drozd, and A. Maende. 2017. Conventional and unconventional petroleum systems of the Delaware Basin. American Association of Petroleum Geologists Search and Discovery Article #10949, 21 p.

Jordan, C.F. 1975. Lower Permian (Wolfcampian) sedimentation in the Orogrande Basin, New Mexico. In Seager, W.R., R.E. Clemons, and J.F. Callender (eds.), *Las Cruces Country.* New Mexico Geological Society 26th Annual Fall Field Conference Guidebook, pp. 109–117.

Keller, G.R., J.M. Hills, and R. Djeddi. 1980. A regional geological and geophysical study of the Delaware Basin, New Mexico and west Texas. In Dickerson, P.W., J.M. Hoffer, and J.F. Callender (eds.), *Trans Pecos Region (West Texas).* New Mexico Geological Society 31st Annual Fall Field Conference Guidebook, pp. 105–111.

KellerLynn, K. 2012. White Sands National Monument: Geologic resources inventory report. Natural Resource Report NPS/NRSS/GRD/NRR—2012/585. National Park Service, Fort Collins, CO, 100 p.

Kerans, C. 2015. Paddling down the Pecos into Epic Geology. Blog, www.jsg.utexas.edu/news/2015/11/paddling-down-the-pecos-into-epic-geology/. Accessed 6 May 2019.

Kerans, C., and X. Janson. 2012. Linking depositional, diagenetic, facies, stratal geometries, and cycle architecture: Examples from Paleozoic carbonate systems of west Texas and southern New Mexico. RCRL spring field course April 2012. Reservoir Characterization Lab, Bureau of Economic Geology, U.T. Austin.

Kohout, M.D. 2010. Big Bend Ranch State Park. http://www.tshaonline.org/handbook/online/articles/gibfk. Accessed 1 December, 2019.

Lehman, T.M. 2014. Paleogeography of interior seaway sands. In Cooper, R.W., and D.A. Cooper (ed.), *Field Guide to Late Cretaceous Geology of the Big Bend Region*. Houston Geological Society Field Guidebook 204G, 94 p.

Lucas, S.G., K. Krainer, L. Corbitt, J. Dibendedtto, and D. Vachard. 2011. The Trans Mountain Road Member, a new stratigraphic unit of the Lower Permian Hueco Group, northern Franklin Mountians, Texas. In Sullivan, M., et al. (eds.), *Fossil Record 3*. New Mexico Museum of Natural History and Science Bulletin 53, pp. 93–109.

McIntosh, W.C., L.L. Kedzie, and J.F. Sutter. 1991. Paleomagnetism and 40Ar/39Ar ages of ignimbrites, Mogollon-Datil volcanic field, southwestern New Mexico. New Mexico Bureau of Mines & Mineral Resources Bulletin 135, 80 p.

Miles, R.W. Handbook of Texas online, Robert W. Miles, "FRANKLIN MOUNTAINS". www.tshaonline.org/handbook/online/articles/rjf14. Accessed 7 May 2019.

National Park Service. 2015. Geologic formations, Guadalupe Mountains National Park, Texas. www.nps.gov/gumo/naturescience/geologicformations.htm. Accessed 7 May 2019.

New Mexico Bureau of Geology and Mineral Resources. 2007–2019. Geologic map of the white sands area. https://geoinfo.nmt.edu/tour/federal/monuments/whitesands/geology.html.

Redfern, F. 2006–2018. Virtual geologic field trips to Big Bend National Park. http://prism-redfern.org/bbvirtualtrip/elena/elena.html. Accessed 7 May 2019.

Rosales, G. 2017. The Organ Mountains-desert peaks national monument: A geological wonder. *Albuquerque Journal*, May 13.

Scholle, P. 2000. An introduction and virtual geologic field trip to the Permian reef complex, Guadalupe and Delaware Mountains, New Mexico-West Texas. https://geoinfo.nmt.edu/staff/scholle/guadalupe.html. Accessed 7 May 2019.

Scotese, C.R. 2001. *Atlas of Earth History*, Volume 1, Paleogeography. PALEOMAP Project, Arlington, TX, 52 pp.

Seager, W.R. 1981. Geology of Organ Mountains and southern San Andres Mountains, New Mexico. New Mexico Bureau of Mines and Mineral Resources, Memoir 36, 98 p.

SEPM. 2013. The geology of the Upper Permian, Permian Basin. Introduction. www.sepmstrata.org/page.aspx?pageid=136.

Shale Experts. Permian Basin overview. www.shaleexperts.com/plays/permian-basin/Overview. Accessed 7 May 2019.

Standen, A., S. Finch, C.R. Williams, R. Lee-Brand, and P. Kirby. 2009. Capitan reef complex structure and stratigraphy. Texas Water Board Contract # 0804830794, 71 p.

Stieb, M. 2015. The geologic wonders of the Guadalupe Mountains National Park. San Antonio Current, September 29. www.sacurrent.com/sanantonio/the-geologic-wonders-of-the-guadalupe-mountains-national-park/Content?oid=2473965. Accessed 7 May 2019.

Sutton, L. 2014. Permian Basin geology: The Midland vs the Delaware Basin Part 2. DI Blog, https://info.drillinginfo.com/blog/permian-basin-geology-midland-vs-delaware-basins/. Accessed 7 May 2019.

Turner, K.J., M.E. Berry, W.R. Page, T.M. Lehman, R.G. Bohannon, R.B. Scott, D.P. Miggins, J.R. Budahn, R.W. Cooper, B.J. Drenth, E.D. Anderson, and V.S. Williams. 2011. Big Bend National Park, Texas. U.S. Geological Survey Scientific Investigations Map 3142, 1:75,000.

US Energy Information Administration. 2018. Permian Basin Wolfcamp Shale Play, geology review. www.eia.gov/maps/pdf/PermianBasin_Wolfcamp_EIAReport_Oct2018.pdf. Accessed 7 May 2019.

U.S. Geological Survey and Bureau of Economic Geology. Explore Texas geology. Online geologic map of Texas. https://txpub.usgs.gov/txgeology/. Accessed 7 May 2019.

Wikipedia. Delaware Basin. https://en.wikipedia.org/wiki/Delaware_Basin. Accessed 7 May 2019.

Wikipedia. Shafter. https://commons.wikimedia.org/wiki/File:Shafter_Presidio_Mine_Cross_Section.png. Accessed 8 April 2019.

Wikipedia. Val Verde Basin. https://en.wikipedia.org/wiki/Val_Verde_Basin. Accessed 7 May 2019.

Woodruff, C.M. Jr., E.W. Collins, E.C. Potter, and R.G. Loucks. 2017. *Canyon Dam Spillway Gorge and Natural Bridge Caverns, Geologic Excursions in the Balcones Fault Zone, Central Texas*. Bureau of Economic Geology Guidebook 29. The University of Texas at Austin, Austin, TX, 56 p.

6 La Popa Basin and Sierra Madre Oriental

Monterrey-Saltillo-Linares Loop, Nuevo León and Coahuila, Mexico

View east toward the overturned north limb of Anticlinal de Los Muertos (Los Muertos Anticline) towering over a fogbank at Monterrey.

OVERVIEW

Ah, the Sierra Madre, famous on the big screen and in legend. Our last but not least transect in North America will take us from the Gulf of Mexico borderlands across the Sierra Madre Oriental (SMO) Fold and Thrust Belt. We will see in outcrop strata equivalent to those in the subsurface along the Gulf of Mexico both onshore and offshore. Starting in the La Popa Basin, we examine giant Jurassic salt-cored uplifts and salt welds and their relationship to platform limestones. Moving to the Monterrey Salient, a bulge in the mountain range south and west of Monterrey, we encounter gigantic Laramide-age folds developed on a 'weak,' evaporite (gypsum) detachment surface. These folds are often overturned on both limbs. Deformation continues into the foreland of the fold-thrust belt, where we see folded sediments shed during mountain-building, as well as delta and deep-water turbidite (submarine landslide) deposits whose sediments derived from the uplifting mountains. The detachment surface itself is exposed in a railroad cut at the mountain front. We enter the Sierra Madre and drive to the Potosí Uplift where we encounter the gypsum-rich detachment zone in the hinterland (back side, or source area) of the fold-thrust belt, as well as sediments deposited during the early, opening stage of Gulf of Mexico rifting. A final drive across the Sierra Madre Thrust Belt in Santa Rosa Canyon reveals structural styles that develop when platform margin to deep-water deposits are thrust on 'strong' shale detachments. We end with a visit to a magnificent waterfall and travertine mound, and stop at a cave linked to a legend of lost treasure in the Sierra Madre.

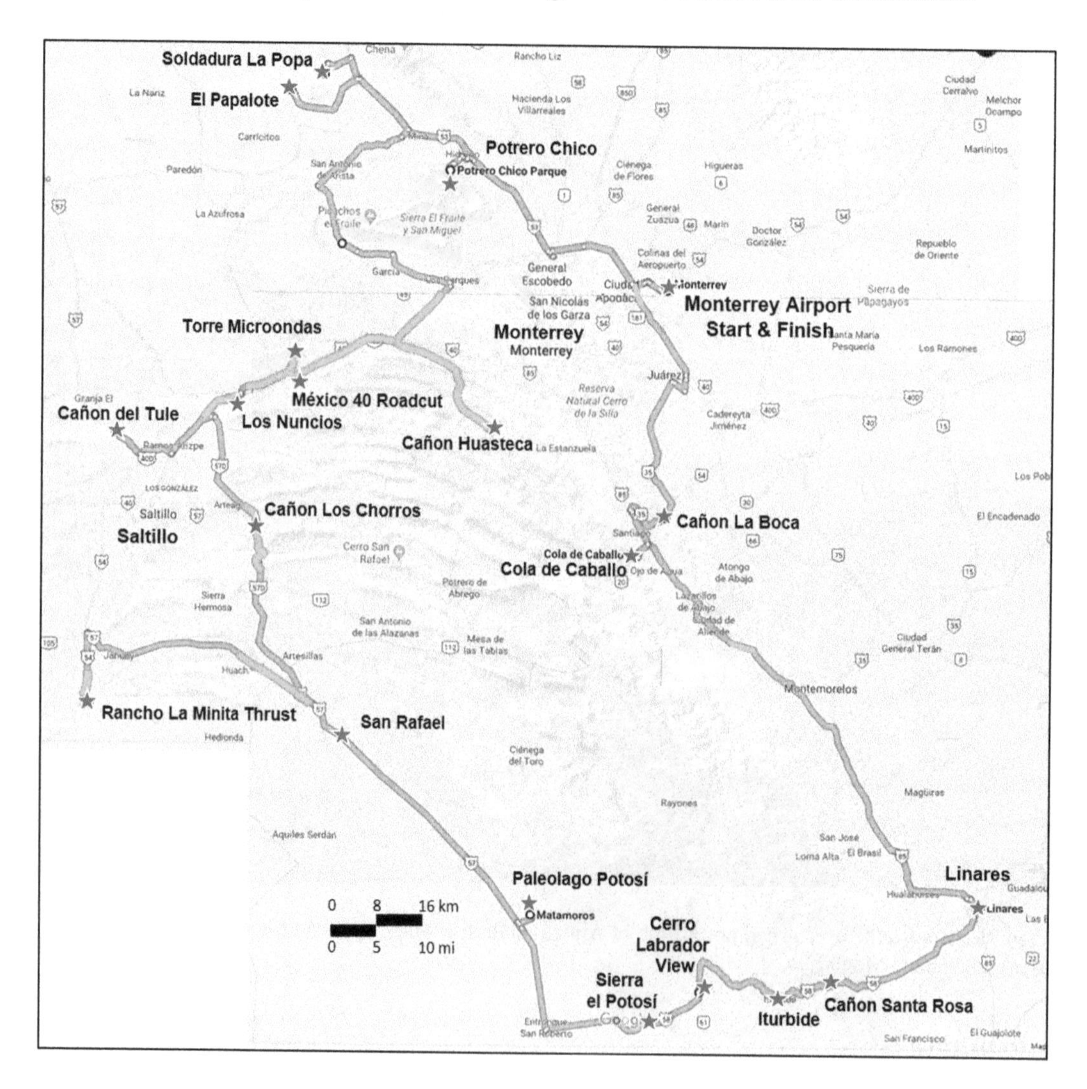

ITINERARY

Begin - Monterrey Airport

1 Potrero Chico Diapir

  Stop 1.1 Entrance to Potrero Chico

  Stop 1.2 Diapir Core

Stop 2 Soldadura La Popa

Stop 3 El Papalote Diapir
4 Cañon Huasteca
Stop 4.1 Anticlinal San Blas
Stop 4.2 Core of Anticlinal de Los Muertos
Stop 4.3 Parque la Huasteca
Stop 5 México 40 Roadcut
Stop 6 Torre Microondas
Stop 7 Los Nuncios
8 Cañon del Tule
Stop 8.1 Difunta Formation Delta and Shoreline
Stop 8.2 Difunta Formation Turbidites
9 Anticlinal Los Chorros
Stop 9.1 North Limb
Stop 9.2 Tunnel Curve
Stop 9.3 South Limb
Side Trip 1 Rancho La Minita Thrust
Stop 10 San Rafael View
Side Trip 2 Paleolago Potosí, Ejido Catarino Rodriguez
11 Sierra El Potosí
Stop 11.1 Sierra El Potosí Uplift
Stop 11.2 San Pablo Outcrop
Stop 11.3 San Pablo Roadcut
Stop 12 Cerro Labrador View
Stop 13 Iturbide Anticline
14 Santa Rosa Canyon
Stop 14.1 Folded Agua Nueva Formation
Stop 14.2 Tamaulipas Superior
Stop 14.3 Anticlinal Pinitos Ebanito
Stop 14.4 San Felipe Formation Synclinal Box Fold
Stop 14.5 Frontal Anticline View
Stop 15 Sierra Madre Oriental Mountain Front View
Stop 16 Cascada Cola de Caballo
17 La Boca Canyon
Stop 17.1 Calera
Stop 17.2 Cueva Murciélagos
Stop 17.3 El Bañito
End - Monterrey Airport

OBSERVATIONS ON DRIVING IN MEXICO

If you take your car into Mexico, be sure you have Mexican driver's insurance, get a tourist visa, and get a temporary vehicle import permit (good for 6 months) when you cross the border. The permit requires a deposit ($400 US in 2018) that will be refunded when you exit Mexico.

Driving in Mexico is not for the faint of heart. Traffic laws are more-often-than-not taken as suggestions by native drivers; there are monster speed bumps (*topes*), sometimes unmarked; there are potholes (*baches*) that can swallow a car. Most advise that you not drive at night and stay on main roads due to narcotics cultivation in some rural, out-of-the-way locations. *Cuotas* (toll roads) are generally better maintained than *Libres* (free roads). Size matters—the largest vehicle will take the right-of-way. Aggressive drivers have been known to pass going uphill on blind curves. There is a reason for the large number of shrines along the roadside.

If you can, let a Mexican national do the driving. They will know the local rules and customs. Carry paper maps, because the Maps app on your phone may be out of service range. When lost, use what is locally known as Mexican GPS and ask the nearest person: "*Disculpe, señor/señora, ¿cómo llegamos a...?*" (excuse me, sir/ma'm, how do we get to ….?).

BEGIN—MONTERREY

Monterrey is the capital and largest city of the state of Nuevo León, and the third largest city in Mexico with a population of around 4.7 million people in 2015. Monterrey is considered the city with the best quality of life in Mexico and is the commercial center of northern Mexico.

The city was founded in 1596 next to a spring called *Ojos de Agua de Santa Lucia*, where the Museum of Mexican History and Santa Lucía river walk are now. An expedition of 13 families led by conquistador Diego de Montemayor named it *Ciudad Metropolitana de Nuestra Señora de Monterrey* (Metropolitan City of Our Lady of Monterrey) in honor of the wife of Gaspar de Zúñiga, fifth Count of Monterrey.

After the War of Mexican Independence (1810–1821), Monterrey became a key economic center for northern Mexico because of its ties to Europe (through Tampico), the United States (through San Antonio), and Mexico City. In 1824, the Spanish province became the Mexican state of Nuevo León, and Monterrey became the capital.

In 1846, the earliest significant battle of the Mexican-American War took place there. Mexican forces eventually surrendered, but only after successfully repelling several U.S. attacks.

By the end of the 19th century, Monterrey was linked to the rest of Mexico by railroad, which spurred industrialization. A steel plant, the Fundidora de Fierro y Acero de Monterrey, and brewery, Cervecería Cuauthemoc, were founded. By 1900, it was one of the world's largest steel-producing centers and continues to be a vital hub in North American trade into the 21st century.

Geologically, Monterrey lies at the boundary between the La Popa Basin and Sierra Madre Oriental. We will first visit the foreland basin to investigate wonderfully exposed salt-cored uplifts and salt welds analogous to oil field structures seen in the subsurface of the Gulf of Mexico.

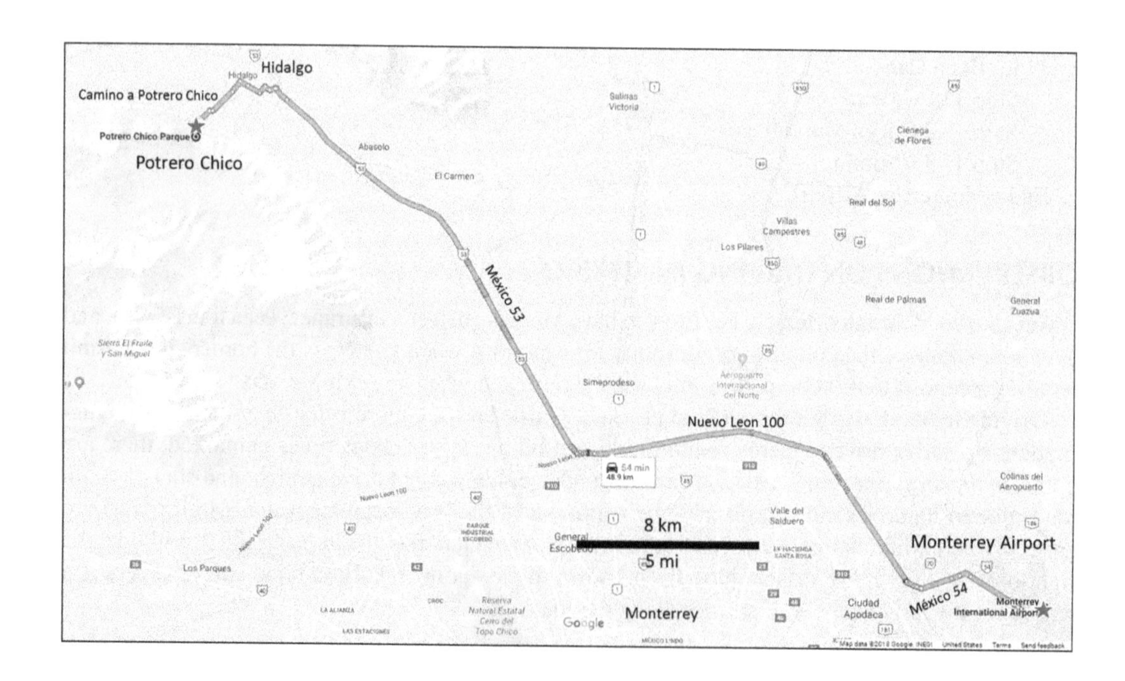

Monterrey airport to Potrero Chico: *Leave Monterrey airport and turn left (west) on México 54; turn right (northwest) onto Nuevo Leon 100 and drive to México 53*; turn right (north) on *México 53* and drive to *Niños Heroes in Hidalgo; turn left (west) and take Niños Heroes to Porfirio Diaz; turn left (southwest) on Porfirio Diaz and drive to Guadalupe G. Lozano; turn right (northwest) on Guadalupe Lozano and drive to Gral. Francisco Villa; turn left (southwest) on Gral. Francisco Villa, which becomes Potrero Chico; continue on Potrero Chico until it merges with Antiguo Camino a Potrero Chico; follow signs to* **Stop 1.1, Parque Recreativo Potrero Chico** *(25.949032, −100.476218), a total of 48.9 km (30.4 mi; 54 min).*

LA POPA BASIN

The La Popa Basin developed east of the Coahuila block, a basement uplift resulting from Triassic-Jurassic opening of the Gulf of Mexico. Jurassic salt and gypsum (evaporites) were deposited by evaporating seawater when the basin was temporarily closed off during the initial opening (rift stage) of the Gulf of Mexico and Atlantic Ocean. These units are overlain by as much as 7 km (23,000 ft) of rock ranging in age from Early Cretaceous to Paleogene (145 to perhaps 39 Ma). The lower 3 km (10,000 ft) is mainly limestone with minor shales, whereas the upper 4 km is primarily shales and sandstones. The transition between the two indicates the beginning of uplift and erosion of the SMO to the south and west.

Diapirism, the slow, upward movement of salt and gypsum, is a result of low-density evaporites moving, under the influence of gravity and squeezing, upward into higher-density sedimentary rocks (kind of like the upward moving low-density blobs in a lava lamp). Movement began shortly after deposition and influenced sedimentation of later units in the basin. Sedimentary layers in this basin, influenced by the growth of salt domes, are known as "halokinetic sequences" (Giles and Lawton, 2002).

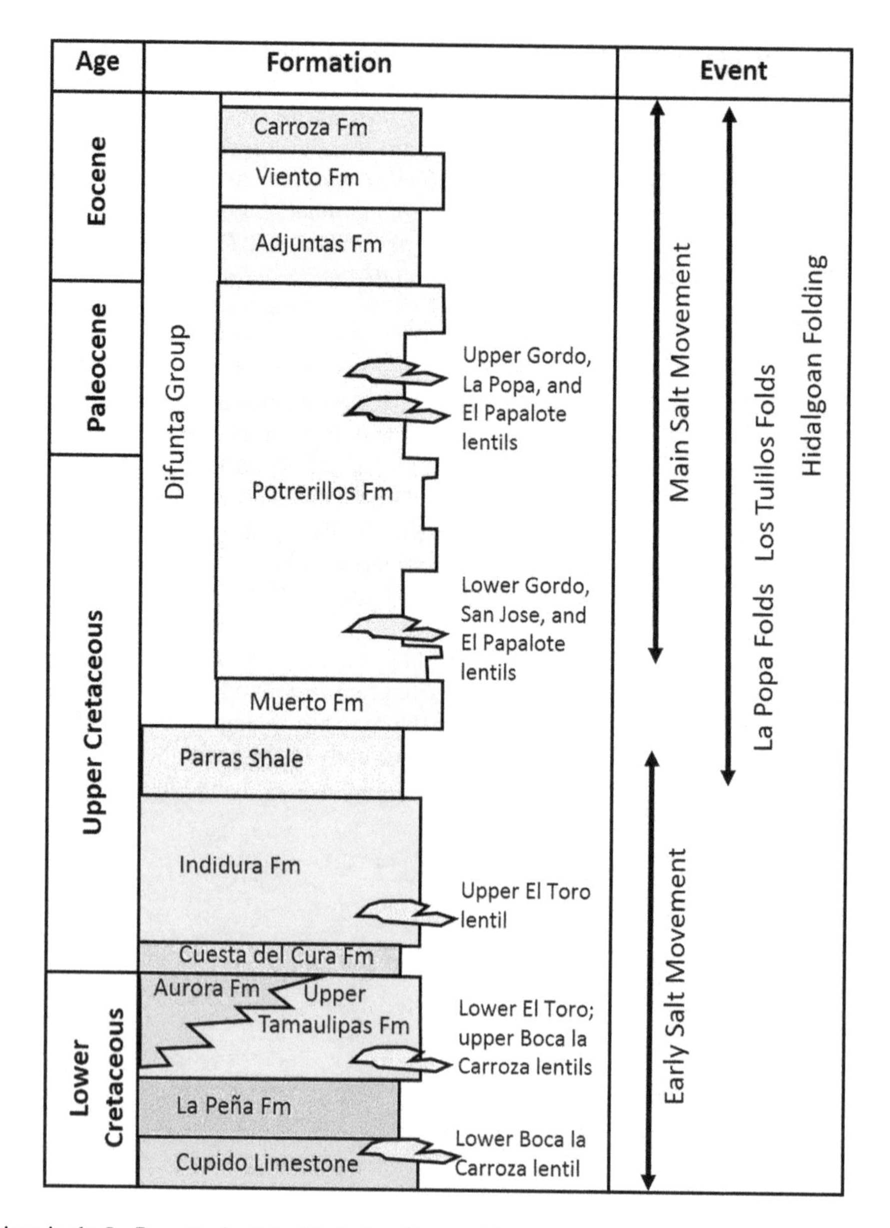

Formations in the La Popa Basin. (Modified after Gray and Lawton, 2011.)

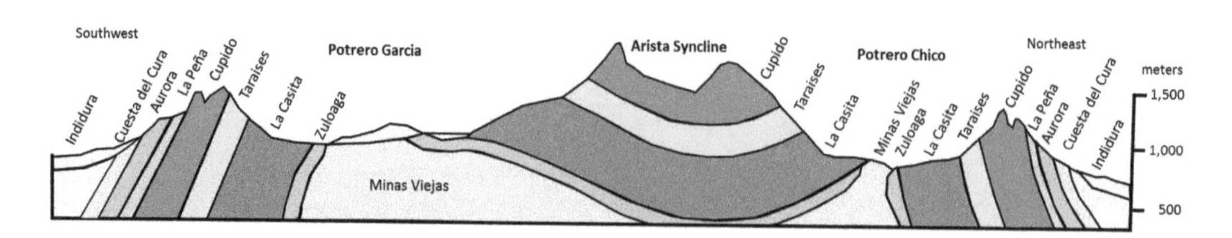

Cross section through Potreros Garcia and Chico. (Modified after Weidie and Ward, 1987.)

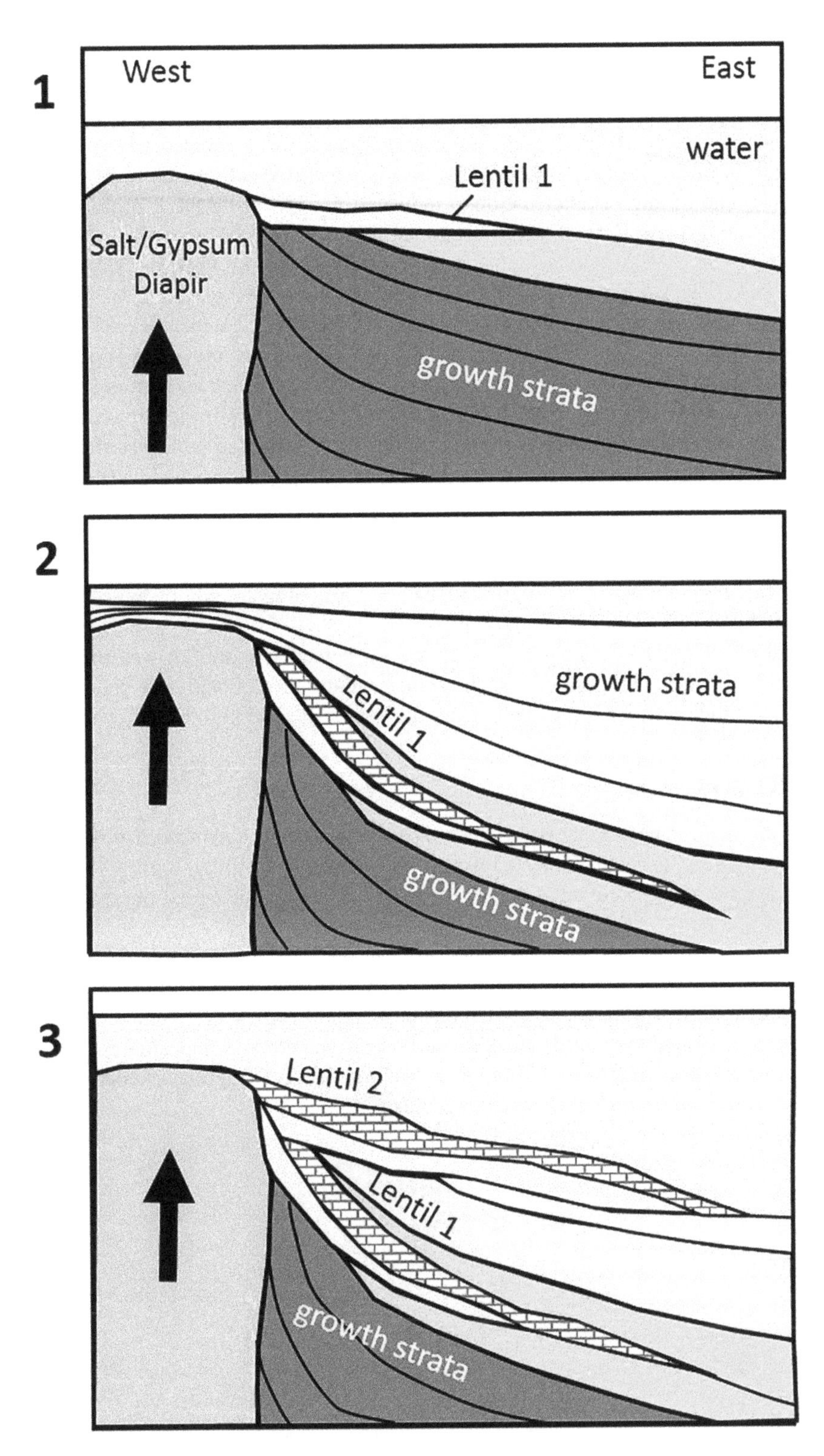

Upper Cretaceous to Early Paleogene development of El Papalote Diapir. 1) Diapir breaches the sea bottom and forms a dome. Lenticular reefs here, or "lentils," grew on the slightly elevated areas above the domes. 2) Sediments bury the dome as it continues to rise. 3) Repeat of phase 1 creates a second lentil. (Modified from Giles and Lawton, 2002.)

Northeast-directed shortening and folding related to the formation of the SMO began to overprint earlier salt deformation in the La Popa Basin during latest Cretaceous (Maastrichtian; 72–66 Ma) time (Eguiluz de Antuñano, 2017). The deformation in the SMO is similar in style to the 'thin-skinned' Sevier Orogeny of the United States and Canada, but occurred later in time, more in line with the Laramide Orogeny (80–35 Ma) and is generally called Laramide here (Gray and Lawton, 2011). The timing of deformation is indicated mainly by thinning of sedimentary units. The Maastrichtian Muerto Formation, for example, thins over the El Gordo salt dome, but retains a constant thickness over the nearby northwest-oriented El Gordo Anticline. Thus, uplift of the diapir came before deposition of the Muerto Formation, and movement on the anticline was later. Anticlines and synclines in the La Popa Basin are up to 8 km (5 mi) wide, 60 km (36 mi) long, and 1 km (3,300 ft) high. These structures formed as detachment folds above the Jurassic evaporites. We saw similar salt-cored folds in the Paradox Basin of Utah, where the salt was Pennsylvanian in age (Volume 1, Tour 3). Salt movement in the La Popa Basin, as determined by sediment thickness variations, initiated in the Lower Cretaceous (Aptian?) and continued until just after early Eocene time (about 40 Ma).

Resources

Throughout this region, there is mining of limestone for aggregate, and of gypsum for sheetrock and agriculture.

A handful of oil wells have been drilled in the La Popa Basin and all were dry. Sampling of organic matter indicates that the source rocks are over-mature for oil and have generated gas in the Sabinas Basin to the north. This may be a result of deep burial and high heat flow in the basin.

1 POTRERO CHICO

A *potrero* is an enclosed pasture. Potrero Chico means a small enclosed pasture. It would be difficult to describe *this* potrero as small. The enclosing walls are 400 m (1,300 ft) high. The central pasture is the result of salt and gypsum being dissolved out of the core of the mountain.

STOP 1.1 ENTRANCE TO POTRERO CHICO

You are between vertical walls of Lower Cretaceous Cupido Limestone, a challenging destination for climbers from around the world. The Cupido Formation here was deposited in a reef and fore-reef environment. Rudist reefs can be found, as well as the reef fragments broken by waves and storms that accumulate in front of the reef (breccia).

Vertical walls of Cupido surround Potrero Chico. They are a favorite of climbers like these.

Rudist reef in the Cupido Formation limestone, entrance to Potrero Chico.

Entrance to Core of Potrero Chico*: Drive 1.6 km (1 mi; 2 min) west into Potrero Chico to* ***Stop 1.2, Core of Potrero Chico*** *Diapir (25.949032, −100.476218). Walk northwest about 200 m (650 ft) to gypsum outcrops.*

STOP 1.2 DIAPIR CORE

Jurassic Minas Viejas Formation (162–169 Ma) gypsum outcrops in the center of the valley. The Minas Viejas correlates to the Louann Salt of the northern Gulf coast (Wilson et al., 1984). There are flow structures (oriented linear mineral grains) and bits of red-purple shale inclusions in the gypsum. The PEMEX Los Ramones #1 well east of Monterrey encountered 1,067 m (3,500 ft) of evaporites (salt and gypsum; Weidie and Ward, 1987). The total thickness of salt has been estimated at over 2 km (1.2 mi) in the basin.

Uniquely in northern Mexico, the La Popa Basin (sometimes considered the southern part of the Sabinas Basin) contains salt/gypsum domes (diapirs). Potrero Chico is one of several diapirs in the area. Others include Potrero Garcia, Potrero Grande, El Gordo, El Papalote, and La Popa. The diapirs, roughly circular uplifts cored by salt and gypsum, began to flow and uplift the overlying rocks in Late Aptian (Early Cretaceous, 125–113 Ma) time, as the Cupido Formation limestone was still being deposited. In fact, the lenticular reefs here, called "lentils," grew on the slightly elevated areas above the early salt domes. The latest movement on these domes was in early Eocene time (56–40 Ma).

Outcrop of Minas Viejas gypsum and salt layers in the core of Potrero Chico.

Above the salt and gypsum is the 550 m (1,800 ft) thick Zuloaga Limestone, equivalent to the Smackover of the U.S. Gulf Coast. This is overlain by the largely covered La Casita Formation limestone (Cotton Valley equivalent, 1,000 m or 3,000 ft thick), the 457 m (1,500 ft) thick Taraises Formation limestone (time equivalent to the Hosston on the Gulf Coast), and the 610 m (2,000 ft) thick Cupido Limestone (Sligo equivalent) that forms the vertical walls of the potrero.

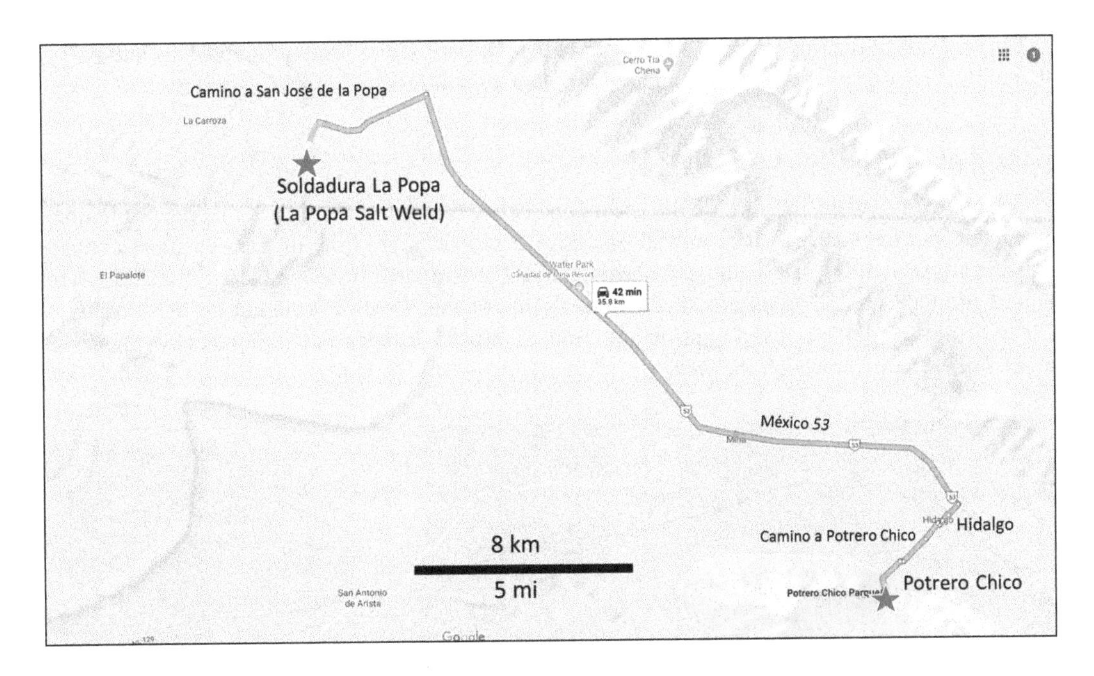

Potrero Chico to Soldadura La Popa: *Return to Carretera Monclova/Hidalgo-Monterrey/ México 53 and turn left; drive northwest on México 53 to road on left to San José de la Popa; immediately after (west of) the track to "Los Lirios," turn left onto an unimproved track and drive southwest; head for the gap in the ridge ahead; drive through the ridge and pull over at* ***Stop 2, Soldadura La Popa*** *(26.091334, −100.692186), a total of 35.8 km (21.5 mi; 42 min).*

Google Earth satellite image of the unimproved track to Soldadura La Popa. Imagery © 2019 CNES/Airbus; Landsat/Copernicus; Maxar Technologies.

The east side of the ridge is near the contact of the upper Cupido Formation limestone and Parras Shale. The Cupido here contains abundant *Gryphaea sp*, a Jurassic-Cretaceous oyster.

STOP 2 SOLDADURA LA POPA

A "soldadura" is a weld, in this case the shear zone between two bodies of rock that once contained rising salt, but the salt has since moved on and the two sides are now in apparent fault contact. The northwest-trending salt weld puts a near-vertical Cupido-aged limestone lentil (northeast side) against the Paleocene-Eocene crossbedded sandstone of the Viento Formation on the southwest. The weld zone contains veins of gypsum and dead oil (black bitumen), indicating that hydrocarbons at one time migrated through the weld.

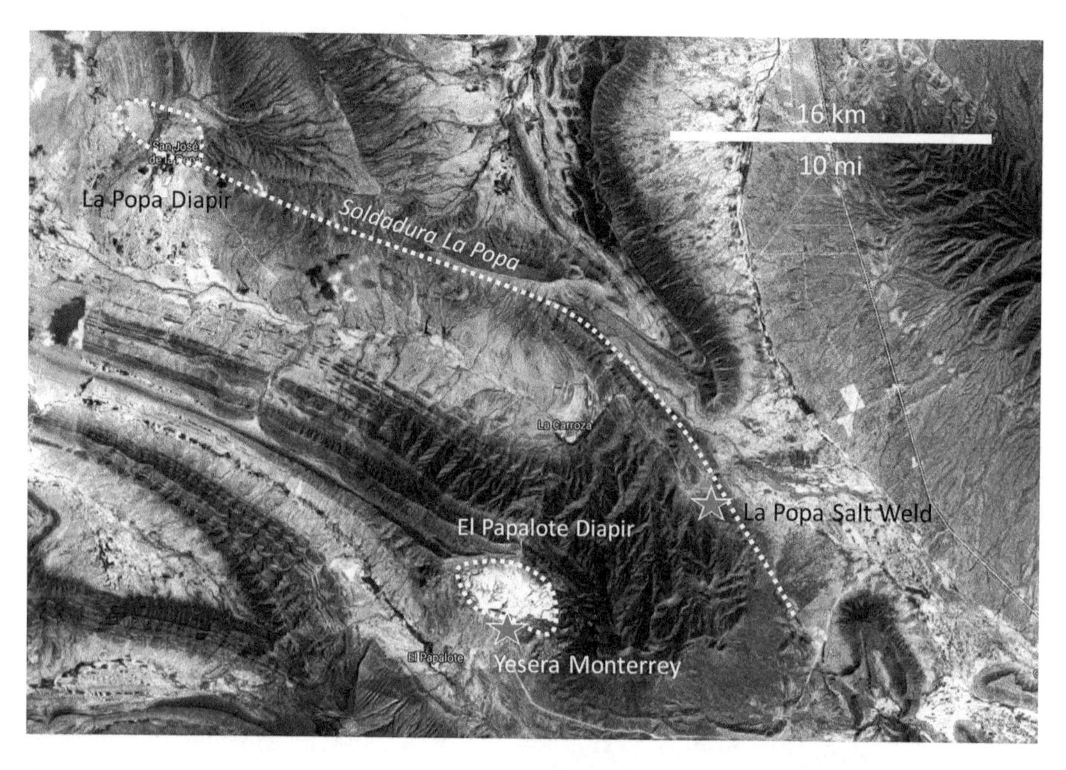

Google Earth satellite image of the Soldadura La Popa and El Papalote stops. Imagery © 2019 CNES/Airbus; Landsat/Copernicus; Maxar Technologies.

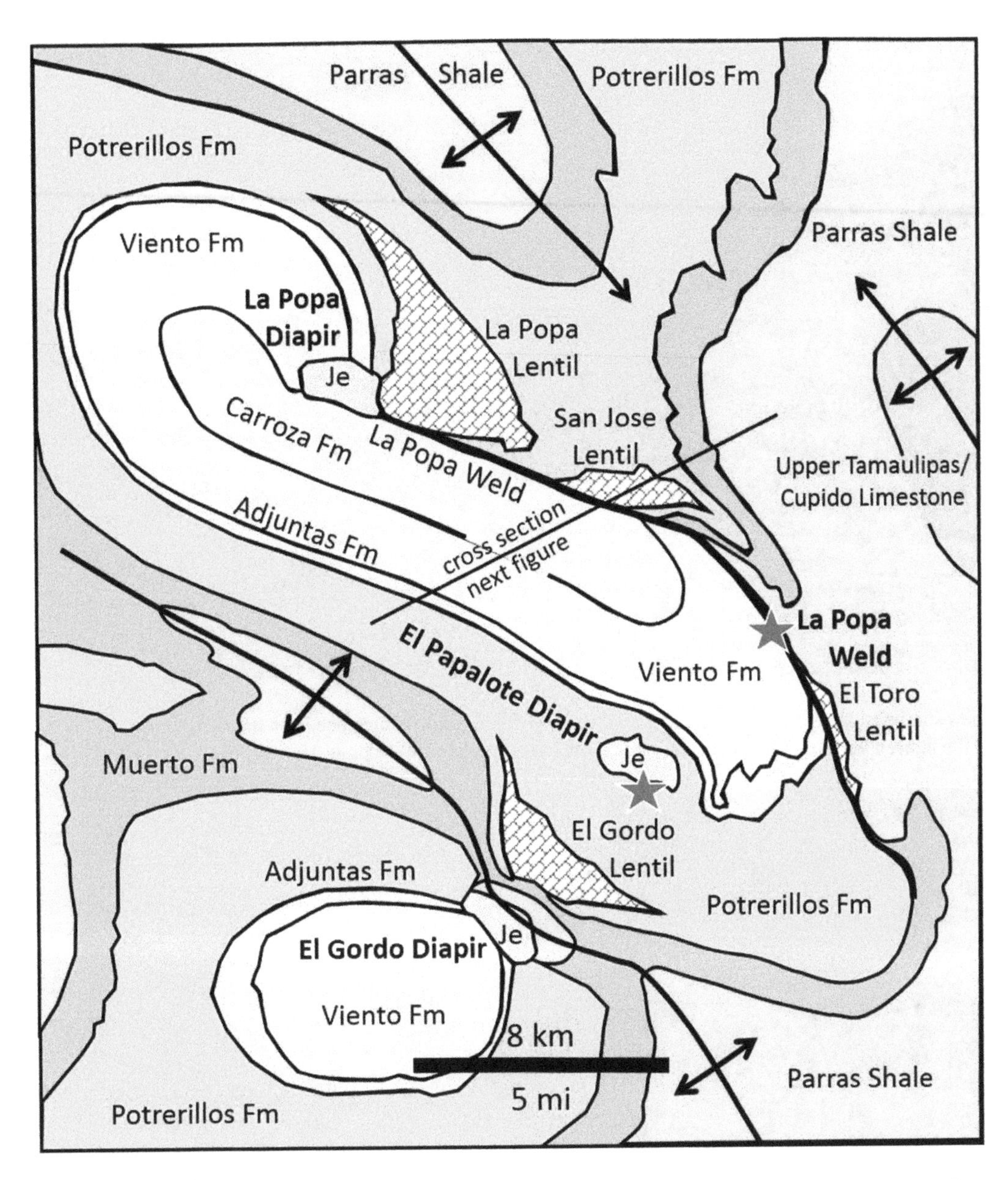

Simplified geologic map of the La Popa Basin. Je = Jurassic evaporite (salt and/or gypsum). Stars indicate stops. (Modified after Giles and Lawton, 2002.)

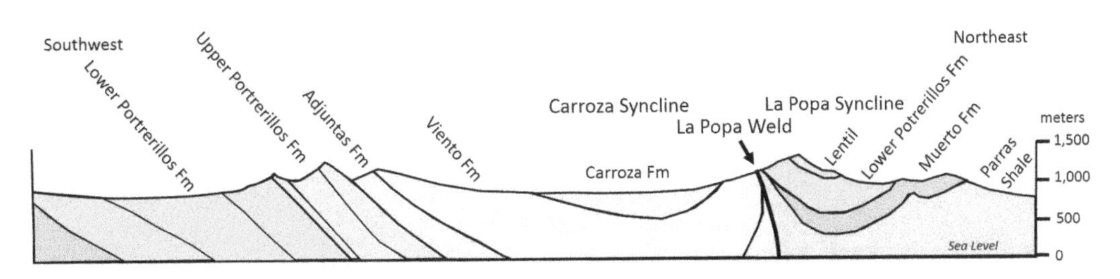

Cross section through the Soldadura near Stop 2. See geologic map for section location. (Modified after Smith et al., 2012.)

View southeast along the La Popa salt weld. Near-vertical Cupido Limestone is on the northeast side (left side of the figure); Tertiary Viento Formation is on the southwest (right) side.

Close-up of the Soldadura La Popa salt weld showing near-vertical shear fabric (broken, elongated, and aligned rock fragments). View southeast.

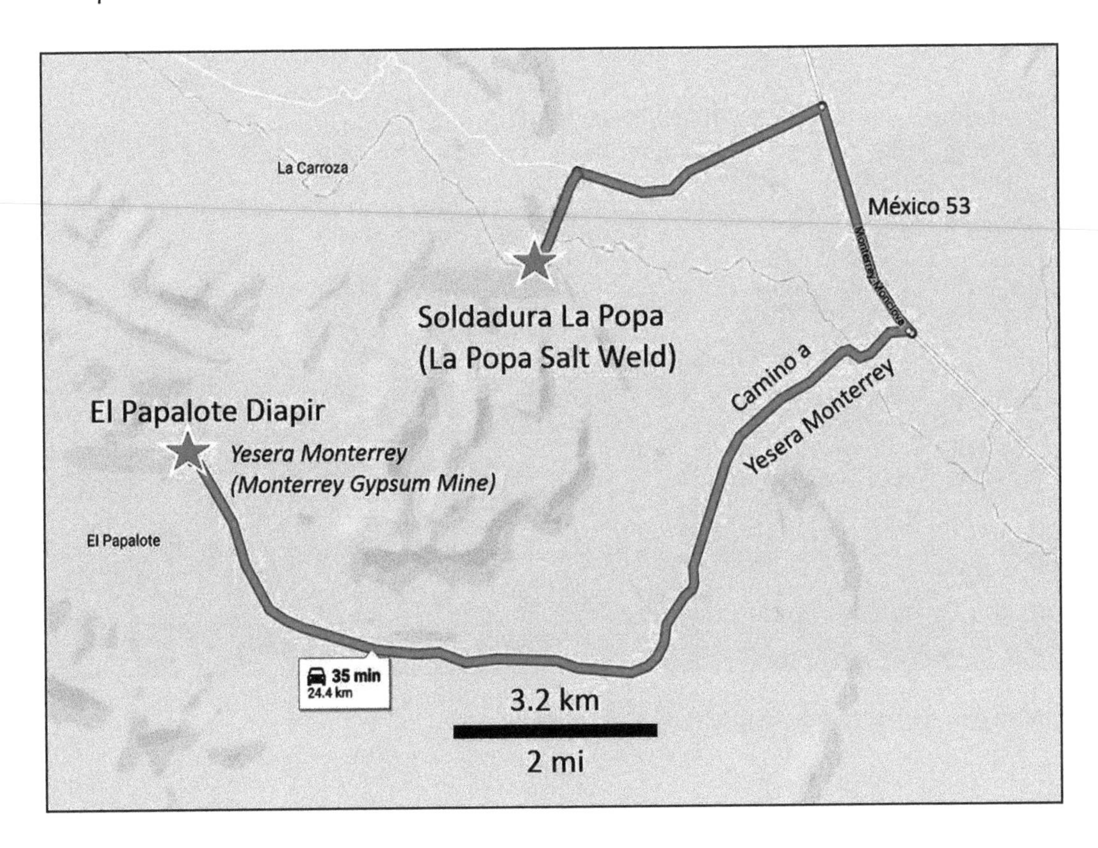

Soldadura La Popa to El Papalote Diapir: *Return to México 53 and turn right (south); drive 3.55 km (2.2 mi) to the curve in the highway and unmarked turnoff to the west (graded dirt road; no sign); turn right (west) onto the unnamed road and drive to Yesera Monterrey (open-pit gypsum mine) at* ***Stop 3, El Papalote Diapir*** *(26.066681, −100.749020), a total of 24.4 km (14.6 mi; 35 min). Stop at gate and request permission to enter the mine.*

STOP 3 EL PAPALOTE DIAPIR

Gypsum is being actively mined at this diapir. You can see flow structures, folding, and limestone inclusions within the gypsum. Layers adjacent to the diapir show thinning and intense deformation. Patch reefs ("lentils") that grew on the rising salt dome can be seen outside the mine. Typically, these reefs grow at the high point, are split into two by the rising diapir, and are rotated to the side as the dome continues to rise, and new reefs grow on the high point during the next period of stability (Vega and Lawton, 2011).

Lower Cretaceous reef lentils adjacent to the El Papalote Diapir.

Limestone inclusion encased in gypsum of the El Papalote Diapir, Yesera Monterrey mine.

Flow-banded gypsum in the El Papalote Diapir, Yesera Monterrey mine.

As you leave the Yesera, the west-inclined ridges north of the road contain the Cretaceous-Tertiary (KT) boundary just below the ridge top.

View north to the K-T boundary just east of Yesera Monterrey.

Visit

Mine administration prefers advance notice and favors student groups. They require that you sign a safety waiver and use a guide while in the mine.

Address: Yesera Monterrey, Serafín Peña 938 Sur, Centro, 64000 Monterrey, N.L., Mexico.
Phone: +52 (81)8345.1200 or +52 81 8345 1122
Email: infoyesera@gpromax.com
Website: www.yeseramonterrey.com/

El Papalote to San Blas View: *Return to México 53 and turn right (south); just before the town of Mina turn right (west) on the road to San Antonio de Arista; continue west on paved road till see* *sign to "Garcia 19 km"; turn left (southeast) and take Camino A Icamole to Road 36 in Garcia; turn right (south) on Road 36 to México 65; turn right (west) on México 65 and drive southwest, south, and southeast to México 40/Eje Metropolitano 910; turn right (southwest) onto México 40 and drive to Nuevo Leon 100 (Anillo Periférico de Monterrey); turn right (southwest) on NL 100 and drive to México 40; turn left (east) on México 40/Carretera Monterrey-Saltillo and drive toward Santa Catarina; bear right (southeast) onto Av Manuel Ordoñez/Road 30; drive to Zaragoza and turn right (south); drive until Zaragoza becomes Cuajuco; at Av las Garza turn left (east) for one block; at Calle Miguel Hidalgo/Road 3/Av Miguel Aleman turn right (south) and drive on Road 20/ Av Dr Ignacio Morones Prieto into Parque La Huasteca; continue on the main gravel road to* ***Stop 4.1, San Blas View*** *(25.586872, −100.424998), for a total of 112 km (69.6 mi; 2 h 14 min).*

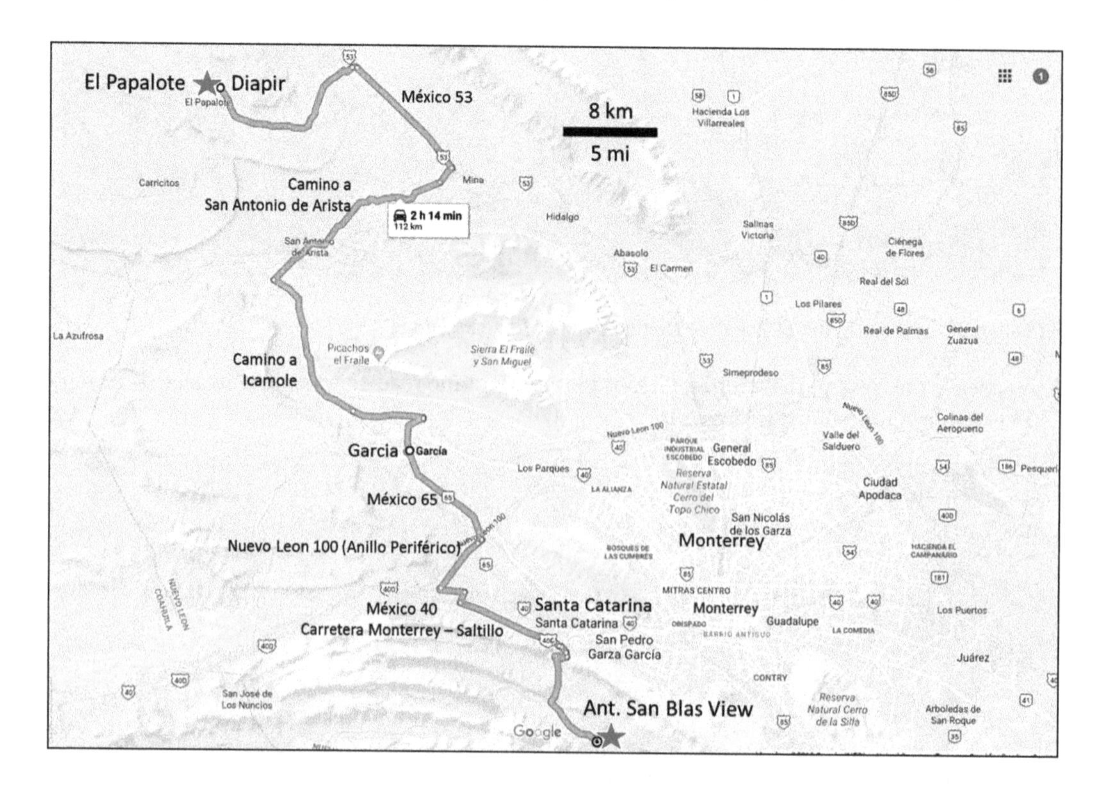

SIERRA MADRE ORIENTAL

We will now enter the Sierra Madre Oriental. The SMO foreland north of Monterrey consists of the La Popa Basin, the Coahuila Uplift to the northwest, and the Picachos Arch to the northeast. East of Monterrey lies the Magiscatzin Basin, a north-northwest-trending trough that connects the La Popa Basin with the Tampico-Misantla Basin to the southeast. The Magiscatzin Basin is bounded by the Tamaulipas Arch on the east (Lehman et al., 2000). These basins (valleys) and uplifts (ranges) developed during the Triassic-Jurassic opening of the Gulf of Mexico. Basement rock, exposed in some of the uplifts, is Paleozoic granite and late Paleozoic (Pennsylvanian-Permian Ouachita-Marathon Orogeny) metamorphic rocks. We last saw these rocks (not metamorphosed) in the Marathon and Solitario uplifts of west Texas.

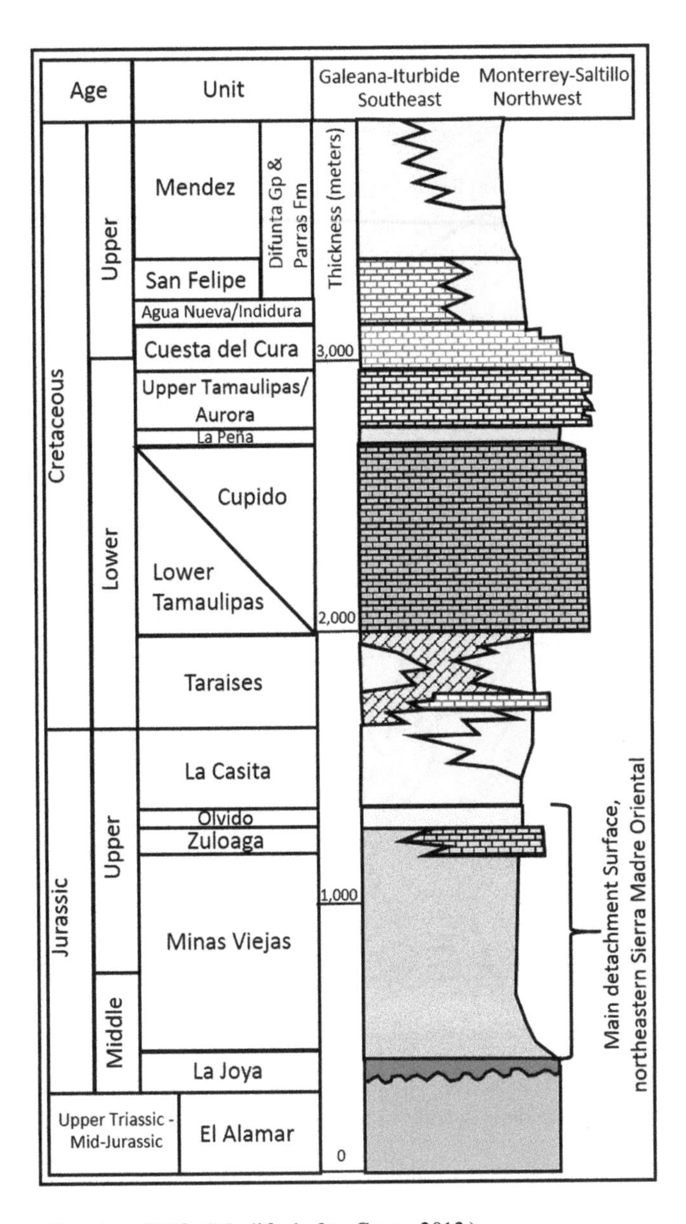

Stratigraphy of the northeastern SMO. (Modified after Cross, 2012.)

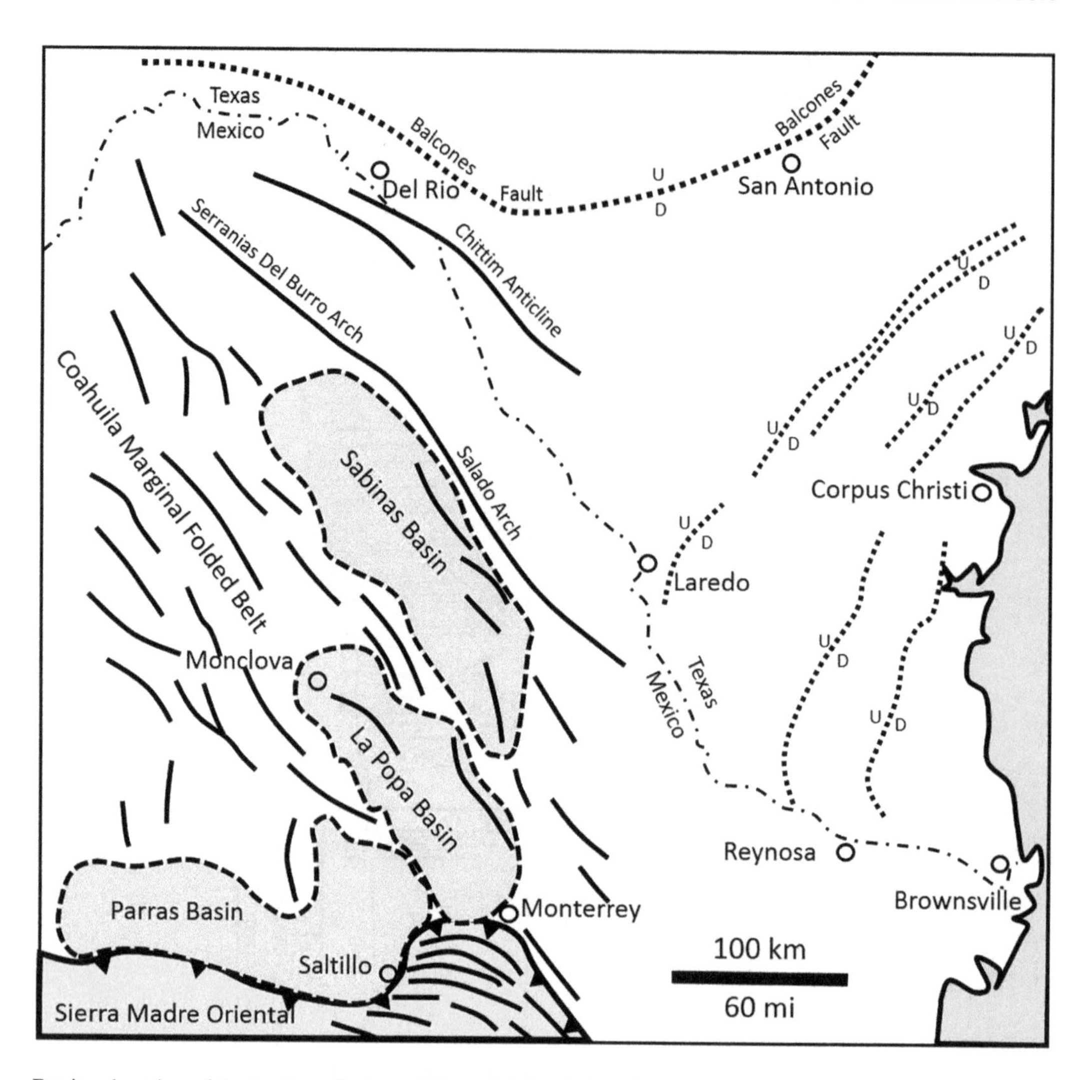

Regional setting of the La Popa Basin and Sierra Madre Oriental. Solid lines represent anticlines and uplifts; dotted lines are normal faults (D = down side; U = up side). (Modified after Fischer and Jackson, 1999.)

The basins fronting the Sierra contain up to 7 km (23,000 ft) of sediments ranging in age from Late Triassic to Eocene. Many of these units can also be seen in the SMO and correlate to formations found in the subsurface along the U.S. Gulf Coast. The Late Triassic-Early Jurassic section is primarily redbed sandstone and shale with some volcanics. In this area, the units include the Upper Triassic El Alamar and Lower-Middle Jurassic La Boca formations, equivalent to the Gulf Coast Eagle Mills Formation. These are overlain by Upper Jurassic evaporites (salt and gypsum): the La Joya, Olvido, and Minas Viejas formations reflect initial marine flooding and are equivalent to the Louann Salt of the Gulf Coast. Evaporites of the Olvido Formation contain salt and gypsum north of Monterrey and mainly gypsum south of there. These evaporites form the main detachment surface in the Monterrey Salient of the SMO. Carbonate ramp and nearshore terrigenous (land-derived) deposits of the Zuloaga, La Joya, and La Gloria formations (Late Jurassic) surround the Coahuila peninsula uplifted block. This succession is overlain by Late Jurassic (Kimmeridgian-Tithonian) shale and sandstone of the La Casita Formation that is equivalent to the Cotton Valley-Buckner strata along the Gulf Coast. In latest Jurassic time, La Casita deep basin limestone was deposited across the La Popa-Sabinas Basin and graded laterally into shales and sandstones around the Coahuila peninsula.

As the influx of sand and shale from the Coahuila peninsula decreased in the Lower Cretaceous, a carbonate platform developed around the uplift. The Taraises Formation open-shelf carbonate was followed by the Cupido Formation, massive reef limestones. Both the Taraises and Cupido changed laterally into the basinal Lower Tamaulipas Formation limestone. The Cupido is equivalent to the Sligo of the Gulf Coast. The Cupido platform limestones were drowned by marine shale of the La Peña Formation (Pearsall equivalent), which thickens south and west off the platform. Above this is the basinal limestone of the Cuesta del Cura/Aurora formations, equivalent to the tidal flat and lagoonal Glen Rose-Fredricksburg section in Texas. Upper Cretaceous strata include interbedded deep-water limestone and shale of the Agua Nueva-San Felipe (Eagle Ford and Austin Chalk), and Parras Shale. Shale becomes much more common up-section. The uppermost Cretaceous is represented by a marine shale (Mendez Formation) or interbedded shale and sandstone (Difunta Group foreland facies) around the margins of the Coahuila Uplift. These are time equivalent to the Taylor-Navarro section in Texas. These units are capped by local Paleocene-Eocene sandstone and shale units topped by an unconformity related to the Laramide-age uplift of the SMO. The youngest unit is recent alluvial material.

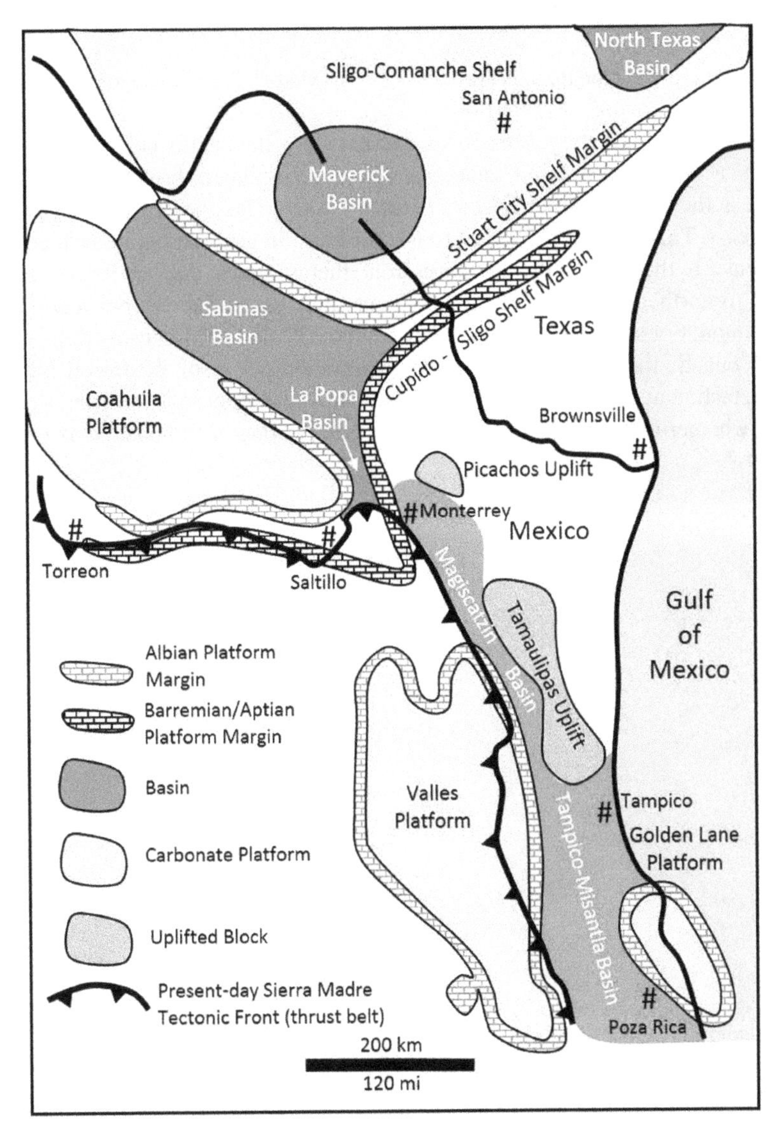

Lower Cretaceous paleogeography, Sierra Madre area. (Modified after Lehman et al., 1999.)

Deformation in the SMO consists of long, east-west to north-south-oriented Laramide-age folds. In the Monterrey Salient, those folds turn a corner from east-west to north-south and are detached above Jurassic gypsum.

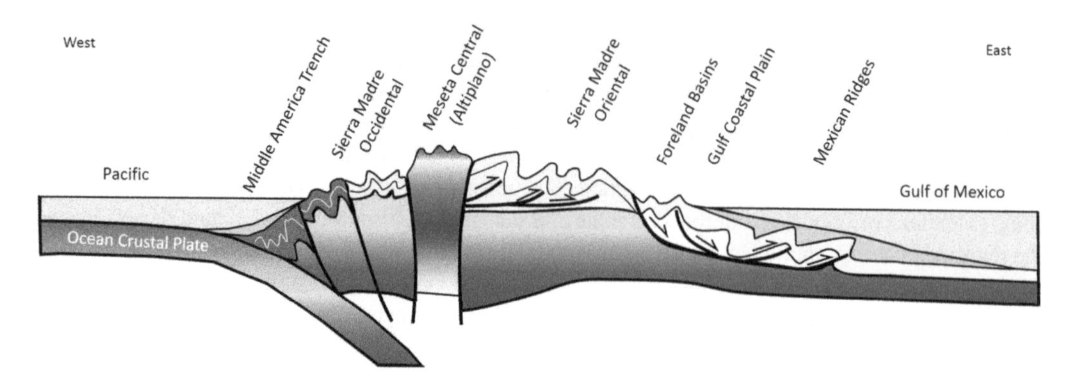

Schematic west-east cross section through northern Mexico. (Modified after Longoria, 1998.)

Probably as a result of convergence and subduction along the Pacific coast of Mexico, uplift and thrusting in the SMO began in late Cretaceous time and continued through Eocene: the youngest unit deformed is the Maastrichtian-Eocene Difunta Group. The age of thrusting gets younger to the north and east. Thrusting was directed to the northeast in general, but in each area, shortening was perpendicular to the trend of the mountain front, that is, almost due north between Torreon and Saltillo, north to northeast in the Monterrey Salient, and toward the east between Monterrey and Linares. The amount of shortening is estimated around 33% in the Monterrey Salient and between 40% and 50% outside the salient. The style of deformation is a result of several factors including whether the detachment surface is weak (low friction as in evaporites) or strong (high friction as in shale), and whether units in a given area are thick and strong or thin and weak (the mechanical stratigraphy).

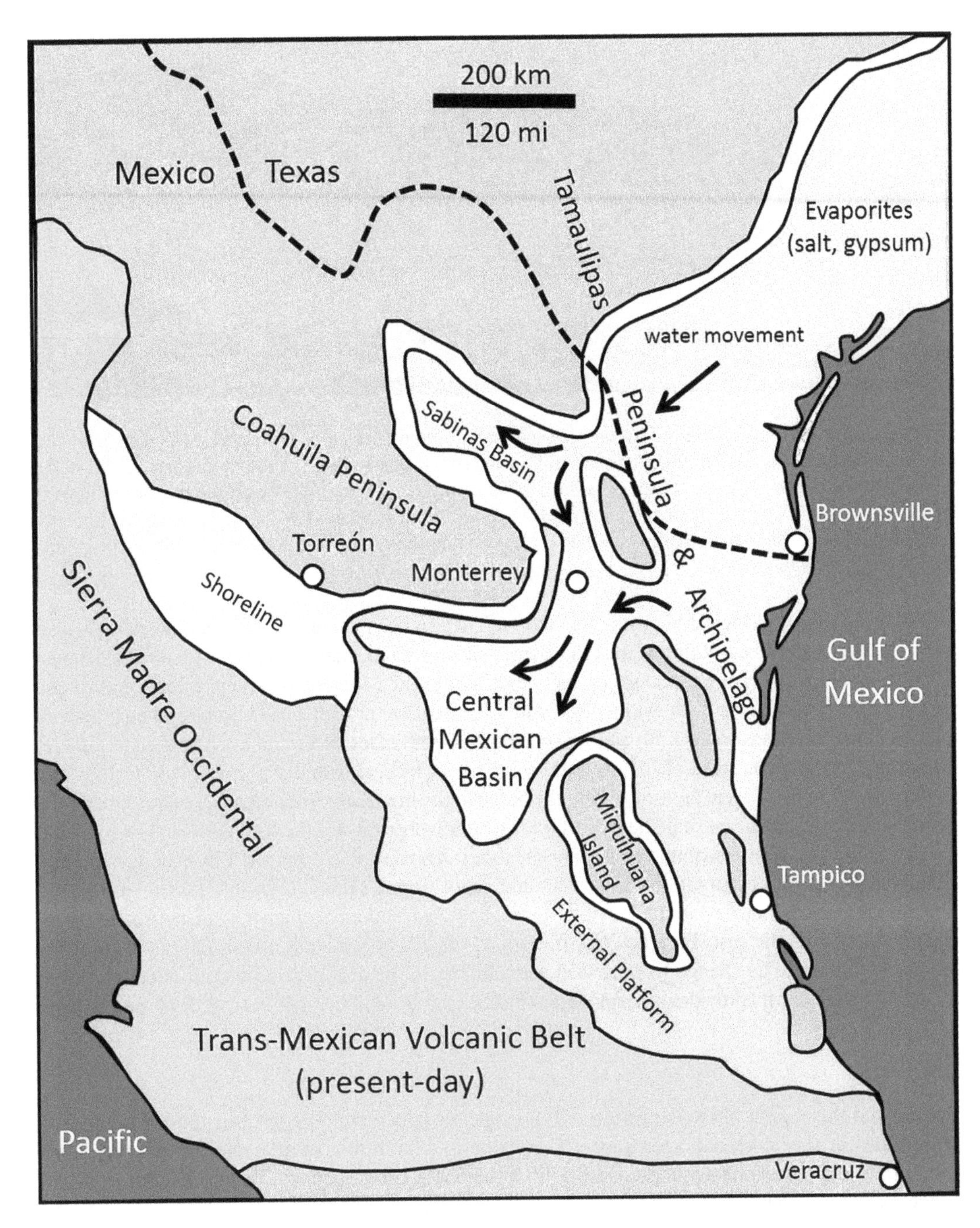

Oxfordian (Upper Jurassic) paleogeography and distribution of evaporites, including the Minas Viejas, Zuloaga, and Olvido formations. Arrows show movement of seawater to areas of evaporation. (Modified after Eguiluz de Antuñano et al., 2000.)

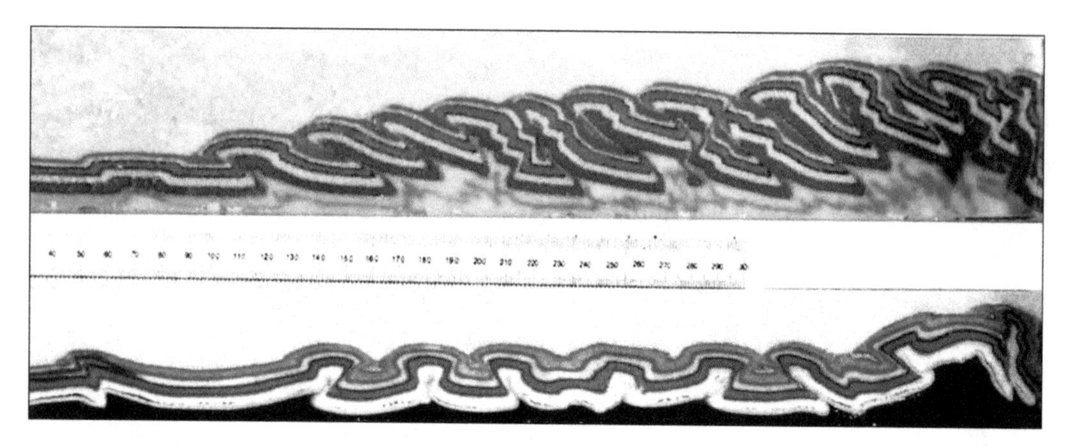

Sandbox models of thrust styles with strong (top) and weak (bottom) detachment surfaces. Strong detachments (e.g., on shale horizons) tend to form stacked, imbricated thrust sheets. Folds in the Monterrey Salient of the Sierra Madre Oriental are detached on gypsum, a weak surface, and form folds that resemble those shown in the bottom model. They are characterized by few faults to the surface and folds with no preferred vergence (inclination); that is, they can be overturned on both sides (fan folds). (From Rowan et al., 2004; experiments carried out by B. Vendeville.)

The initiation of folding and uplift in the SMO led to a switch from mainly limestone-shale deposition in the Jurassic and early Cretaceous to mainly sandstone and shale eroded from Late Cretaceous-Paleogene mountains rising to the south and west. Within the Monterrey Salient, a Jurassic salt and gypsum detachment surface allowed folding with large vertical or overturned limbs and no preferred fold vergence: the overturned limb can be on either or both sides of the fold. Thrusts that come to the surface are rare. Because of the weak detachment surface, deformation propagated farther into the foreland than in adjacent areas, forming the bulge in the mountain front. Outside of the salient, the main detachments occur on shale beds, and folds tend to verge toward the foreland (they lean toward the north or east). The distribution of evaporites versus shales was controlled by paleogeography: closed basins allowed evaporation of seawater and deposition of salt and gypsum/anhydrite.

According to Eguiluz de Antuñano (1991), Upper Jurassic evaporites pinch out against ancient basement highs both to the east (Picachos Uplift, Tamaulipas Arch) and west (Coahuila Platform) of the Monterrey Salient. The change from a weak detachment in the anhydrite to a sticky detachment in shale around the uplifts provides an explanation for the curvature of the folds in the Monterrey Salient.

Resources

Throughout this region, there is mining of limestone for concrete and aggregate, and local mining of gypsum for sheetrock and agriculture. Travertine is mined on a small scale for building stone and sculptures. Unfortunately, there is no gold in the SMO. Small deposits of silver, lead, zinc, and copper have been found in the far southern part of the range.

To date, the SMO has not had any oil or gas discoveries. The only hydrocarbons found in the SMO were sampled from a water well at the pueblo of San Antonio, about 25 km southeast of Los Chorros Canyon. The gas was 86% methane, 13.6% carbon dioxide, and 0.36% ethane. It came from fractures in the Parras Formation in the core of a syncline. The source was never determined (Eguiluz de Antuñano, 1992). Deep burial and high heat flow again suggest that any hydrocarbons encountered would probably be gas.

The Tampico-Misantla Basin, a foreland basin located just east of the SMO, is a major oil-producing area (Peterson, 1985). Oil was discovered at Ebano in 1904, and with the Golden Lane (Faja de Oro) discovery in 1908, this basin quickly became a world-class producer. Perhaps the greatest well of all time, the Cerro Azul #4, was drilled to a depth of 534 m (1,752 ft) during 1915–1916 and bottomed in Lower Cretaceous cavernous limestone. As a result of high pressure, the well blew out with a 183 m (600 ft) high gusher that flowed at the incredible rate of 41,472 m^3/day (260,850 barrels/day; Blickwede and Rosenfeld, 2011). The Mexican Petroleum Company, controlled by California oilman Edward Doheny, drilled the well. When the well blew out, the sound could be heard 26 km (16 mi) away. Over the next 14 years, Cerro Azul #4 produced over 9.06 million m^3 (57 million barrels) of oil. Doheny formed the Pan American Petroleum and Transport Company; the Mexican portion would later become PEMEX. By 1921, based on production from the Tampico-Misantla Basin, Mexico was the second largest oil producer in the world, with an annual production of 30.7 million m^3 (193 million barrels). The Poza Rica Field was discovered in the basin in 1930: it produces oil from high-porosity Cretaceous reef limestone, and is one of the largest stratigraphic traps in the world (Guzmán, 2013).

The gusher at Cerro Azul no. 4. https://commons.wikimedia.org/wiki/File:MEDI_D099_The_gusher_of_the_Cerro_Azul_oil_well.jpg.

The Ebano-Pánuco Trend in the basin produces oil from Upper Jurassic and Cretaceous fractured limestone. Production is located on high blocks uplifted during the Laramide Orogeny. In the Tuxpan Platform, production is from Lower Cretaceous fringe reefs and platform carbonates at the Golden Lane and El Abra trends, and limestone breccia in the Poza Rica Trend. West of the Poza Rica Trend and close to the mountain front is the Chicontepec Trend (Cruz-Helú and Meneses-Rocha, 1998). Oil in this fairway is trapped in Paleocene-Eocene turbidites (deep-water landslide deposits).

Mesozoic and Tertiary strata have produced 1.03 billion m^3 (6.5 billion barrels) of oil equivalent as of 2012 (Padilla y Sánchez, 2014). Few discoveries have been made in this basin since the 1970s, and production has fallen off sharply. Out of 225 oil and gas fields in the basin, only 120 are producing. PEMEX and the Mexican government are working to reverse the slide in production. As of 2012, the U.S. Geological Survey estimates 874 million m^3 (5.5 billion barrels) of undiscovered conventional oil in the basin in all reservoirs and 402 billion m^3 (14.2 trillion ft^3) of conventional gas (USGS, 2012).

4 CAÑON HUASTECA

Driving from Santa Catarina south into Cañon Huasteca, you are entering the frontal folds of the Monterrey Salient of the SMO. These folds, developed in Upper Jurassic and Lower Cretaceous limestones, are all detached above a Jurassic anhydrite (gypsum) layer.

STOP 4.1 ANTICLINAL SAN BLAS

This stop is just north of Anticlinal San Blas (San Blas Anticline), which is developed primarily in Lower Cretaceous Cupido Formation limestone (Vokes, 1963). The folds here are enormous and appear to be kink folds with long, relatively straight flanks separated by sharp hinges.

There is some debate as to whether these folds developed by "hinge migration" or "buckling and rigid rotation." In hinge migration, all the layers in the vertical limbs of the fold passed through a 90° bend and would show the resulting fracturing. In buckling and rotation, the hinges, once developed, remain in a fixed location and the fold limbs rotate as relatively straight panels from horizontal to vertical between the hinges. The consensus is leaning toward buckling and rigid rotation because the fold limbs do not show features (such as intense fracturing) associated with having moved through a fold hinge.

View west to Anticlinal San Blas from Santa Catarina River (dry). The fold is outlined by Cupido Formation limestone (dashed lines) rising 850 m (2,800 ft) above the valley floor.

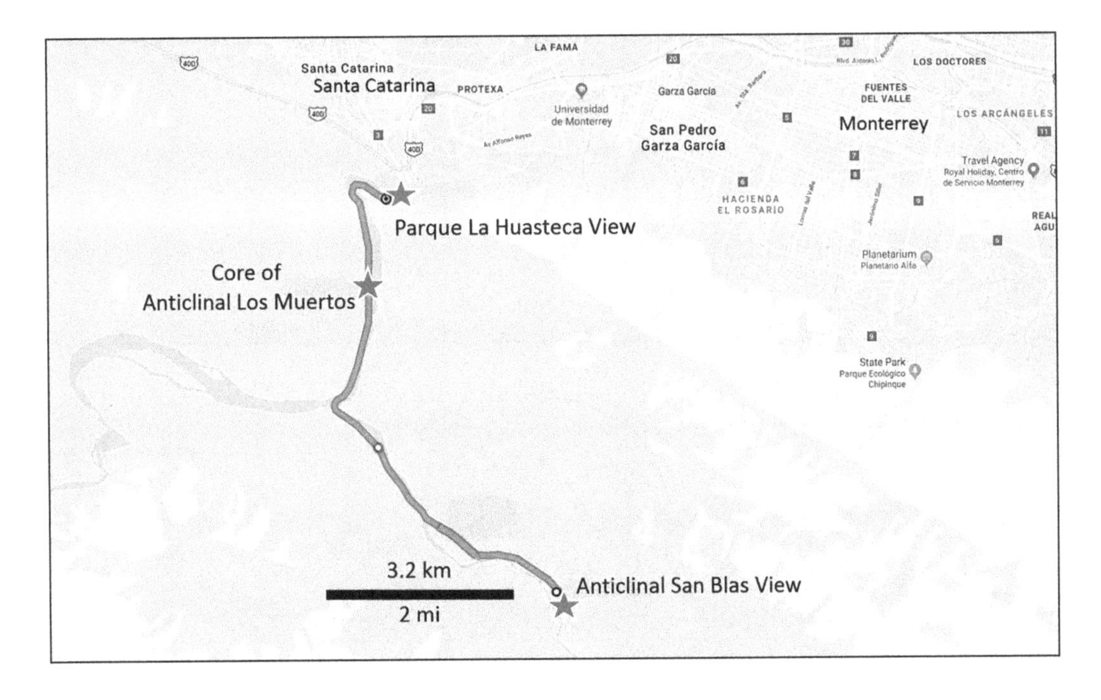

Anticlinal San Blas to Anticlinal de Los Muertos: *Return north on the gravel road 7.6 km (4.7 mi; 13 min) to* ***Stop 4.2, Core Anticlinal de Los Muertos*** *(25.634501, −100.457914) just north of the village of Nogales.*

STOP 4.2 CORE OF ANTICLINAL DE LOS MUERTOS

The east-west valley here lies in the eroded core of this huge anticline. The core contains outcrops of Upper Jurassic Zuloaga Limestone, and the covered areas consist of La Casita Formation shale and Taraises Formation shaley limestone. The near-vertical walls are 1,000 m (3,000 ft) of Lower Cretaceous Cupido Formation limestone. The base of the Cupido is a thick-layered reef limestone. Los Muertos Anticline, like the others in the Monterrey Salient, is detached on Jurassic gypsum in the shallow subsurface.

View west along the core of Anticlinal de Los Muertos.

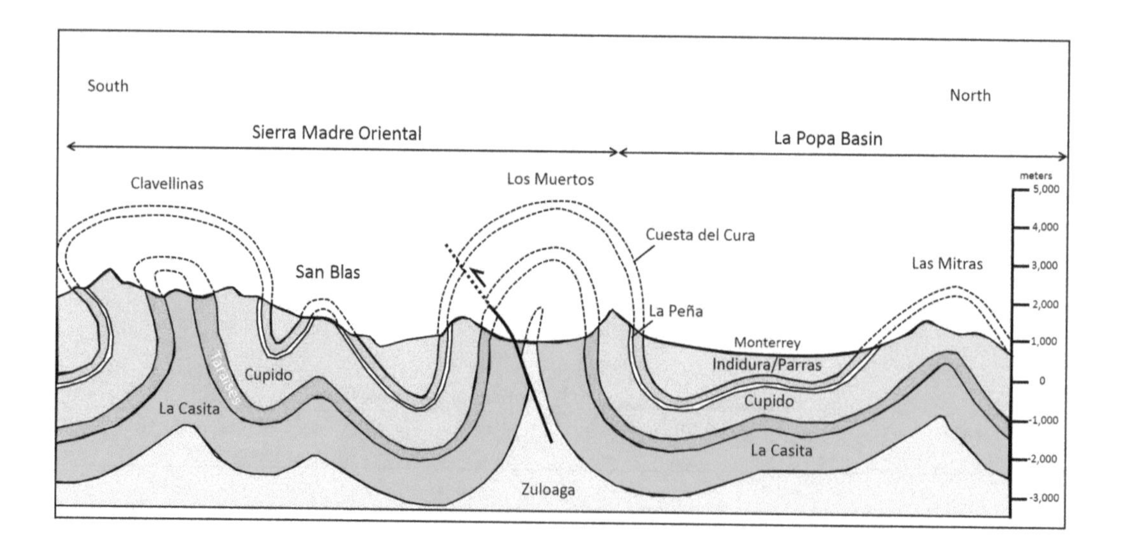

Cross section through the frontal folds of the Monterrey Salient, Sierra Madre Oriental, just west of Monterrey. (Modified after South Texas Geological Society, 1999.)

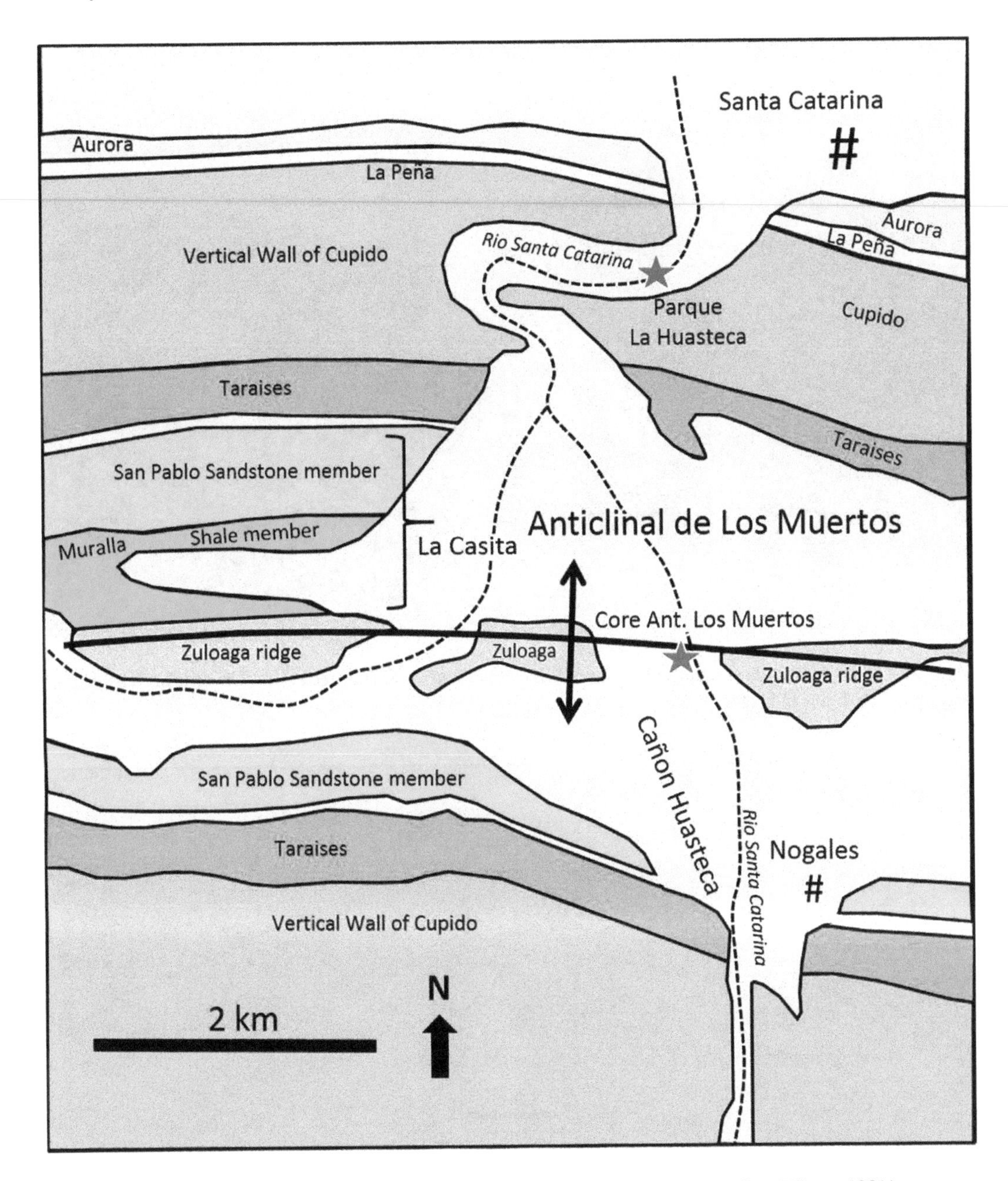

Geologic map of Anticlinal de Los Muertos and Parque la Huasteca. (Modified after Wilson, 1981.)

Anticlinal de Los Muertos to Parque La Huasteca: *Continue driving north 2.5 km (1.6 mi; 5 min) to* ***Stop 4.3, Parque La Huasteca*** *(25.648265, −100.454869).*

STOP 4.3 PARQUE LA HUASTECA

This stop allows us to view the near-vertical north flank of Anticlinal de Los Muertos. Looking west, you can see impressive caverns that developed when groundwater was actively flowing through the limestone.

View west at vertical walls of Cupido Formation limestone, Parque la Huasteca view. Caverns are visible half way up the cliff, left of center.

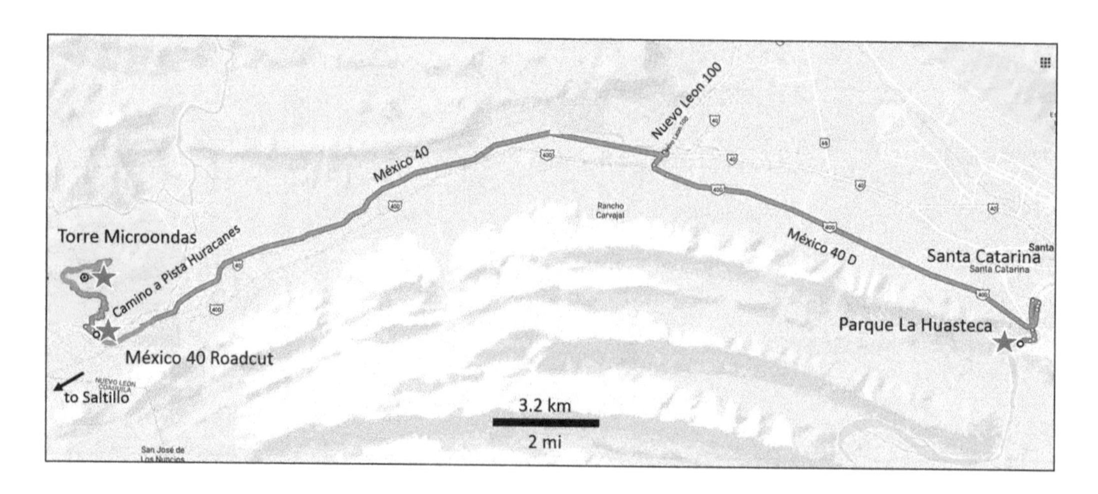

Parque la Huasteca to México 40 Roadcut: *Return north on Av. Dr. Ignacio Morones Prieto to Av. Miguel Aleman; bear left onto Miguel Aleman and drive north to Benito Juarez; bear right onto Benito Juarez and drive north to Av. Montana; turn right (east) on Av. Montana and drive to Autopista Monterrey-Saltillo/México 40 D Cuota (toll road); turn right (south) on México 40 D and drive south and west to Nuevo Leon 100/Anillo Periférico de Monterrey; turn right (north) on Nuevo Leon 100 and take it to Carretera Monterrey-Saltillo Libre/México 40; turn right to go west on México 40 and drive to pullout on the right by shrines before a large roadcut at* ***Stop 5, México 40 Roadcut*** *(25.648394, −100.745380), a total of 43 km (26.7 mi; 54 min).*

STOP 5 MÉXICO 40 ROADCUT

Stop 5 is in lightly metamorphosed shales of the Upper Cretaceous Parras Shale. Note the well-developed "pencil cleavage." Pencil cleavage here refers to the slender, pencil-shaped rock shards created by intersecting bedding and axial planar cleavage, a set of fractures developed more-or-less parallel to the axial plane of folding. The cleavage suggests a fair degree of burial since the end Cretaceous.

Parras shale in México 40 roadcut.

Parras Shale shards exhibit well-developed pencil cleavage. Centimeter scale shown.

México 40 Roadcut to Torre Microondas: *Drive west on México 40 for 0.4 km (0.24 mi; 1 min) to unmarked turnoff on the right just beyond roadcut; turn right onto dirt road to PEMEX Station 7 and Pista Huracanes; drive 225 m (728 ft) east to unmarked road on the left just before Pista Huracanes; turn left and drive on a rock-paved road to* ***Stop 6, Torre Microondas*** *(Microwave Tower: 25.663961, −100.749169) on top of the mountain, a total of 6.6 km (4 mi; 25 min).*

STOP 6 TORRE MICROONDAS

From this vantage point at 1,620 m (5,315 ft) elevation, you can see the sweeping east-to-west arc of Anticlinal de Los Muertos, the frontal fold of the Monterrey Salient to the south, as well as folding in the Upper Cretaceous to Paleocene Difunta Group to the west and north of this hill. The Los Muertos Anticline here is a box fold characterized by long, straight flanks and abrupt changes in dip at fold hinges. The fold is overturned toward the north in Cortinas Canyon southeast of here and in Los Nuncios Canyon southwest of here. Folding in the Difunta Group shales and sandstones has no apparent vergence; that is, it is more-or-less symmetrical and the fold axis is not inclined in any particular direction.

View west from the microwave tower to open folds in the Difunta Group.

View southeast from the microwave tower to the overturned north limb of Anticlinal de Los Muertos at Cortinas Canyon. Cliffs are Lower Cretaceous Cupido Limestone. A small parasitic fold is enlarged on the right.

View southwest from microwave tower to the overturned north flank of Anticlinal de Los Muertos at Los Nuncios Canyon. A light fog obscures the vista.

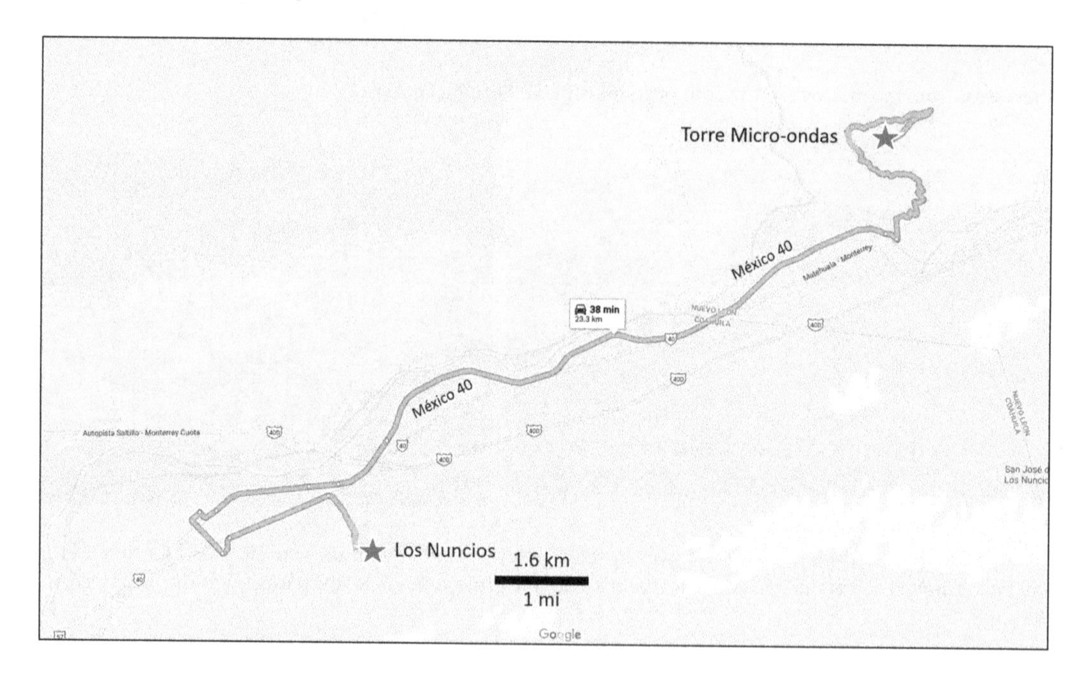

Torre Microondas to Los Nuncios: *Return to México 40 and turn right (west); drive west on México 40 for 12.9km (8 mi) to Cementera exit on right; exit México 40 to Holcim Planta Ramos Arizpe; take first left turn (southwest), then left again (southeast) to the underpass beneath México 40; drive to the onramp for eastbound México 40 and turn left (east) onto México 40; drive 1.46km (0.9 mi) east to turnout on the right (truck stop with blue-pink-green buildings marked "Baños" just before México 40 splits into Monterrey Libre and Monterrey Cuota); turn right (south) and drive*

*1.3 km (0.8 mi) on a graded track that runs beneath power lines to a gravel quarry at **Stop 7, Los Nuncios** (25.604421, −100.835815), for a total of 23.3 km (14 mi; 38 min).*

STOP 7 LOS NUNCIOS

Los Nuncios Canyon cuts into the north limb of Anticlinal de Los Muertos. You are looking at a Lower Cretaceous succession of Cupido limestone (main cliff-forming unit) and the overlying La Peña and Aurora formations. Mapping shows that Anticlinal de Los Muertos changes from overturned on both limbs (to the east) to just north-verging at the western end (Higuera-Díaz et al., 2005).

View east at north flank of Anticlinal de Los Muertos from a gravel quarry in Los Nuncios Canyon.

Looking west, you can see a small, tight subsidiary fold that is overturned to the north and broadens out to the west.

Looking west from Los Nuncios Canyon at a subsidiary fold on the north flank of Anticlinal de Los Muertos.

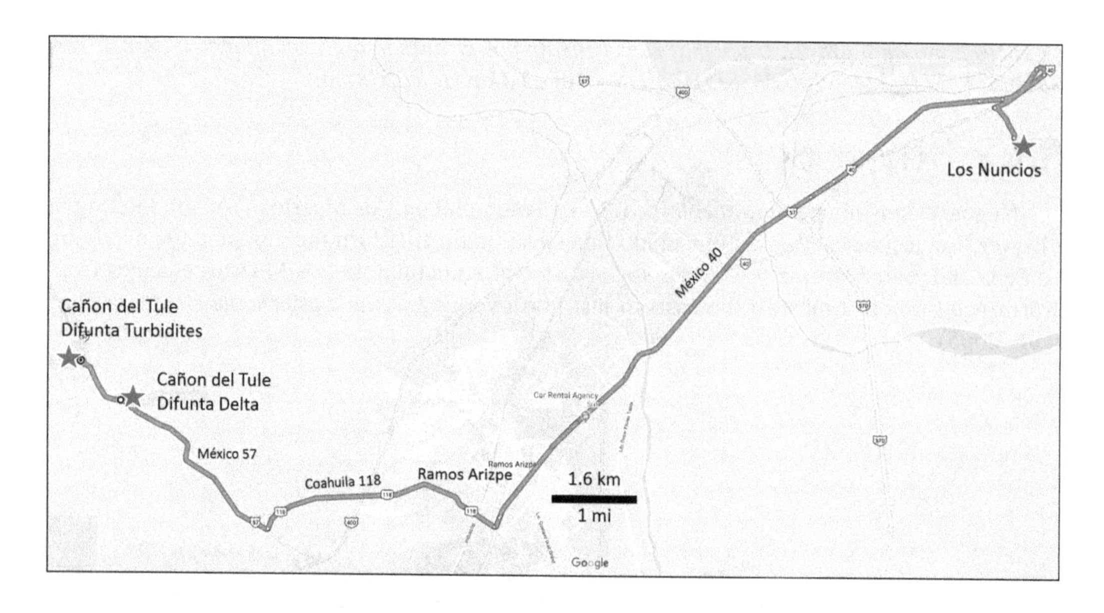

Los Nuncios to Cañon del Tule: *Return to México 40 and turn right (east); go straight (or slight left) to Monterrey Libre and take first turnoff on the left to get to México 40 west to Saltillo; take México 40 west to Ramos Arizpe; take exit on the right toward Los Pinos on Coahuila 118; drive west on Coahuila 118 to Castaños-Saltillo/México 57; turn right (northwest) onto México 57 and drive to pullout on the right after the sign to "San Joaquin las Vacas" and before the "no passing" sign. This is* ***Stop 8.1, Cañon del Tule*** *delta and shoreline outcrops (25.553143, −101.036960), a total of 30.4 km (18.9 mi; 30 min).*

8 CAÑON DEL TULE

Cañon del Tule contains excellent outcrops of the Upper Cretaceous to Eocene Difunta Group that records a deepening sea and sedimentation derived from the rising Sierra Madre to the southeast and southwest (Weidie, 1961).

STOP 8.1 DIFUNTA FORMATION DELTA AND SHORELINE

The red and green shales along the side of the road are in the Cerro Huerta Formation of the Late Cretaceous Difunta Group. The shale was deposited on a delta plain at or near sea level. As you walk around the curve to the north, the delta changes to beach sandstone, indicating that sea levels were rising. The PEMEX Encinas-1 well near here penetrated 4,000 m (13,100 ft) of Difunta and Parras shale and never got out of it.

North-inclined delta succession of the Difunta Group exposed in México 57 roadcut.

Cañon del Tule Difunta Delta to Difunta Turbidites: *Drive 1.2 km (0.7 mi; 1 min) north on México 57 to dirt track on the left at the end of the guard rail. Pull off on the left and park off the road at* ***Stop 8.2, Cañon del Tule Difunta Turbidites*** *(25.560800, −101. 045833). Be careful, this is a dangerous curve! Walk no more than 280 m (900 ft) to outcrops in the riverbed.*

STOP 8.2 DIFUNTA FORMATION TURBIDITES

As sea levels rose, sediments could only reach this area by way of turbidity flows, that is, a slurry of sand and mud that flowed downslope after accumulating near the coast to the point of becoming unstable. Each time there was a sediment flow, the larger grains would settle out first, followed by progressively finer material. This led to a series of coarse-to-fine packages of sandstone-siltstone-shale that is evident in the river bottom and in the outcrop across the valley. Scour marks at the base of each sandstone bed indicate the original flow direction was from west-southwest to east-northeast.

The sandstones and shales of the Difunta Group record the rise of the SMO in Late Cretaceous time.

Difunta turbidite outcrop, Cañon del Tule.

Alternating turbidite sequences of sandstone to shale. A five peso coin (2.5 cm dia.) is shown for scale.

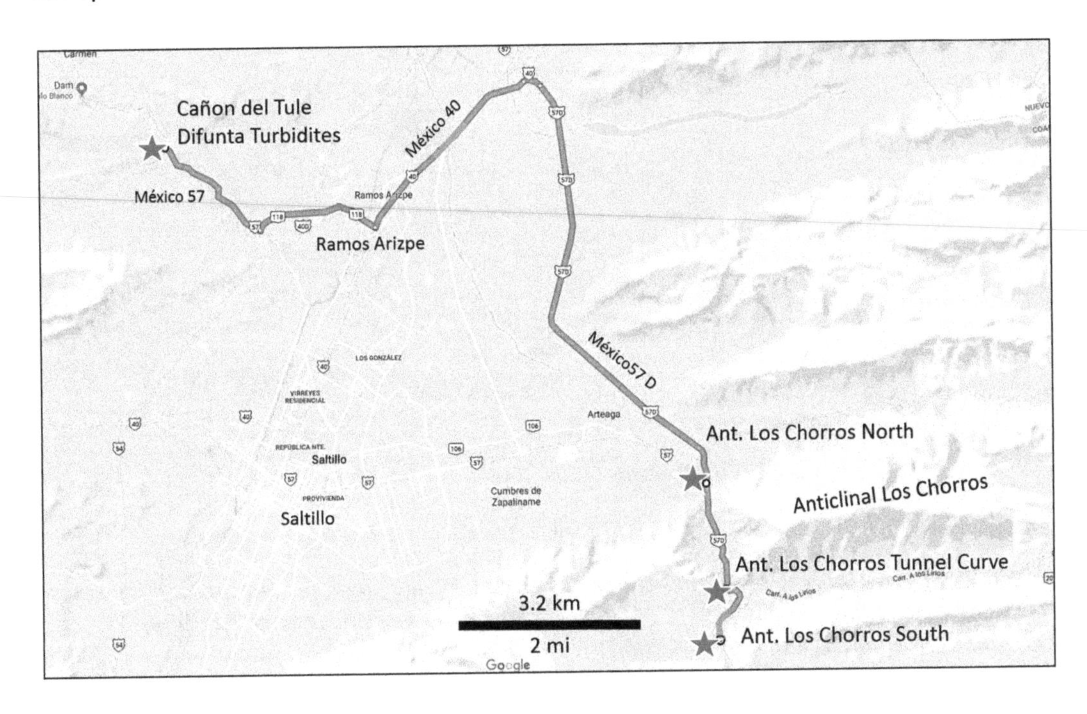

Cañon del Tule to Anticlinal Los Chorros North: *Return southeast on México 57 to Coahuila 118; turn left (northeast) onto Coahuila 118 and drive to Carretera Monterrey-Saltillo/México 40; turn left (northeast) onto México 40 and drive to Matehuala Cuota (toll road) exit on the right; merge onto México 57 D and drive south to the junction with México 57. Just past this junction there is a PEMEX station on the right. This is* ***Stop 9.1, Anticlinal Los Chorros North*** *(25.427408, −100.805555), for a total of 42.9 km (26.7 mi; 37 min).*

9 ANTICLINAL LOS CHORROS

The following three stops traverse Anticlinal Los Chorros by means of a canyon that cuts through it. You will see the overturned north and south limbs of this "fan fold" anticline, and intense deformation in a weak, thin-bedded unit within the fold.

This box fold has fairly straight inclined layers that abruptly change dip at pronounced hinges in the thick, strong platform limestone of the Cupido Formation. Stylolites, zigzag suture lines indicating where the limestone has been dissolved by pressure solution, are common and may have helped accommodate deformation in these beds (Gray et al., 1997). Interbedded shales show bedding plane slip, and thin, weak limestone layers can be highly deformed.

STOP 9.1 NORTH LIMB

From this vantage point, you can see the overturned north limb of Anticlinal Los Chorros. The Lower Cretaceous Cupido limestone is nearly flat along the crest of the fold and then goes vertical to overturned across a sharp, well-defined hinge.

View east from PEMEX station toward the overturned north limb of Los Chorros Anticline.

Anticlinal Los Chorros to Tunnel Curve: *Continue driving south on Matehuala-Monterrey/ México 57 D to pullout on the right at the highway tunnel curve. This is* ***Stop 9.2, Tunnel Curve, Anticlinal Los Chorros*** *(25.386215, −100.797014), a total of 4.9 km (3 mi; 6 min).*

STOP 9.2 TUNNEL CURVE

As you drive to the next stop, notice how the bedding goes from inclined to the north to nearly flat to inclined to the south. In the core of the anticline are calcareous shale, siltstone, and sandstone of the Upper Jurassic-Lower Cretaceous La Casita (and Carbonera?) Formation.

At this stop, beds of the Lower Cretaceous Cupido-La Peña -Tamaulipas Superior/Aurora-Cuesta del Cura succession (equivalent to the Hosston/Sligo to Buda series along the Gulf Coast) are inclined 60°–70° south. The weak, thin-bedded limestone-shale sequence of the Cuesta del Cura is intensely folded.

Tight folding in the weak, thin-bedded Upper Cretaceous Cuesta del Cura Formation shale- sandstone sequence. Chain link covers the outcrop to prevent falling rocks.

Tunnel Curve to South Limb, Anticlinal Los Chorros: *Continue driving south on Matehuala-Monterrey/México 57 D to Los Lirios/El Diamante exit on the right; take the exit and pull over on the right at the restaurant. This is* ***Stop 9.3, South Limb view, Anticlinal Los Chorros*** *(25.377370, −100.795767), a total of 1.8 km (1.1 mi; 2 min).*

STOP 9.3 SOUTH LIMB

Both the north and south flanks of Anticlinal Los Chorros are overturned. Units seen on the overturned south flank include the Lower Cretaceous Cupido-La Peña-Tamaulipas Superior formations.

Physical modeling suggests that doubly overturned folds with no preferred vergence are indicative of a weak detachment surface (Costa and Vendeville, 2002). In this area that would be the Jurassic Minas Viejas and Olvido anhydrites.

View northeast to the overturned south flank of Anticlinal Los Chorros from Los Lirios.

Side Trip 1—Rancho La Minita Thrust *(requires permission to enter private land): From South Limb Anticlinal Los Chorros, merge back onto Matehuala-Monterrey/México 57 D and continue driving south on México 57 D to Huachichil exit on right; in 3.5 km (2.2 mi) turn right again onto México 57 going west to Chapultepec and Jaguey; continue west to intersection with México 54; use right lane to merge onto México 54/Saltillo-Zacatecas going south; drive to sign for Los Angeles/Rural E-18; turn left (east) onto E-18 and drive 390 m (1,275 ft) to dirt track on left just past concrete structure; turn left and take main track leading north; in 529 m (1,727 ft) the track forks: stay on main track (left fork) to get permission at Rancho La Minita; go right to get to railroad tunnel cut. On main track, continue north 2.66 km (1.65 mi) to house on the right, Rancho la Minita. Return south to the fork and turn left; drive about 180 m (590 ft) to a gate. From the gate, walk the remaining 850 m (2,770 ft) to the railroad tunnel and climb the last few meters to the cut. This is* ***Side Trip 1, Rancho La Minita Thrust*** *(25.160744, −101.089718), a total of 68.25 km (48.4 mi; 1 h) drive and 30 minute walk.*

A few kilometers north of this stop, on the outskirts of Saltillo, lies the little village of Buena Vista, site of the Battle of Buena Vista (February 22–February 23, 1847) during the Mexican-American War. Known in Mexico as the Battle of Angostura, it was one of the largest battles of the war and saw use of artillery by the U.S. Army to beat back a much larger Mexican force. Buena Vista today is a village on the southern outskirts of Saltillo.

After the Battle of Monterrey, Major General Zachary Taylor's army moved on Saltillo. Part of Taylor's force under General Wool advanced to Agua Nueva, south of Saltillo, on December 21, to prepare for an impending attack. In early January, the Mexican leader General Santa Anna got word of this move and made plans to attack. Santa Anna's army reached Encarnación, south of Saltillo, with 15,142 men on February 20.

Taylor moved 4,650 of his men to Agua Nueva, but after a brief encounter with Santa Ana, he withdrew to Angostura, 2 km south of Hacienda San Juan de la Buena Vista. General Wool thought the site excellent for defense since the road passed through a narrow valley crossed by ravines east of the road and arroyos west of the road. Santa Anna advanced and demanded surrender, to which one of Taylor's aides replied, "I beg leave to say that I decline acceding to your request." In 2 days of intense fighting, the Americans, including a young Colonel Jefferson Davis, were driven back to the hacienda. An artillery battery under Captain Braxton Bragg was brought up with orders to "maintain the position at all costs." In a conversation with Taylor, Bragg mentioned he was using single canister shot. Taylor ordered him to "double-shot your guns and give 'em hell, Bragg." This later became a campaign slogan that carried Taylor into the White House. The Mexican attack was repulsed and fighting ended as rain fell. The battle ended in a draw as Santa Ana retreated south and Taylor did not pursue him.

The American wounded were taken to Cerralvo, a town east of Monterrey, where many of them stayed after the war. Apparently, there are a large number of red-headed descendants in that area today. Santa Ana lost his wooden leg in the encounter, and the Americans later traded it for the flag taken by the Mexicans from the Alamo.

Battle of Buena Vista. Hand-colored lithograph (1851) by Adolphe Jean-Baptiste Bayot, after a drawing by Carl Nebel. https://commons.wikimedia.org/wiki/File:Nebel_Mexican_War_03_Battle_of_Buena_Vista_(cropped).jpg

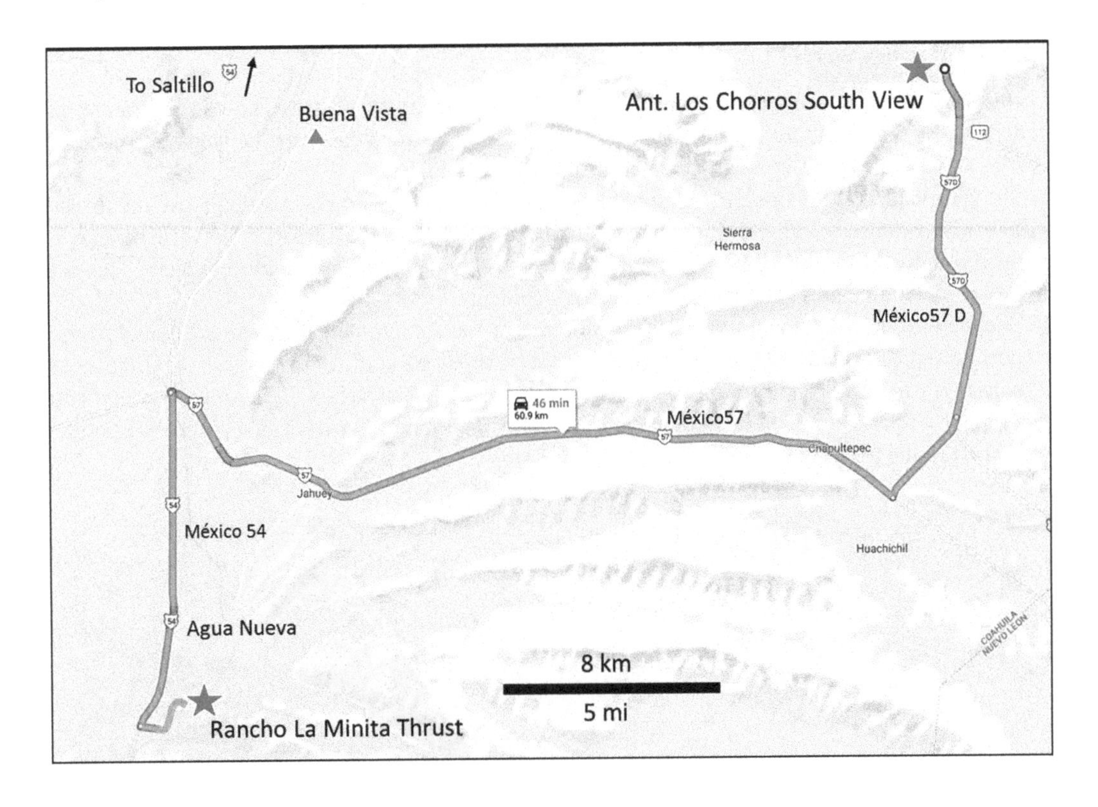

Google Earth image showing access to the Railroad Cut site, Rancho La Minita Thrust (informal name). Imagery © 2019 CNES/Airbus; Maxar Technologies.

SIDE TRIP 1 RANCHO LA MINITA THRUST

At this stop, you can see an example of an emergent thrust at the eastern end of the Transverse Ranges of the Sierra Madre Oriental. The Transverse Ranges trend east-west between Torreon and Saltillo and are the result of north-directed thrusting and folding.

The railroad cut at the tunnel has Jurassic carbonates of the Olvido Formation (above the thrust) pushed northward over Upper Cretaceous Parras Shale (below the thrust). The thrust plane is inclined roughly 30° to the south and contains broken rock in a gypsum/anhydrite matrix. This is a rare example of an emergent thrust in a weak detachment.

The view to the west from this stop is of the Sierra La Catana, which contains Upper Jurassic evaporites and carbonates of the Olvido and La Caja formations at the base, and Lower Cretaceous carbonates of the Taraises through Cuesta del Cura formations at the top. The mountain is thrust over Upper Cretaceous shale of the Parras Formation. The mountain front is a complex isoclinal fold thrust to the north.

Railroad cut looking east. Jurassic Olvido limestone and gypsum are thrust over Cretaceous Parras Shale. The shear zone contains flow-banded, elongated, and aligned rock fragments.

Touching the thrust plane. Railroad cut looking west.

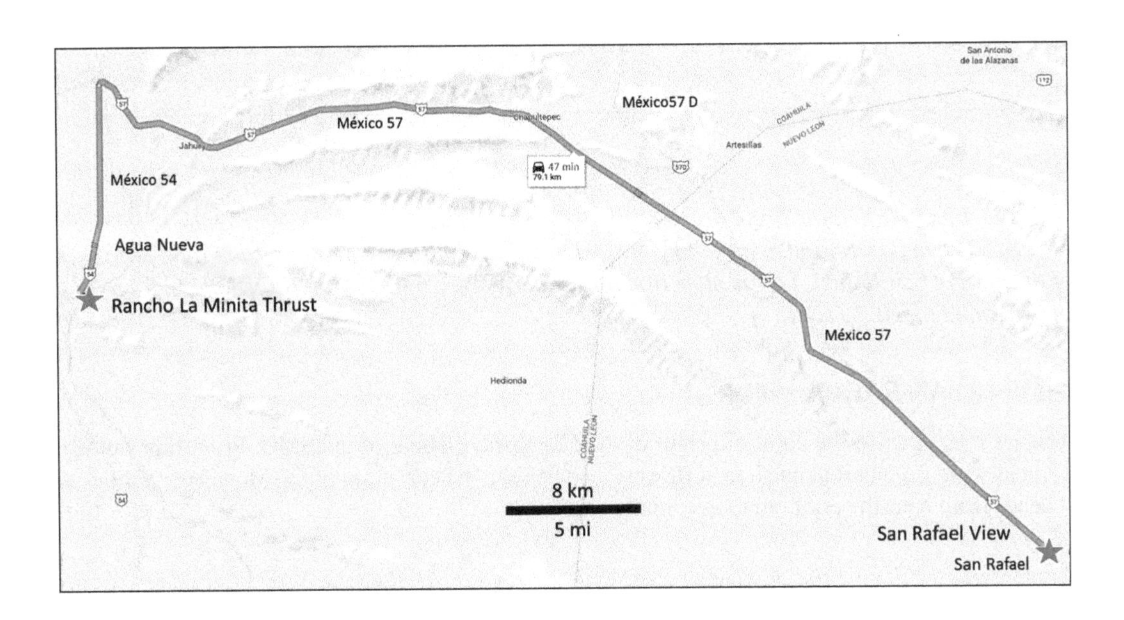

Side Trip 1—Rancho La Minita Thrust to San Rafael: *Return to Rural E-18 and turn right (west); drive to México 54 and turn right (north); take México 54 to turnoff to México 57 on the right; turn right (east) onto México 57 and drive to México 57 D; merge (southeast) onto México 57 D and drive to a PEMEX station on the right just south of the town of San Rafael. This is* ***Stop 10, San Rafael View*** *(25.022724, −100.545942), a total of 79.1 km (41.2 mi; 47 min).*

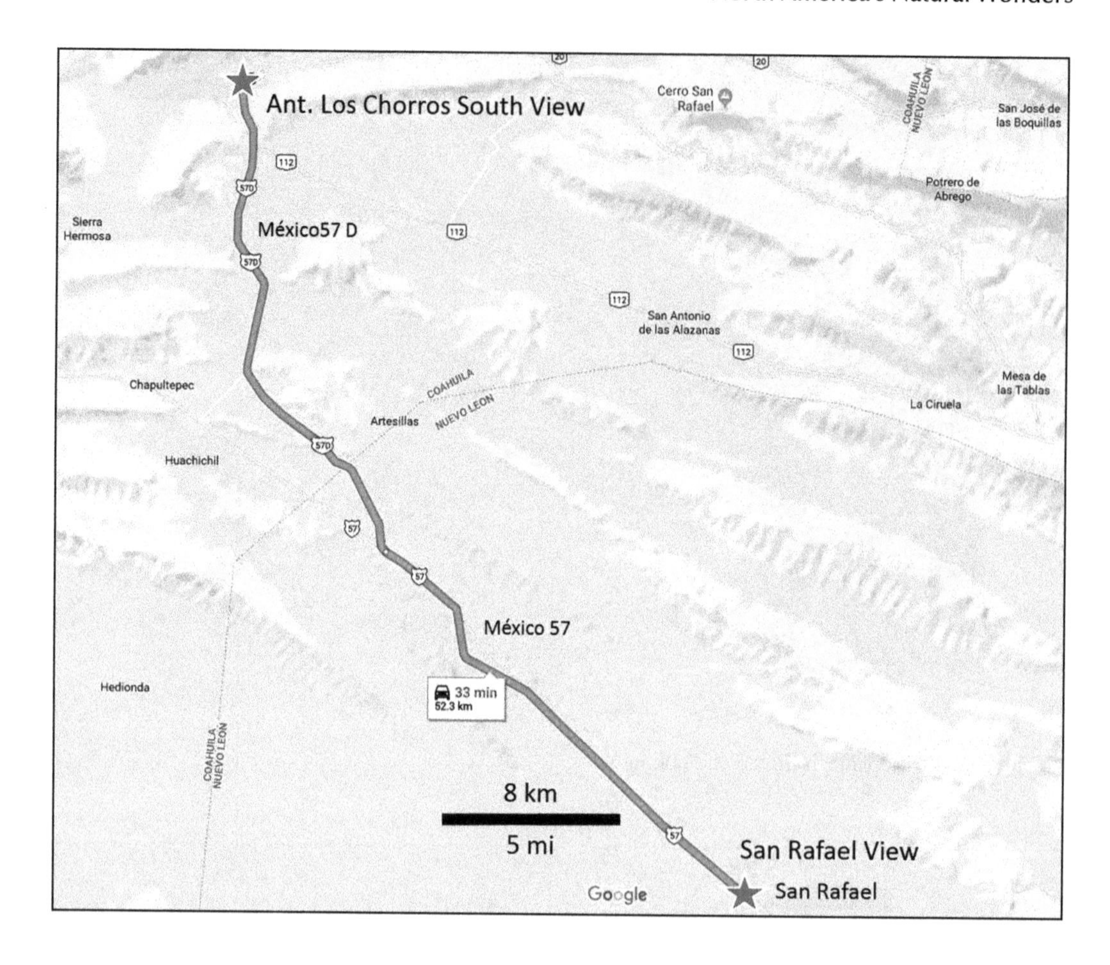

South Limb Anticlinal Los Chorros to San Rafael: *Merge back onto Matehuala-Monterrey/ México 57 D and continue driving south on México 57 D to a PEMEX station on the right just south of the town of San Rafael. This is* ***Stop 10, San Rafael View*** *(25.022724, −100.545942), for a total of 52.3 km (32.5 mi; 33 min).*

STOP 10 SAN RAFAEL VIEW

You are now close to the point of origin of the Monterrey Salient of the SMO. From this vantage point looking from east to north to northwest, you can see the curvature of the Monterrey Salient as it bends from a northwest trend to essentially east-west.

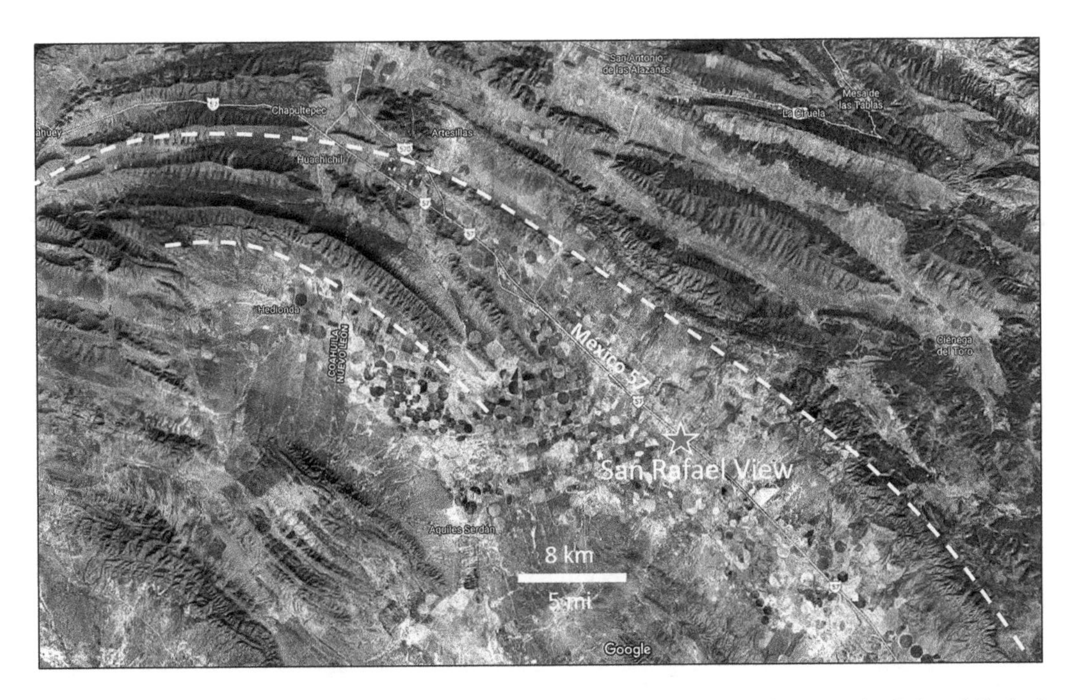

Google Earth satellite image showing curvature of the Monterrey Salient of the Sierra Madre Oriental (dashed lines) around the village of San Rafael. Imagery © 2019 TerraMetrics.

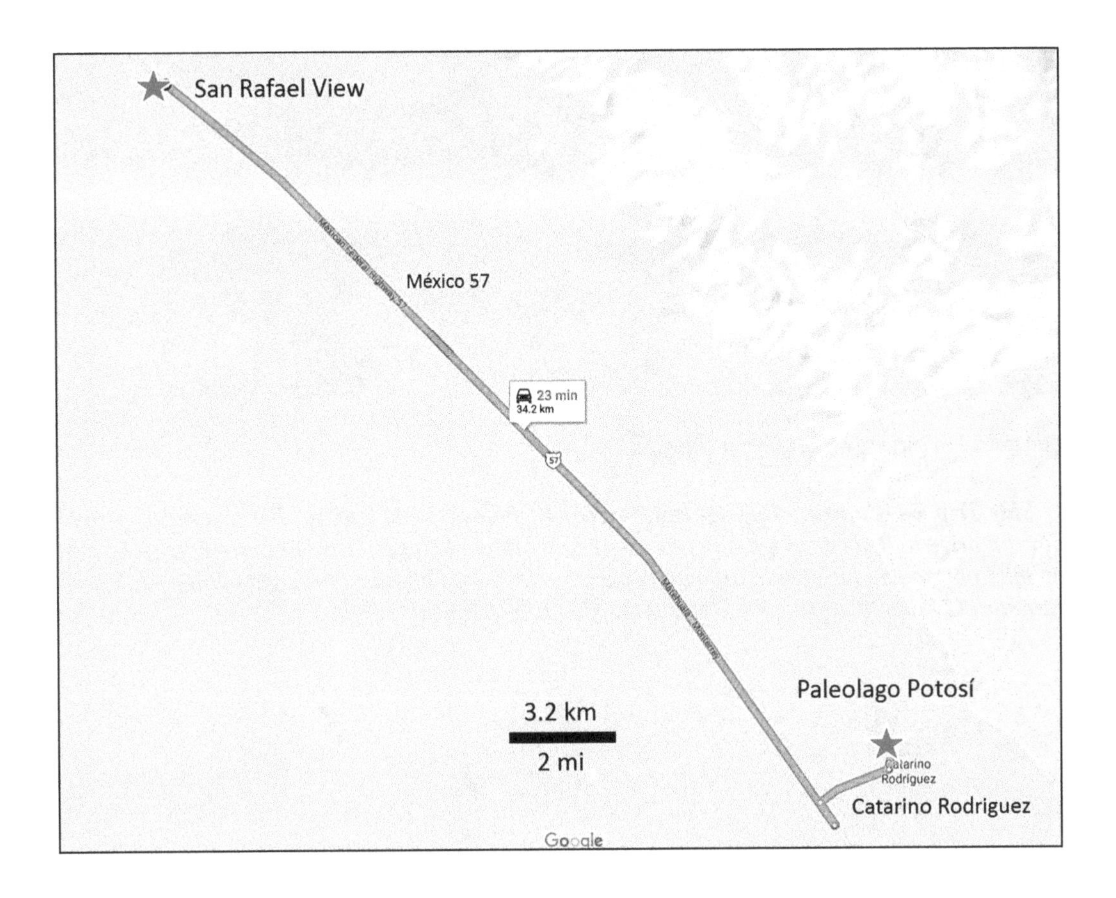

Side Trip 2—San Rafael to Paleolago Potosí: *Continue driving southeast on México 57 to turnoff to El Potosí on the left. Use Retorno to get on northbound lanes and turn right onto road to Ejido Catarino Rodriguez; drive to Plaza Mayor and turn left; drive beyond the church and school. This is* ***Side Trip 2, Paleolago Potosí*** *(24.846539, −100.327264), a total of 34.3 km (21.3 mi; 23 min). Walk about 100 m northwest to where the ground burns ("donde quema la tierra").*

SIDE TRIP 2 PALEOLAGO POTOSÍ, EJIDO CATARINO RODRIGUEZ

The valley here is the site of a Pleistocene lake, Paleolago Potosí. Lake deposits consist of lake muds, swamp deposits, and river deltas. Fossil pollen and algae indicate an overall warming climate here changing from warm humid to warm temperate, with long periods of drought. Among the evidence for this lake is peat that was deposited in bogs around the margins of the lake (Amezcua Torres, 2003). The peat, up to 10 m (33 ft) thick, had been burning in this area for many years, causing the ground to subside and collapse as the peat is reduced to ash or otherwise dewaters and compacts.

Collapsed soil surface where underground peat has burned.

Side Trip 2—Catarino Rodriguez to Sierra El Potosí: *Return to México 57 and turn right (north); drive to Retorno at left and cross over to southbound lanes; drive to Entronque (junction) at San Roberto and turn left (east) onto México 58. Drive to pullout on the right at* ***Stop 11, Sierra El Potosí*** *(24.688241, −100.180573), a total of 33.6 km (20.9 mi; 23 min).*

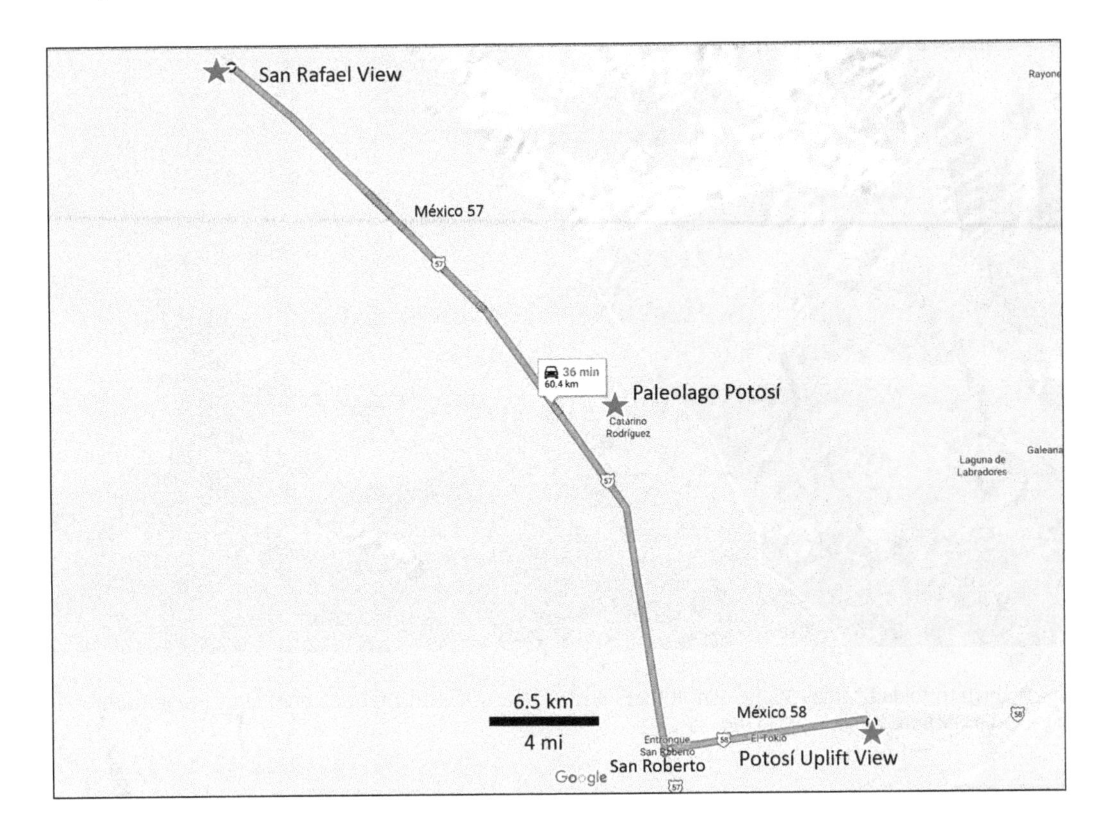

San Rafael to Sierra El Potosí: Continue southeast on México 57; drive to Entronque (junction) at San Roberto and turn left (east) onto México 58. Drive to pullout on the right at ***Stop 11.1, Potosí Uplift*** *(24.688241, −100.180573), a total of 60.4 km (37.5 mi; 36 min).*

11 SIERRA EL POTOSÍ

The Sierra El Potosí is a basement uplift within the otherwise thin-skinned (thrust faulted) SMO. The uplift exposes deformed Jurassic Minas Viejas Formation evaporites (gypsum and anhydrite) that serve as the main detachment for the Sierra Madre Fold-Thrust Belt north of this area. The uplift probably occurred after the main Laramide-age thrusting, as the evaporite layers are themselves highly deformed, probably as a result of folding above the thrust surface. The Minas Viejas here is on the order of 920 m (3,020 ft) thick gypsum interval that contains at least four thick interbedded limestone layers and several thin carbonate layers. These evaporites were deposited during the initial rifting phase of the Gulf of Mexico.

Overall, the uplift is a broad north-south anticlinorium with folding dominant in the west and thrusting dominant on the east flank. Within the uplift, only the detachment interval (Jurassic Minas Viejas) and redbeds below the detachment (Triassic El Alamar and the overlying Jurassic La Joya formations) are exposed.

Some workers have interpreted the Potosí Uplift as originally down-to-the-west, then uplifted (inverted) during the Laramide Orogeny.

STOP 11.1 SIERRA EL POTOSÍ UPLIFT

At this stop, you see deformed Jurassic Minas Viejas Formation that serves as the detachment surface for folding seen in the Monterrey Salient. The fold axes here generally trend northeast-southwest.

View north to folded Minas Viejas Fm in the Sierra el Potosí Uplift. Light-colored units are gypsum-rich layers; darker units contain limestone layers.

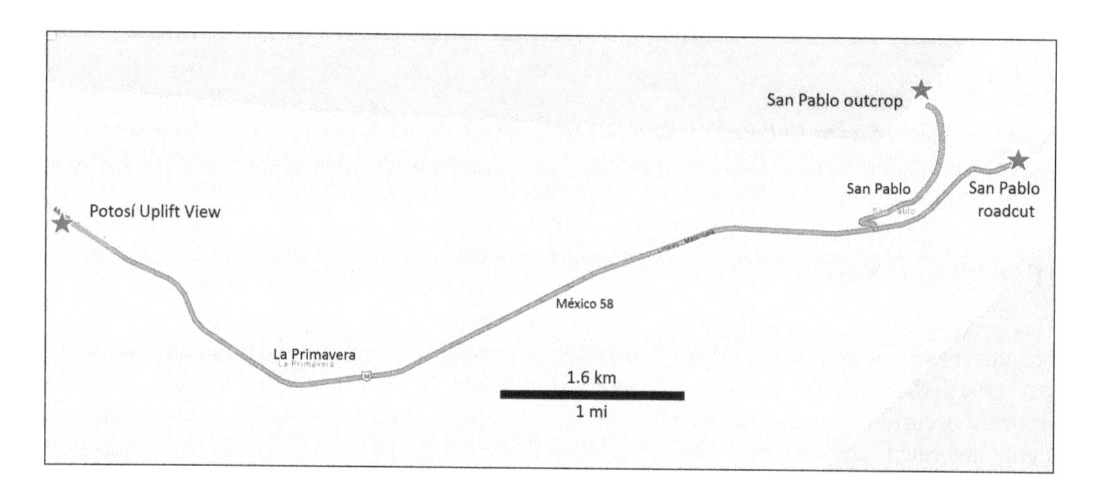

Sierra El Potosí to San Pablo Outcrop: *Continue east on México 58 to unmarked turnoff on the left at km 78 (a small sign says "Mina"); turn left and drive across an arroyo and through the pueblo of San Pablo and continue toward the bright red outcrops to the north,* ***Stop 11.2, San Pablo Outcrop*** *(24.696066, −100.100306), for a total of 10 km (6.2 mi; 9 min).*

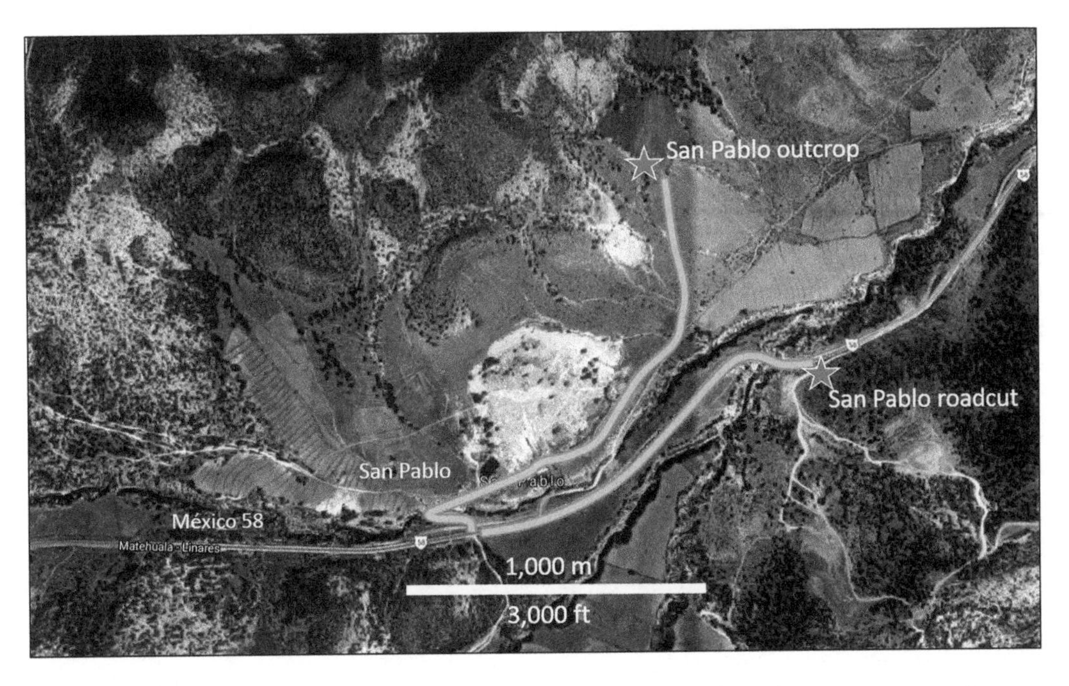

Google Earth satellite image of the San Pablo area stops. Light areas are Jurassic gypsum outcrops. Imagery © 2019 CNES/Airbus; Maxar Technologies.

STOP 11.2 SAN PABLO OUTCROP

This colorful outcrop contains the Triassic El Alamar Formation (equivalent to the Eagle Mills on the U.S. Gulf Coast; formerly referred to as the La Boca Formation or Huizachal Formation) and the overlying Jurassic La Joya and Minas Viejas formations. The orange-to-green El Alamar was deposited in rift basins as alluvial fan, river, and lake deposits of shale and sandstone with interbedded volcanic rocks. The bright red La Joya Formation is a continental, subaerially deposited conglomerate. The Minas Viejas consists of shallow marine evaporites and limestones. These units were deposited during initial rifting of the Gulf of Mexico.

The gypsum is being mined south of the highway, and there are plans to start a mine here as well.

View north to San Pablo outcrop.

San Pablo Outcrop to San Pablo Roadcut: *Return to México 58 and turn left (east); drive 1.5 km (0.9 mi) to pullout on right after the "Curva Peligrosa" (dangerous curve) sign and roadcut, and right after the "Curve to Left" sign. This is* ***Stop 11.3, San Pablo Roadcut*** *(24.691751, −100.093732).*

STOP 11.3 SAN PABLO ROADCUT

From this stop, walk slowly west through the roadcut. You are looking at the Triassic El Alamar Formation, red sandstone and shale deposited in a rift basin at the time of the initial opening of the Gulf of Mexico and Atlantic Oceans. There are channel deposits filled with conglomerate and small offset (0.3–2 m) west-directed (and east-dipping) thrust faults. At the west end of the roadcut, the Minas Viejas gypsum (on the west) is downfaulted against the El Alamar Formation (on the east) across a normal fault.

South side of México 58 roadcut east of San Pablo.

South side of the road at the San Pablo Roadcut. A small west-directed thrust cuts the El Alamar Formation.

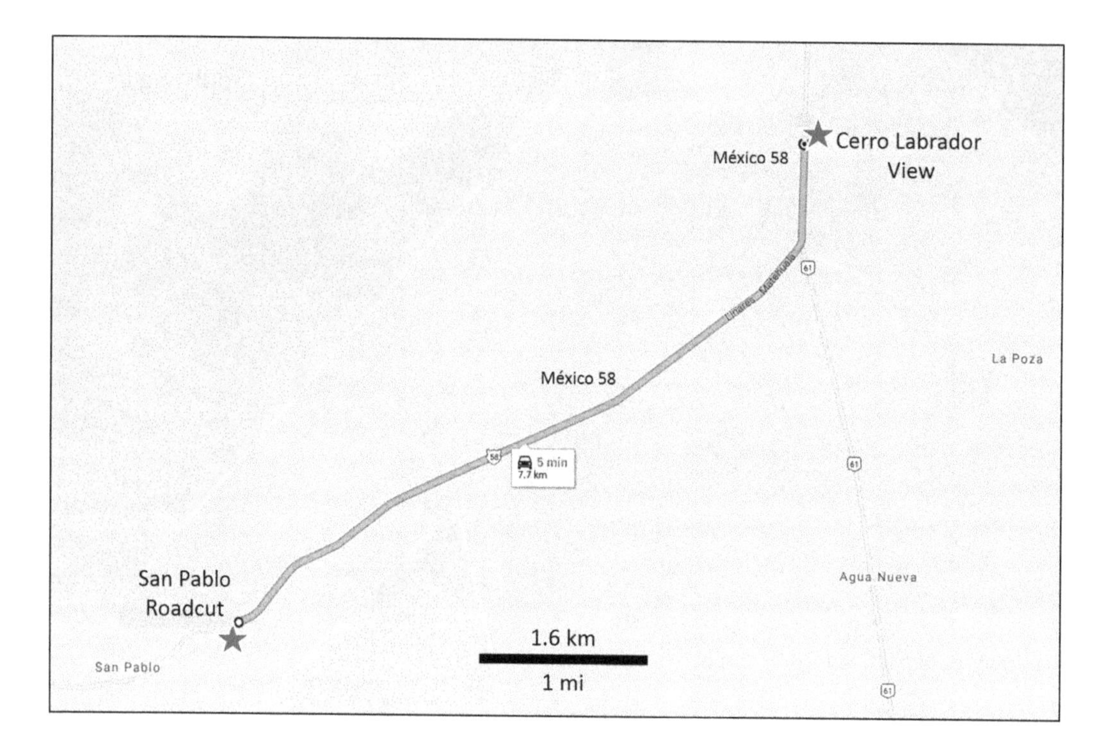

San Pablo Roadcut to Cerro Labrador View: *Continue northeast on México 58 to the intersection with México 61; turn left (north) and continue on México 58 and drive to pullout on right, a total of 7.7 km (4.8 mi; 5 min). This is* ***Stop 12, Cerro Labrador View*** *(24.733131, −100.038768).*

STOP 12 CERRO LABRADOR VIEW

From this stop, you can look north to Cerro Labrador and see vertical beds of the Cupido platform edge. The reef margin starts here, extends north to around Lampazos de Naranjo, and then turns northeast toward Laredo. These massive reef limestones are in stark contrast to the thin, basin limestones of the Tamaulipas Inferior found further east in Santa Rosa Canyon. The peak is part of a large anticline that is overturned to the east.

West of here is the Potosí Uplift. You are looking west at Loma San Pablo (San Pablo Hill), with the light-colored gypsum deformed by thrusting.

East of here is the Sierra Tapias, the back side (oldest deformation) of the folded and thrusted front range of the SMO.

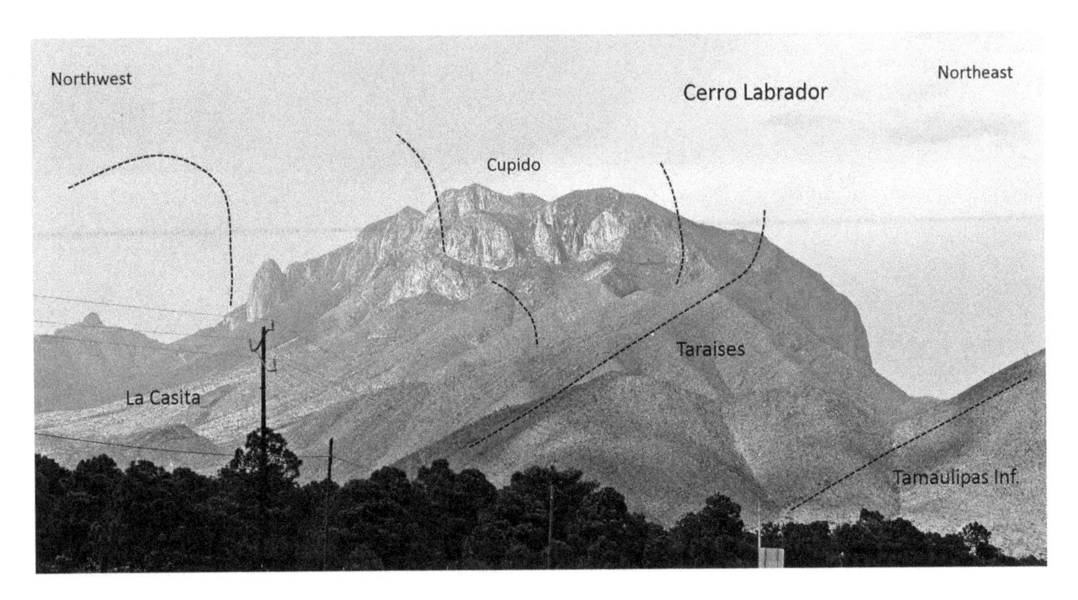

View north to Cerro Labrador and vertical to overturned Lower Cretaceous units. The Cupido is the platform (reef) equivalent of the deeper marine Tamaulipas Inferior limestone seen in Santa Rosa Canyon to the east. The Taraises Formation in the foreground dips west.

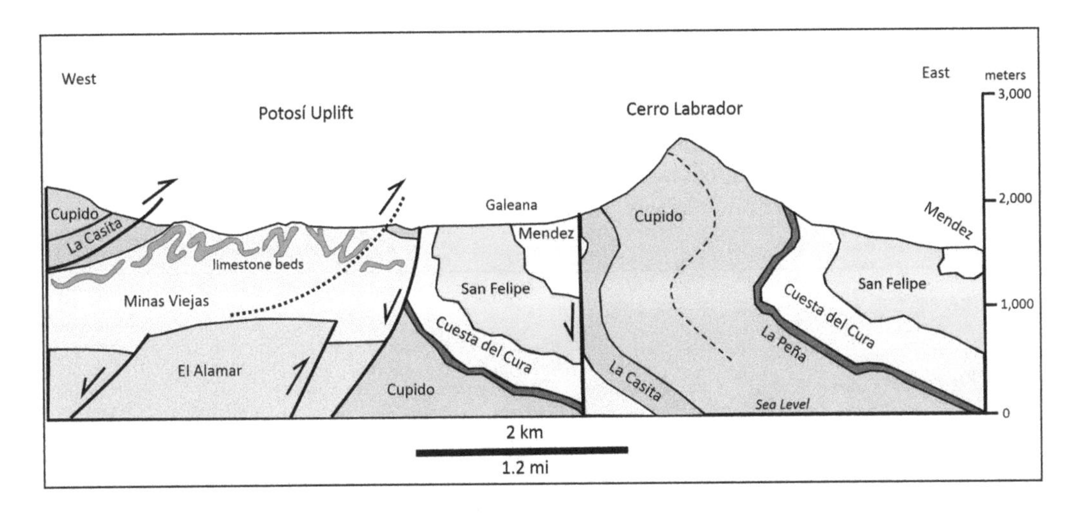

Cross section through Cerro Labrador. (Modified after Cross, 2012.)

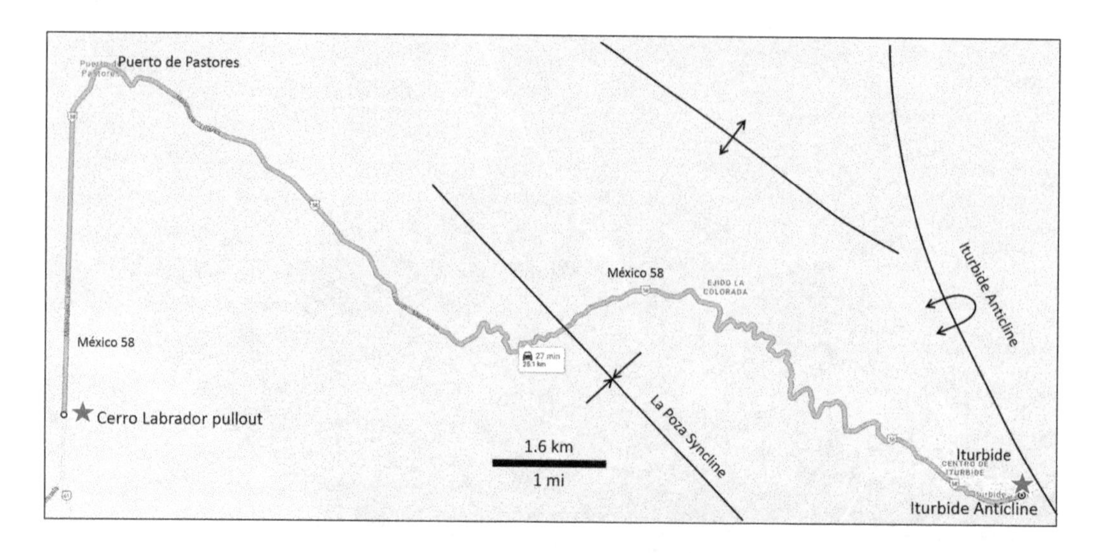

Cerro Labrador View to Anticlinal Iturbide: *Drive north on México 58 to Puerto Pastores, and then continue southeast on México 58 to pullout on the left at Parque el Alamo in Iturbide,* ***Stop 13, Iturbide Anticline****, a total of 25.1 km (15.6 mi; 27 min).*

STOP 13 ANTICLINAL ITURBIDE

The first anticline encountered in this traverse of the SMO is the large Iturbide Anticline. The fold exposes Upper Jurassic La Casita Formation in the core (near the town of Iturbide) and has Lower Cretaceous Tamaulipas Inferior limestone on the flanks. The anticline is overturned to the east (toward the foreland), but is not cut by a thrust. The relatively thin layers of calcareous shale and siltstone in the La Casita are often intensely deformed.

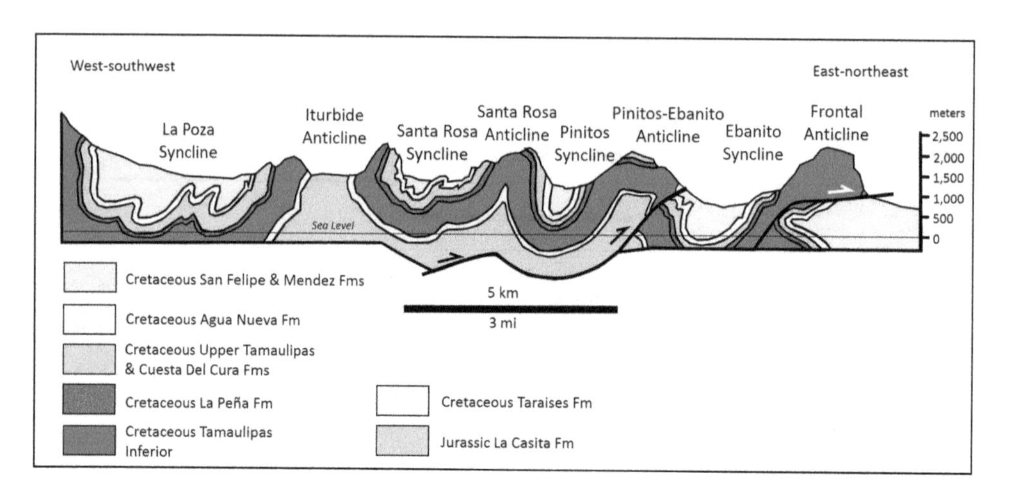

Cross section from Iturbide to the SMO mountain front west of Linares. (Modified after Cross, 2012.)

View north from Parque El Alamo, Iturbide. This is a subsidiary fold on the Iturbide Anticline, a large box fold overturned to the east. The Jurassic La Casita Formation here is vertical to overturned in the core of the anticline.

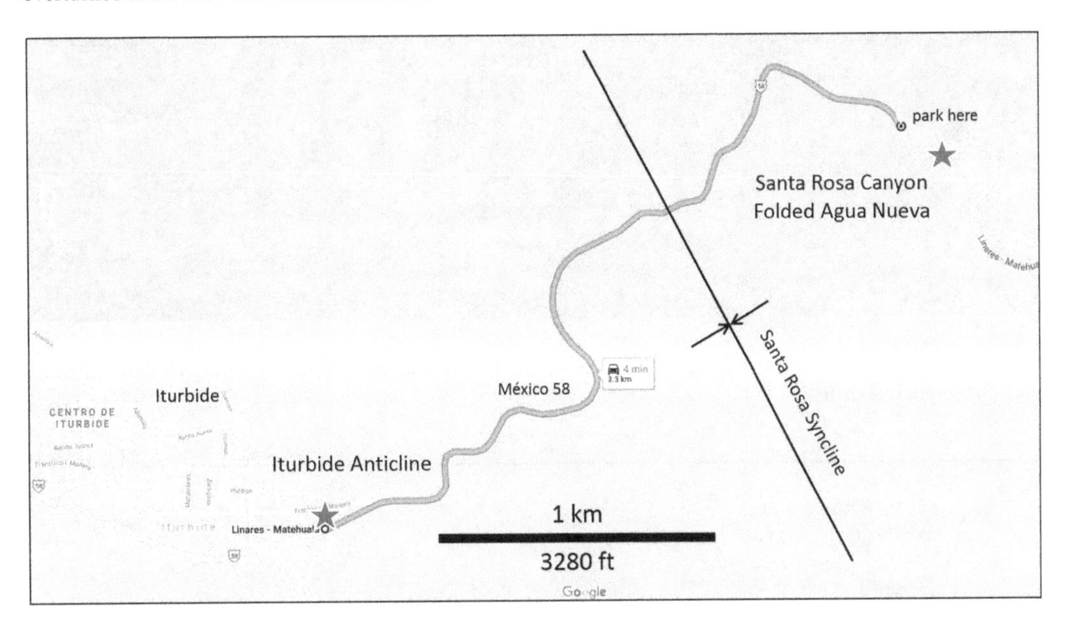

Iturbide Anticline to Santa Rosa Canyon Folded Agua Nueva: *Continue driving east on México 58 to pullout on the left after 3.3 km (2 mi; 4 min). This is* ***Stop 14.1, Santa Rosa Canyon folded Agua Nueva Formation*** *(24.735855, −99.875683).*

14 SANTA ROSA CANYON

Driving east from Iturbide you enter Santa Rosa Canyon. Santa Rosa Canyon cuts through the frontal ranges in the Linares sector of the SMO and provides outcrops and views of the deformation style in this section of the sierra. A series of four anticlines and synclines are overturned to the east. Folds are cut by thrusts that come to the surface, and the main detachments here are thought to be on shale bedding planes. As such, the stronger, sticky shale detachments result in a deformation style similar to that seen in the Appalachian, Idaho-Wyoming, and Canadian thrust belts.

STOP 14.1 FOLDED AGUA NUEVA FORMATION

The first stop in Santa Rosa Canyon is in the Upper Cretaceous Agua Nueva Formation on the east flank of the northwest-trending Santa Rosa Syncline and west flank of the Santa Rosa Anticline. A small thrust appears to truncate bedding in the footwall (below) the thrust. The Santa Rosa Anticline is the second large anticline, overturned to the east, encountered in Santa Rosa Canyon. As there is no major thrust that comes to the surface here, Santa Rosa Anticline is thought to be a fault bend fold. Fault bend folds form over a ramp where a bedding-parallel thrust fault cuts up-section.

View south at folded and thrusted Agua Nueva Formation, east side Santa Rosa Syncline. Dotted line is an out-of-the-syncline thrust.

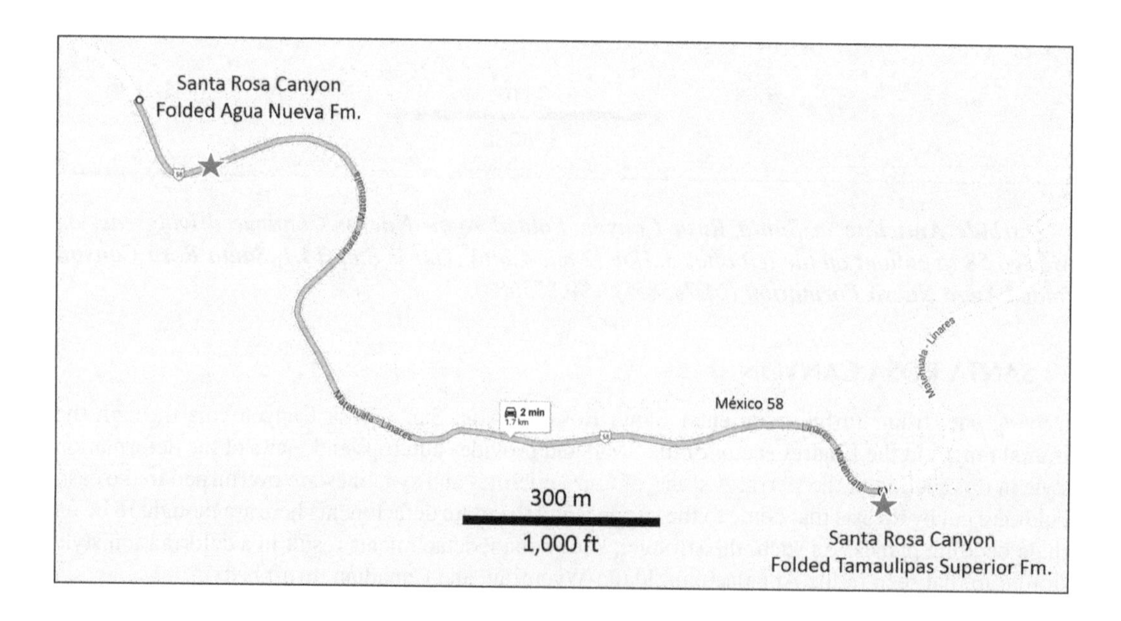

Santa Rosa Canyon folded Agua Nueva to Santa Rosa Canyon Tamaulipas Superior: *Continue driving east on México 58 to pullout on the left, a total of 1.7 km (1.1 mi; 2 min). This is* ***Stop 14.2, Santa Rosa Canyon Tamaulipas Superior*** *(24.730852, −99.864558).*

STOP 14.2 TAMAULIPAS SUPERIOR

At this stop, we examine chevron folds in the thin-bedded Lower Cretaceous Tamaulipas Superior. We are still on the west limb of the east-verging Santa Rosa Anticline. Looking north across the valley, we see long, straight bedding planes that suggest, at least in the stronger, more competent layers, that this is a box or kink fold. Kink folds have straight, planar limbs and abrupt hinge zones where dip changes suddenly. Kink folds form in rocks with strong, competent layers interbedded with weaker layers. The resulting folds are chevron or box-shaped.

Looking southeast from pullout toward intensely folded Lower Cretaceous Tamaulipas Superior Formation.

View north at the west limb, Santa Rosa Anticline. Long, straight west-dipping panels of Tamaulipas Superior indicate a large box fold with abrupt changes in dip at fold hinges.

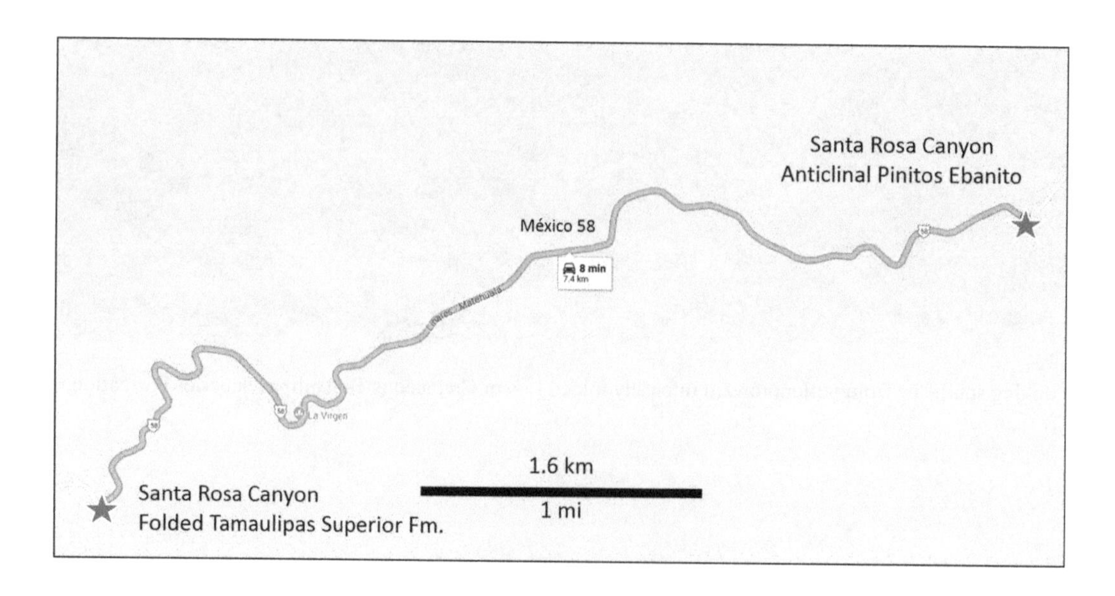

Santa Rosa Canyon Tamaulipas Superior to Anticlinal Pinitos Ebanito: *Continue east on México 58 to pullout on the right, a total of 7.4 km (4.6 mi; 8 min). This is* ***Stop 14.3, Santa Rosa Canyon and Anticlinal Pinitos Ebanito*** *(24.745235, −99.812939).*

STOP 14.3 ANTICLINAL PINITOS EBANITO

This impressive peak is at the leading edge of the Pinitos Ebanito Anticline (Cross, 2012) on the Cabalgadura Secundaria (Second Thrust) of Chávez Cabello et al. (2011). This is a classic fault propagation fold, that is, a fold formed when a propagating thrust fault loses slip and terminates by transferring its displacement to a fold.

A thrust fault puts Lower Cretaceous Tamaulipas Inferior over Tamaulipas Superior at the leading edge of Anticlinal Pinitos Ebanito. View south.

Anticlinal Pinitos Ebanito to San Felipe Box Fold: *Continue driving east on México 58 to pullout on the right immediately across from synclinal fold, a total of 1.6 km (1 mi; 2 min). This is* ***Stop 14.4, San Felipe Box Fold*** *(24.750098, −99.799634).*

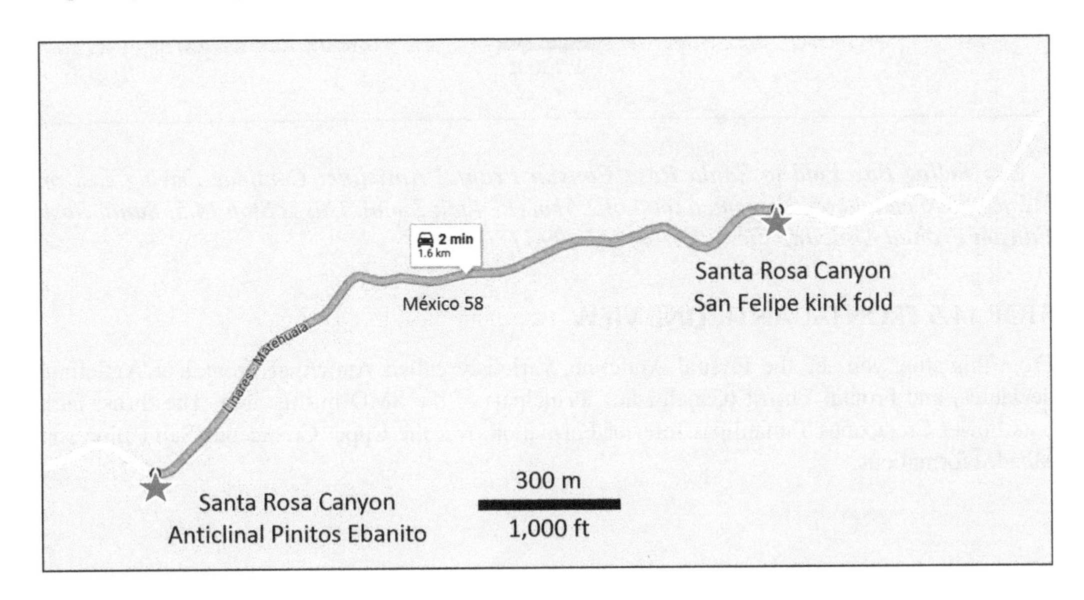

STOP 14.4 SAN FELIPE FORMATION SYNCLINAL BOX FOLD

At this outcrop, we see a box, or kink fold syncline developed in the Upper Cretaceous San Felipe Formation (equivalent to the Austin Chalk on the Texas Gulf Coast).

Panoramic view of a synclinal box fold developed in the Cretaceous San Felipe Formation, north side of road.

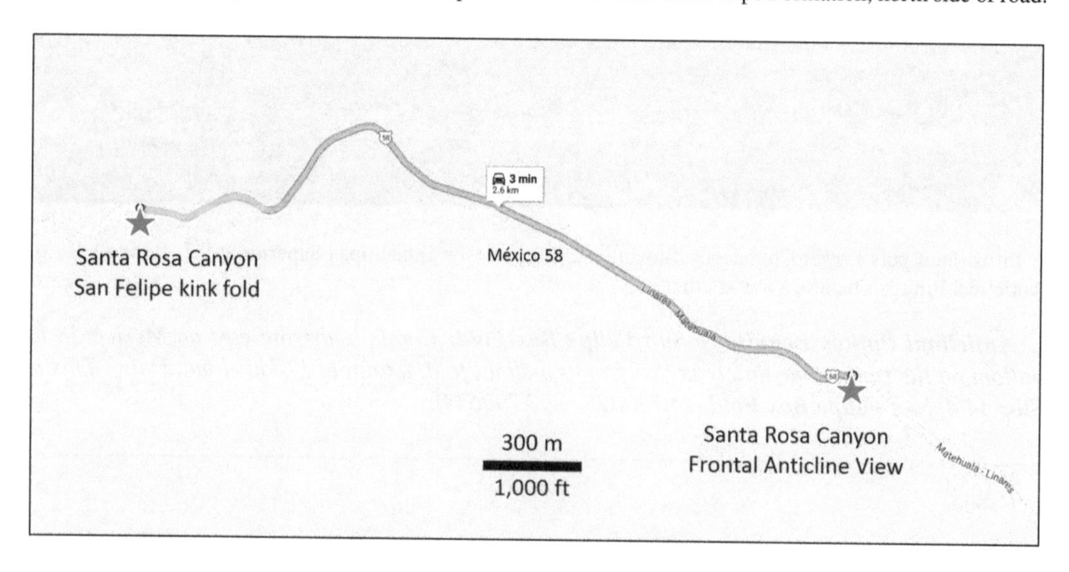

San Felipe Box Fold to Santa Rosa Canyon Frontal Anticline: *Continue driving east on México 58 to pullout on the right, a total of 2.6 km (1.24 mi; 3 min). This is* ***Stop 14.5, Santa Rosa Canyon Frontal Anticline View*** *(24.745722, −99.777197).*

STOP 14.5 FRONTAL ANTICLINE VIEW

From this stop, you see the Frontal Anticline, variously called Anticlinal Frontal or Anticlinal de Jáures, and Frontal Thrust (Cabalgadura Principal) of the SMO in this area. The thrust fault puts Lower Cretaceous Tamaulipas Inferior Formation over the Upper Cretaceous San Felipe and Mendez formations.

View south to the Frontal Anticline in the SMO developed in Lower Cretaceous limestone.

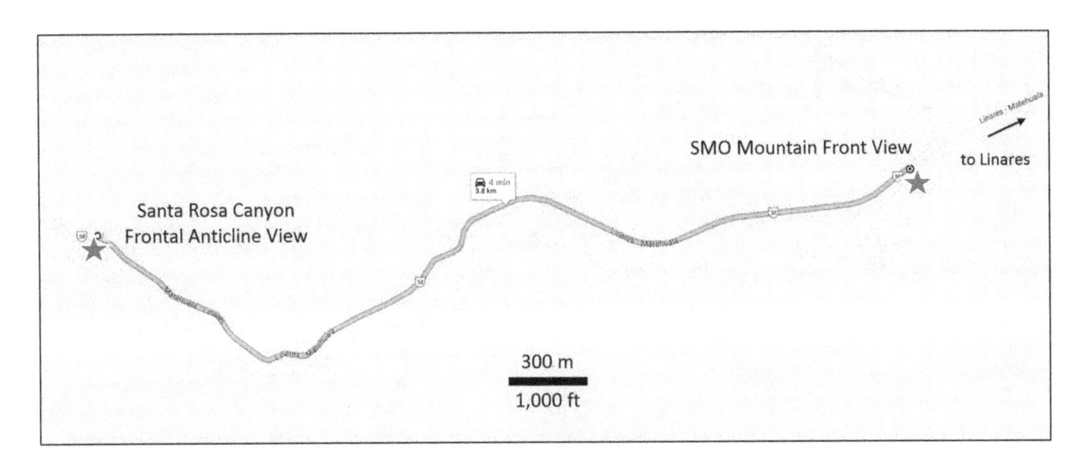

Santa Rosa Canyon Frontal Anticline to Sierra Madre Oriental Mountain Front: *Continue driving east 3.8 km (2.4 mi; 4 min) on México 58 to pullout on the right. This is* ***Stop 15, Sierra Madre Oriental Mountain Front*** *(24.747952, −99.744431).*

STOP 15 SIERRA MADRE ORIENTAL MOUNTAIN FRONT

Looking west you see both the Frontal Anticline and the Frontal Thrust, the Cabalgadura Principal, of the SMO. The thrust carries the east-vergent Frontal Anticline. As we saw at the last stop, the thrust (at the base of the cliffs) puts massive limestone of the Lower Cretaceous Tamaulipas Inferior Formation over Upper Cretaceous shaley limestone of the San Felipe Formation and calcareous shale of the Mendez Formation.

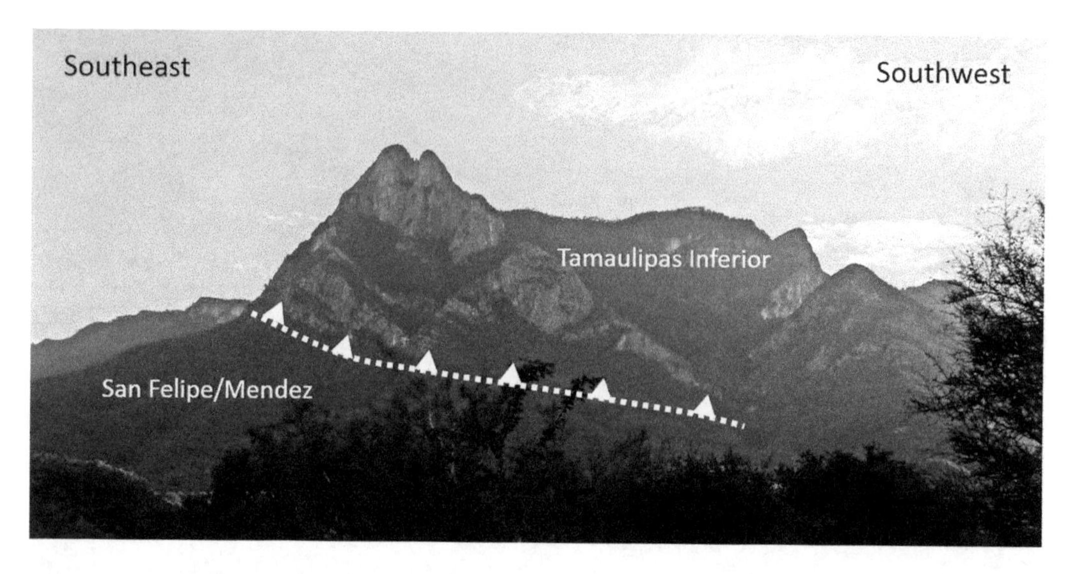

View southwest to the leading edge anticline and thrust of the SMO west of Linares.

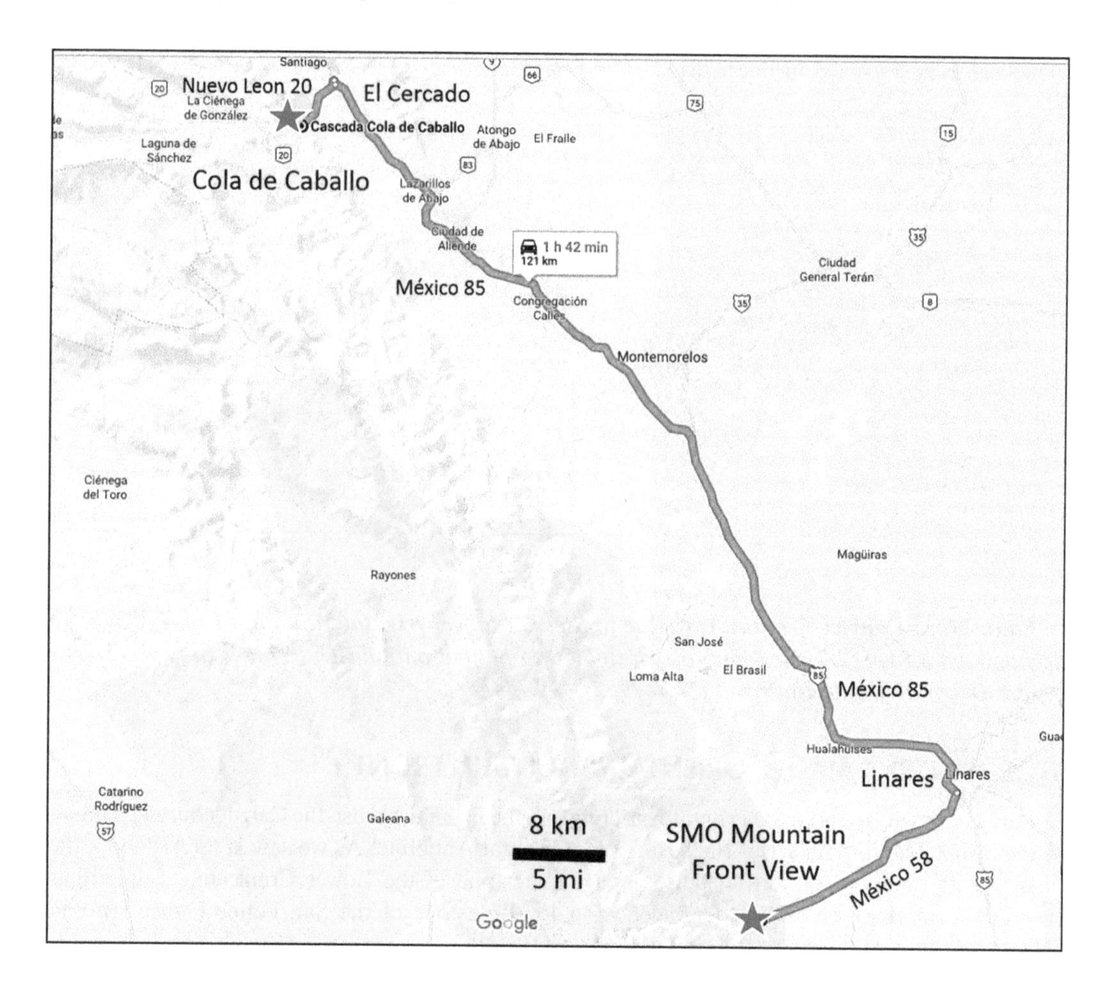

Sierra Madre Oriental Mountain Front to Cascada Cola de Caballo: *Continue driving east on México 58 to México 85 in Linares; turn left (north) on México 85 and drive toward Monterrey till get to El Cercado; exit México 85 at Nuevo Leon 20/Cola de Caballo and drive west on Nueva Leon 20. Follow signs to* ***Stop 16, Cascada Cola de Caballo*** *(25.366267, −100.161469), a total of 121 km (75.2 mi; 1 h 42 min). There is a small fee to enter the park.*

STOP 16 CASCADA COLA DE CABALLO

The scenic Cascada Cola de Caballo (Horsetail Falls) emerges from the Lower Cretaceous Cupido Formation limestone and plunges down a large travertine buildup. The drive to the falls is through Upper Cretaceous San Felipe and Mendez formations.

According to those that have mapped this area (e.g., Longoria, 1998), there is no emergent Frontal Thrust at Cola de Caballo. Thus, it is likely we have re-entered the Monterrey Salient and have moved from thrusting in shale (strong detachment) to detachment folding above a gypsum decollement (weak detachment surface).

The travertine deposited at springs along the mountain front is carved by locals into flower pots and statues and is used as building stone.

Cola de Caballo waterfall and travertine mound.

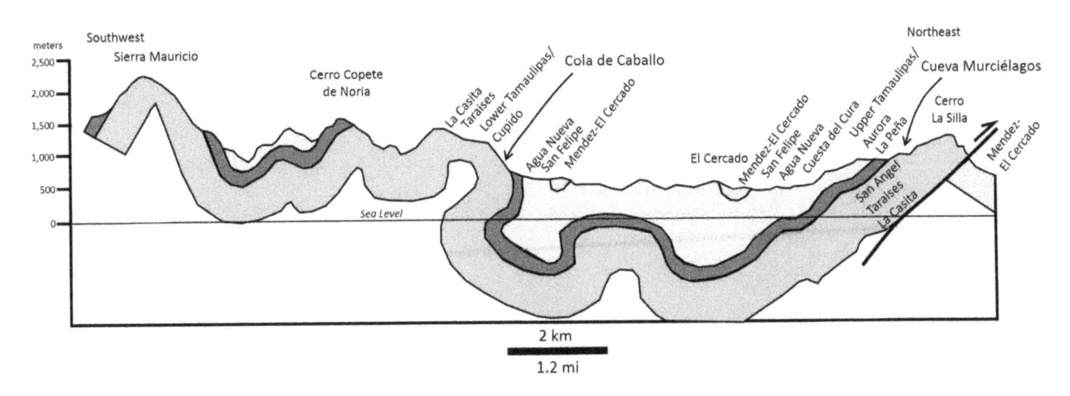

Cross section through Cola de Caballo and the Sierra La Silla. (Modified after Longoria, 1998.)

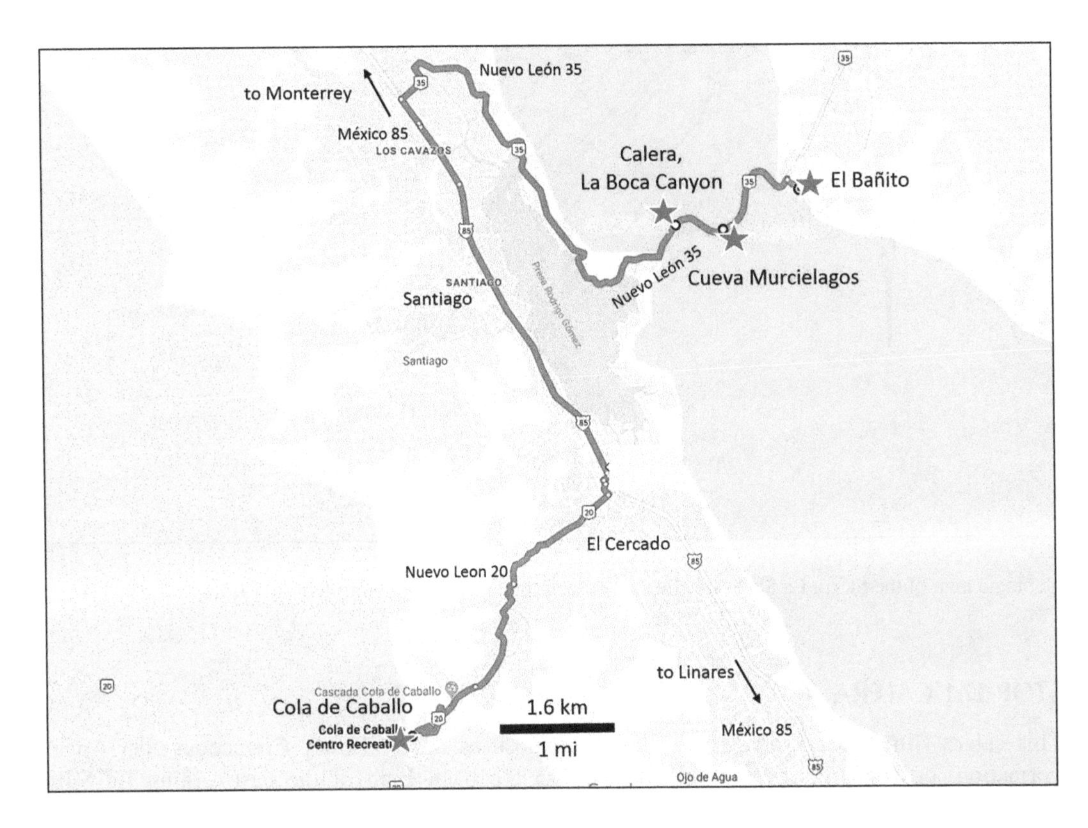

Cola de Caballo to La Boca Canyon: *Drive northeast on Nueva Leon 20/Cola de Caballo to México 85; turn left (north) on México 85 and drive to Nuevo León 35; turn right (east) onto Nuevo León 35 at sign for "Cueva de las Murciélagos" and drive to pullout on the right at the quarry entrance. This is* ***Stop 17.1, Calera, La Boca Canyon*** *(25.434310, −100.121155), a total of 20.2 km (12.5 mi; 33 min).*

17 LA BOCA CANYON

La Boca Canyon cuts across the Sierra La Silla and allows us to see outcrops of the Upper Jurassic through Upper Cretaceous units carried on this east-directed thrust sheet.

As you drive to the quarry (calera), you are going past outcrops of Upper Cretaceous San Felipe Formation (shaley limestone and shale) dipping west about 45° along the shore of Presa

Rodrigo Gómez. Continuing east, you go down-section through the Agua Nueva (shaley limestone), Cuesta del Cura (wavy-bedded limestone and calcareous shale), Tamaulipas Superior (medium to thick-bedded limestone), and La Peña (thin-bedded shaley limestone) formations.

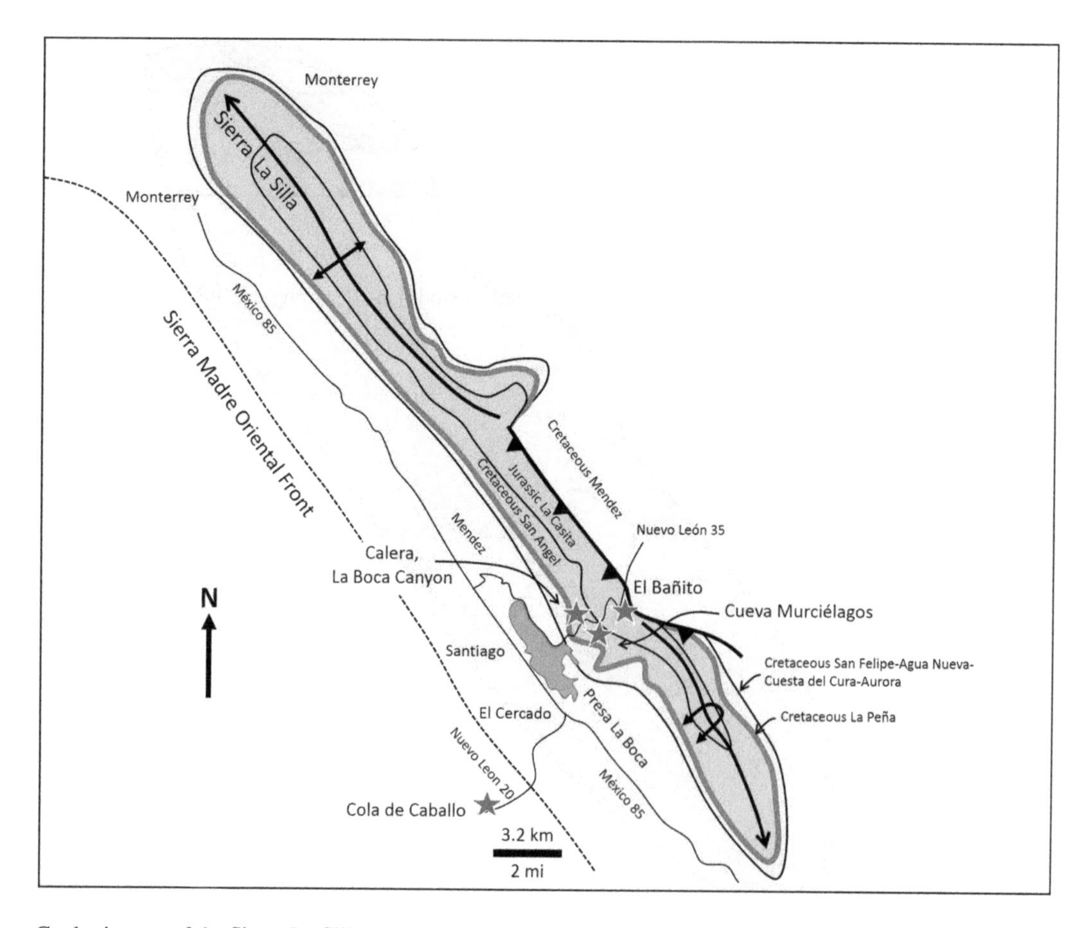

Geologic map of the Sierra La Silla showing stops in La Boca Canyon. (Modified after Longoria, 1998.)

STOP 17.1 CALERA

This calera (lime quarry) provides excellent exposures of the Lower Cretaceous San Angel Formation, or Tamaulipas Inferior limestone on the back limb of the east-verging La Silla Anticline.

***Calera to Cueva Murciélagos:** Continue east on Nuevo León 35 for 550 m (0.34 mi; 1 min) to pullout on the right, **Stop 17.2, Cueva Murciélagos** (25.433701, −100.114073).*

STOP 17.2 CUEVA MURCIÉLAGOS

Cueva Murciélagos (Bat Cave), also known as Cueva La Boca (The Mouth) and Cueva de Agapito Treviño, is a large cavern in the Lower Cretaceous San Angel (or Tamaulipas Inferior) Formation. The entrance to the cave is 30 m (100 ft) wide and 40 m (130 ft) high, and clouds of bats emerge around sunset. You can climb up to the cave entrance. Much of the cave remains unexplored.

Legend has it that the Robin Hood of this area, Agapito Treviño (1829–1854), stole from the rich in this region during the 1800s and hid his treasure in this cave. Eventually he was caught. Before he was shot by a firing squad, he is claimed to have said "Goodbye Monterrey, goodbye friends, forgive me if I hurt you." His treasure has never been found and the legend lives on. Could this be the real 'treasure of the Sierra Madre'?

View south to Cueva Murciélagos (or Cueva La Boca) in the San Angel Formation.

Cueva Murciélagos to El Bañito: *Continue east on Nuevo León 35 to sign for "Camino al Bañito" on the right; turn right (southeast) onto gravel road and drive to* ***Stop 17.3, El Bañito*** *(25.438970, −100.102444), a total of 2.14 km (1.3 mi; 4 min).*

STOP 17.3 EL BAÑITO

At this stop, you have gone down-section into the Upper Jurassic La Casita Formation consisting of calcareous shale and a dark, platy limestone of the Olvido Formation. These are carried just above the northeast-directed Rancho La Noria Thrust, and overlie massive shales of the Upper Cretaceous Mendez and El Cercado formations below the thrust and to the east. The west-dipping units are mapped as overturned here.

The small baths may or may not be open. If you want to swim, please check with the locals before entering La Boca Creek as it has had serious pollution warnings in the past.

We end this tour by returning to the Monterrey International Airport. Most of this drive is over Late Cretaceous shale of the Mendez Formation.

Outcrop of Jurassic La Casita Formation just above the Rancho La Noria Thrust, east side of the Sierra La Silla, El Bañito, La Boca Canyon.

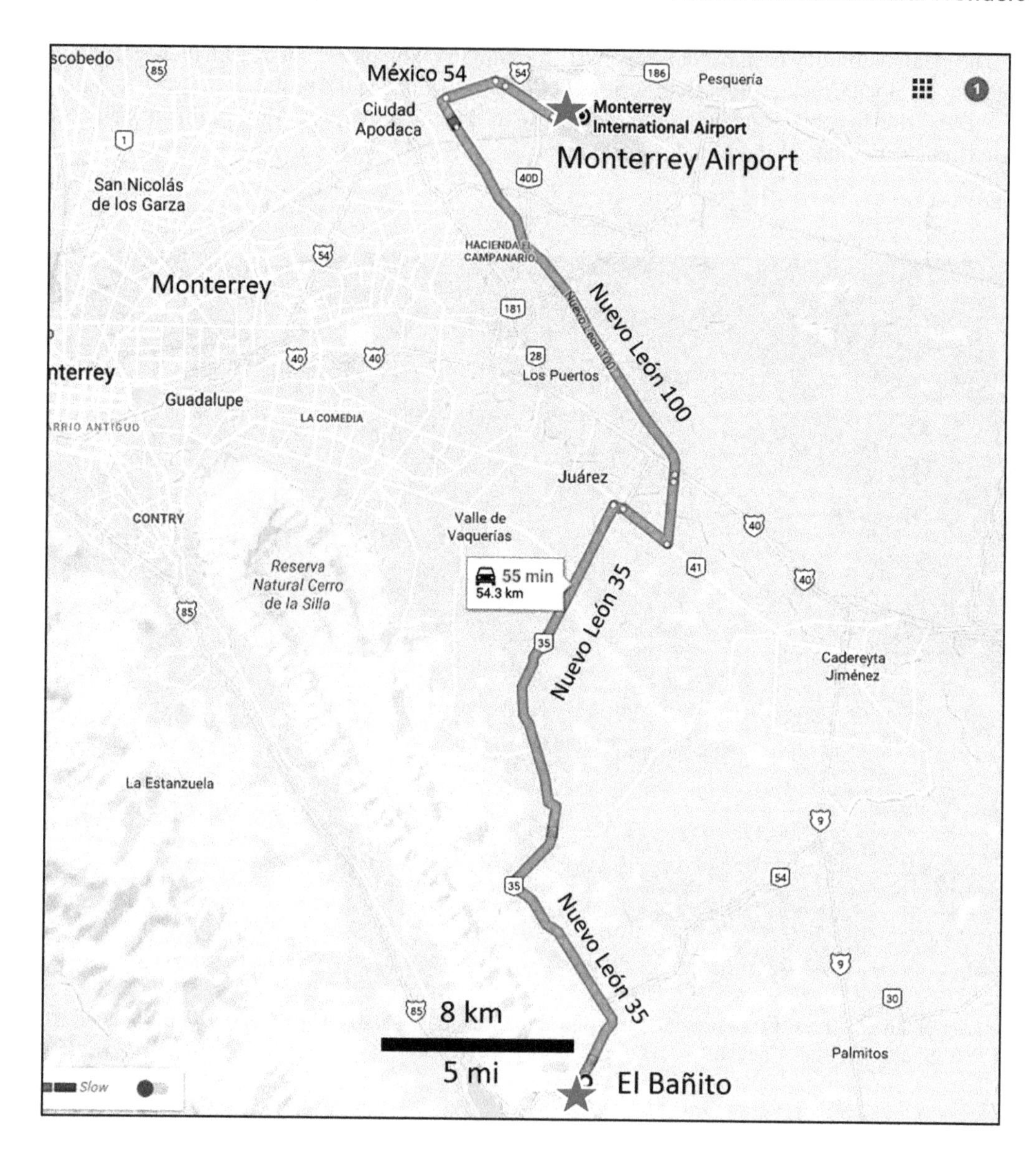

***El Bañito to Aeropuerto Monterrey**: Continue driving north on México 35 to Monterrey; bear right on México 35 to Carretera Reynosa/ México 41 going southeast; turn left (north) onto Nuevo Leon 100/Arco Vial; continue north on Nuevo Leon 100 to México 54; turn right (east) on México 54 and follow signs to the **Aeropuerto Monterrey,** for a total of 54.3 km (33.7 mi; 1 h).*

ACKNOWLEDGMENTS

This trip could not have been accomplished without the advice, suggestions, assistance, and good cheer of Jon Blickwede and Samuel Eguiluz de Antuñano. Their participation is gratefully acknowledged. Thanks also to Josmavic Villarreal C, Yesera Monterrey, for permission to tour their mine at El Papalote Diapir. I also thank (the former) Amoco Production Company and PEMEX for allowing me to work in this wonderful country, my mentor Joshua Rosenfeld, for introducing me to Mexico, and la brigada "Los Leones de PEMEX:" Jose Guadalupe Galicia, Mario Aranda, Salvador Ortuño, Javier Banda, Guillermo Perez-Cruz, Maximino Palma, Francisco Couttulenc, Fernando Ariaga, and Samuel Eguiluz de Antuñano, among others.

REFERENCES AND FURTHER READING

Amezcua Torres, N. 2003. Análisis de la Cuenca Lacustre del Potosí y Sus Peligros Geológicos Asociados a la Materia Orgánica Sedimentaria, Nuevo León, MX. Maestro en Ciencias Geológicas, Universidad Autónoma de Nuevo León, Facultad de Ciencias de la Tierra.

Blickwede, J., and J. Rosenfeld. 2011. The greatest oil well in history? The story of Cerro Azul #4. In Bryce, W.R. (ed.), *International Symposium on the History of the Oil Industry*, April 29th–May 1st, 2010, Lafayette, LA. Oil-Industry History, v. 11, no. 1, 2010, pp. 191–195.

Chávez Cabello, G.C., J.A. Torres Ramos, N.D. Porras Vázquez, T. Cossio Torres, and J.J. Aranda Gómez. 2011. Structural evolution of the tectonic from of the Sierra Madre Oriental at the Santa Rosa canyon, Linares, N.L. *Boletín De La Sociedad Geológica Mexicana*, v. 63, no. 2, pp. 253–270.

Costa, E., and B.C. Vendeville. 2002. Experimental insights on the geometry and kinematics of fold-and-thrust belts above weak, viscous evaporitic décollement. *Journal of Structural Geology*, v. 24, pp. 1729–1739.

Cross, G.E. 2012. Evaporite deformation in the Sierra Madre Oriental, northeastern Mexico: Décollement kinematics in an evaporite-detached thin-skinned fold belt. *PhD Dissertation*, University of Texas, Austin, TX, 575 p.

Cruz-Helu, P., and J.J. Meneses-Rocha. 1998. PEMEX plots ambitious E&D spending increase. *Oil & Gas Journal*, v. 96. pp. 86–88.

Eguiluz de Antuñano, S. 1991. Interpretación geológica y geofísica de la curvature de Monterrey, en el noreste de México. *Revista Inginieria Petrolera*, v. 31, pp. 25–39.

Eguiluz de Antuñano, S. 1992. Presencia de Hidrocarburos en el Municipio de Arteaga. *Asociación Mexicana de Geólogos Petroleros*, v. 42, no. 1, pp. 1–8.

Eguiluz de Antuñano, S. 2017. Exploración Petrolera en el Frente Plegado Laramide de la Sierra Madre Oriental. CiENCiA UANL No. 85, pp. 21–26.

Eguiluz de Antuñano, S., M. Aranda-García, and R. Marrett. 2000. Tectónica de la Sierra Madre Oriental. *Boletín de la Sociedad Geológica Mexicana*, v. LIII, pp. 1–26.

Fischer, M.P., and P.B. Jackson. 1999. Stratigraphic controls on deformation patterns in fault-related folds: A detachment fold example from the Sierra Madre Oriental, northeast Mexico. *Journal of Structural Geology*, v. 21, pp. 613–633.

Giles, K.A., and T.F. Lawton. 2002. Halokinetic sequence stratigraphy adjacent to the El Papalote Diapir, northeastern Mexico. *American Association of Petroleum Geologists Bulletin*, v. 86, no. 5, pp. 823–840.

Gray, G.G., S. Eguiluz de Antuñano, R.J. Chuchla, and D.A. Yurewicz. 1997. Structural evolution of the Saltillo-Monterrey corridor, Sierra Madre Oriental: Applications to exploration challenges in fold-thrust belts – A field guidebook. *1997 AAPG/AMGP International Research Symposium: Oil and Gas Exploration and Production in Thrust Belts*, pp. 1–20.

Gray, G.G., and T.F. Lawton, 2011. New constraints on timing of Hidalgoan (Laramide) deformation in the Parras and La Popa basins, NE Mexico. *Boletín de la Sociedad Geológica Mexicana*, v. 63, no. 2, pp. 333–343.

Guzmán, A.E. 2013. Petroleum history of Mexico: How it got to where it is today. AAPG Search and Discovery Article 10530, 27 p.

Higuera-Díaz, C., M.P. Fischer, and M.S. Wilkerson. 2005. Geometry and kinematics of the Nuncios detachment fold complex: Implications for lithotectonics in northeastern Mexico. *Tectonics*, v. 24, TC4010, 19 p.

Lehman, C., D.A. Osleger, and I.P. Montañez. 2000. Sequence stratigraphy of Lower Cretaceous (Barremian-Albian) carbonate platforms of northeastern Mexico: regional and global correlations. *Journal of Sedimentary Research*, v. 70, no. 2, pp. 373–391.

Lehman, C., D.A. Osleger, I.P. Montañez, W. Sliter, A. Arnaud-Vanneau, and J. Banner. 1999. Evolution of Cupido and Coahuila carbonate platforms, Early Cretaceous, northeastern Mexico. *GSA Bulletin*, v. 111, no. 7, pp. 1010–1029.

Longoria, J.F. 1998. The Mesozoic of the Mexican Cordillera in Nuevo Leon, NE Mexico. In Longoria, J.F., P.R. Krutak, and M.A. Gamper (eds.), *Geologic Studies in Nuevo Leon, Mexico*. Sociedad Mexicana de Paleontologia, Monterrey, NL Mexico, A. C. Special Publication, July 5, 1998, 117 p.

Padilla y Sánchez, R.J. 1985. Las Estructuras de la Curvatura de Monterrey, Estados de Coahuila, Nuevo Leon, Zacatecas y San Luis Potosi. *Univ. Nat. Autón. México,* Revista, Inst. Geología, v. 6, no. 1, pp. 1–20.

Padilla y Sánchez, R.J. 2014. Tectonics of Eastern Mexico – Gulf of Mexico and its Hydrocarbon Potential. AAPG Search and Discovery Article #10622, AAPG Annual Convention, Houston. 54 p.

Peterson, J.A. 1985. Petroleum geology and resources of Northeastern Mexico. U.S. Geological Survey Circular 943, 38 p.

Rowan, M.G., F.J. Peel, and B.C. Vendeville. 2004. Gravity-driven fold belts on passive margins. In McClay, K.R. (ed.), *Thrust Tectonics and Hydrocarbon Systems.* American Association of Petroleum Geologists Memoir 82, Tusla, OK, pp. 157–182.

Smith, A.P., Fischer, M.P., and Evans, M.A. 2012. Fracture-controlled palaeohydrology of a secondary salt weld, La Popa Basin, NE Mexico. In Alsop, G.I., S.G. Archer, A.J. Hartley, N.T. Grant, and R. Hodgkinson (eds), *Salt Tectonics, Sediments and Prospectivity.* Geological Society, London, Special Publications, 363, pp. 107–130.

South Texas Geological Society. 1999. Stratigraphy and structure of the Jurassic and Cretaceous platform and basin systems of the Sierra Madre Oriental - A field book and related papers. *Annual Meeting of the AAPG and SEPM*, San Antonio, TX.

U.S. Geological Survey. 2012. Assessment of undiscovered, conventional oil and gas resources of Mexico, Guatemala, and Belize, 2012. World Petroleum Resources Project, 4 p.

Vega, F.J., and Lawton, T.F. 2011. Upper Jurassic (Lower Kimmeridgian-Olvido) carbonate strata from the La Popa Basin diapirs, NE Mexico. *Boletín de la Sociedad Geológica Mexicana*, v. 63, no. 2, pp. 313–321

Vokes, H.E. 1963. Geologic map of the Huasteca Canyon area, Nuevo Leon, Mexico. *Tulane Studies in Geology*, v. 1, no 4, pp. 127–148.

Weidie, A.E. Jr. 1961. The stratigraphy and structure of the Parras Basin, Coahuila and Nuevo Leon, Mexico. *LSU Historical Dissertations and Theses*, 710 p.

Weidie, A.E. Jr., and W.C. Ward. 1987. *A Field Guide to Jurassic and Lower Cretaceous Stratigraphy of northeastern Mexico.* New Orleans Geological Society, New Orleans, LA, 75 p.

Wilson, J.L. 1981. Lower Cretaceous stratigraphy in the Monterrey-Saltillo area. In Smith, C.I. (ed.), *Lower Cretaceous Stratigraphy and Structure, northern Mexico.* West Texas Geological Society Publication, Midland, TX, pp. 81–74.

Wilson, J.L., W.C. Ward, and J.M. Finneran. 1984. *A Field Guide to Upper Jurassic and Lower Cretaceous Carbonate Platform and Basin Systems, Monterrey-Saltillo Area, Northeast Mexico.* Gulf Coast Section, Society of Economic Paleontologists and Mineralogists, Houston, TX. 42 p.

The Final Word

I hope you have enjoyed these travels across some of the most geologically interesting and scenically spectacular landscapes of North America. If you also learned a few things, all the better. If you liked these trips, stick around. The next in this series will visit several World Heritage Sites and other geologic landmarks in Latin America. After that, who knows, the whole world lies before us. There are so many places to see. Let's get going!

Index

C

D

E

H

I

J

K

L

M

N

O

P

Q

R

S

T

U

V

W

X

Y

Z

Lightning Source UK Ltd.
Milton Keynes UK
UKHW051829291022
411168UK00011B/383